Principles
of Statics
and Strength
of Materials

Principles of Statics and Strength of Materials

ENGINEERING

Andrew L. Simon
University of Akron

David A. Ross
University of Akron

wcb

Wm. C. Brown Co. Publishers
Dubuque, Iowa

wcb
group

Wm. C. Brown
Chairman of the Board

Mark C. Falb
Executive Vice President

wcb

Wm. C. Brown Company Publishers
College Division

Lawrence E. Cremer
President

David Wm. Smith
Vice President, Marketing

E.F. Jogerst
Vice President, Cost Analyst

David A. Corona
Assistant Vice President,
Production Development and Design

James L. Romig
Executive Editor

Marcia H. Stout
Marketing Manager

Janis M. Machala
Director of Marketing Research

Marilyn A. Phelps
Manager of Design

William A. Moss
Production Editorial Manager

Mary M. Heller
Visual Research Manager

Book Team

Robert B. Stern
Editor

Anthony L. Saizon
Designer

Barbara Rowe Day
Production Editor

Mavis M. Oeth
Permissions Editor

Stellitano Associates, *Illustrators*

To Rebekah, Anthony, and Andrew

Contents

Preface

This book was designed for technological, architectural, or engineering programs. It offers a practical and combined treatment of statics and mechanics of solids. The material may be covered in one semester or in two quarters. The sole prerequisite is the sound knowledge of algebra and trigonometry.

The authors believe that a firm grasp of the basic concepts is essential for practical applications. For this reason the usual dogmatic presentation of more involved concepts is avoided. For instance, the chapter on indeterminate beams is preceded by a detailed treatment of the deformation analysis of beams, which is the basis of indeterminate analysis. That chapter, in turn, starts out with a careful explanation of the geometry of a curved line. Of course, the instructor has the choice of treating these subjects simply by the method of superposition. To aid this, an extensive table of beam reaction and deformation formulas is included in Appendix B. Some of the worked examples use symbolic rather than numerical notations. The reason for this is to show the reader how the tabulated formulas came about. Such approach will give the students confidence in their acquired knowledge.

Another feature rarely found in texts at this level is the use of three-dimensional statics. Some instructors may skip this section without loss of understanding of later chapters. Throughout the book the concept of free body equilibrium is given great emphasis.

In the experience of the authors students at all levels consistently encounter difficulties in the understanding of the procedure of constructing shear and moment diagrams, even if they are familiar with calculus. For this reason, shear and moment diagrams are introduced in separate chapters. Only after the students gain a firm knowledge of how shear diagrams are constructed from loading diagrams are they taught how moment diagrams are made from loading diagrams. They will then be introduced to the method of making a moment diagram from a shear diagram.

A separate chapter is devoted to the geometric properties of cross-sectional areas. The computation of centroids, moments of areas, polar moments of inertia, and partial moments of areas are explained in numerous examples worked in a tabular manner.

Wherever involved geometric explanations may tend to confuse the reader, such as in the derivation of Mohr's circle or in the case of beam deformations, the geometric considerations are separated and explained before they are applied to the mechanics problems at hand. The reader has the choice to skip the mathematical fundamentals and proceed with the applications.

An interesting aspect of this text is the consistent attempt to teach by way of examples. A large number of problems, worked out in detail, are interspersed with explanations. Another interesting feature of the book is the inclusion of generalized problems for which each individual student may be assigned a different basic variable. The generalized solution for these problems is available in the solutions manual. Each chapter contains a good selection of exercise problems. Solutions for these are included in the solutions manual.

U.S. Customary units and S.I. units are treated equally throughout. Approximately 50 percent of the worked examples as well as the problems are in S.I. units. Selected standard steel tables are included in Appendix A in U.S. Customary units.

In preparing this text the authors have become greatly indebted to many who have made it possible. The authors acknowledge a great debt to their wives, whose patience and helpfulness have made this endeavor what it is. The members of the staff at William C. Brown Company Publishers have been most helpful and cooperative, particularly the editor, Robert B. Stern.

The authors wish to express their sincere appreciation to the reviewers of the manuscript, Steven Ridenour, Temple University; Kimberly H. Gillett, Ferris State College; Roger H. Nelson, Washington State University; Lawrence J. Wolf, University of Houston; Marvin Heifetz, Milwaukee School of Engineering; Eino W. Fagerlund, Wentworth Institute of Technology; Dale King, Western Michigan University; and Minnie Pritchard, University of Akron. Their detailed comments and their meticulous, line-by-line review of the initial text have greatly contributed to its improvement. The authors would also like to thank Richard Yen, for his helpful suggestions, and Agnes Stitz for carefully typing the manuscript.

David A. Ross   Andrew L. Simon

Principles
of Statics
and Strength
of Materials

Basic Concepts 1

1.1 Introduction

The study of mechanics is a very old endeavor. Among the earliest students of mechanics was Archimedes (287–212 B.C.) to whom is attributed the discovery of how a lever works and also the discovery of the principle of buoyancy. Most of the principles of mechanics that we now study were formulated in the sixteenth, seventeenth, and eighteenth centuries by such scientists as Stevinus (1548–1620), Galileo (1564–1642), and Newton (1642–1727).

Mechanics may be defined as the *science of predicting the conditions of rest or motion of bodies under the action of forces.* There are two branches of mechanics that are concerned with rigid bodies exclusively. These two branches are *statics*, which deals with systems at rest, and *dynamics*, in which movement occurs in the system. In this study we will concern ourselves exclusively with the study of static systems. In other words, the systems we will be studying are all completely stationary. Another branch of mechanics concerns the study of *deformable bodies*. This study is often referred to as the study of the *mechanics of materials*. Once we have learned to analyze systems at rest (i.e., the subject matter of statics), we will continue our investigations by considering some of the basic principles of the mechanics of materials. It is well to remember that the two branches—statics and the mechanics of materials—address different questions. The study of statics enables us to answer questions like: What forces are acting on a structure? Why doesn't a piece of machinery, or a complete building, move? What sorts of forces are acting inside particular beams, trusses, and columns? The study of the mechanics of materials, on the other hand, enables us to answer questions like: How strong do I have to make a beam (or a column) so that it will not collapse? How much can I expect a shaft to twist in a piece of machinery? How much will a beam deflect when I walk across it?

In practice one typically needs to check the design of a part or build something to perform a particular task. The task might require the building of a steam boiler, an engine shaft, a crane, or even a complete building. The range of functions of a designer is thus very diverse. However, the basic principles of mechanics relate to a great variety of particular applications. In this study the principles of mechanics are applied to a number of common situations in which engineers and technologists are involved, and in this way an understanding of *practical mechanics* may be gained.

1.2 Fundamentals of Mechanics

In the sixteenth century Galileo Galilei dropped steel balls of various sizes off the Leaning Tower of Pisa, measuring their time of arrival. From the unexpected result that they all arrived at the ground together, regardless of their weight, he concluded that they all accelerated at the same rate. In further experiments, he allowed the steel balls to roll down inclined planes of different slopes—at speeds low enough that he could measure the time intervals with some precision—and derived a value of the acceleration reasonably close to the value accepted today: 32.2 feet per second per second or 9.81 meters per second per second. The nature of this *gravitational acceleration* puzzles mankind even today. We can measure it, we apply it all the time in technical and scientific computation, but we still do not know how and why it acts. It does not behave like other known things: light, radiation, heat, or whatever. It acts from millions of miles away, as effectively across empty space as through the heaviest objects.

Modern mechanics originated with Newton. It probably is only a myth that he got his inspiration when, sitting under a tree, he was hit on the head by a falling apple. In any case, he postulated that a force is a product of mass and acceleration:

(1.1)
$$F = m \times a$$

The force F could be measured on a spring scale, and since nothing was moving, the only acceleration that was present was the gravitational acceleration, g. Rearranging the equation by dividing through with g gives the formula for *mass*:

(1.2)
$$\text{mass} = \frac{\text{force}}{\text{gravitational acceleration}} = \frac{F}{g}$$

The mass per unit volume is called the *density* of a substance:

$$\text{density} = \frac{\text{mass}}{\text{volume}} = \rho$$

The Greek letter ρ (rho) is universally used to signify density.

Equation 1.1 is referred to as *Newton's second law of motion*. Two other laws are equally important. The *first law* states that a body at rest will remain at rest, or a body moving at a constant velocity will remain moving with the same velocity unless a force acts on it. The *third law* states that for every action (force) there is an equal but opposite reaction.

1.3 Units

Most of the quantities used by engineers in calculations have dimensions that are expressed in particular units, and thus a standard system of units is necessary in order that various professionals can communicate with one another. Until quite recently there has not been a single system of units that has been agreed to and used internationally. There are two systems of units that merit our attention at this time—the *International System of Units* and the *United States Customary Units*. Each of these systems needs to be understood by the modern practitioner.

International System of Units (S.I. units)

This system was adopted at an international convention in France in 1872, and since that time most nations in the world have adopted some version of it. The basic quantities in the S.I. system are *mass* (for which the unit of the kilogram is chosen), *length* (for which the standard is the meter), and *time* (of which the second is the standard unit). These basic units and some common derived units are shown in Table 1.1. Derived units are those that may be defined in terms of the three basic units. For example, the S.I. unit of force—the newton—is defined as follows: *A force of one newton must be applied to a mass of one kilogram in order to accelerate it with an acceleration of one meter per second squared.* In

Table 1.1 Common S.I. Units

Quantity	Unit Name	Unit Symbol
Basic Quantities		
Length	meter	m
Mass	kilogram	kg
Time	second	s
Derived Quantities		
Frequency	hertz	Hz
Force	newton	N
Energy	joule	J
Power	watt	W
Pressure or stress	pascal	Pa

this way the unit of force is independent of the location on earth, or, for that matter, independent of any particular location. For that reason, the S.I. system of units is known as an *absolute* system of units.

U.S. Customary Units

This system of units is extensively and traditionally used in the United States. Although it is in many respects similar to the British Imperial system of units, it is distinct in that it relies on different standard measures. The basic quantities of the U.S. Customary system of units are force (for which the standard is the pound), length (for which the standard is the foot), and time (for which the standard is the second). The problem with this system is with the unit of force. The *standard pound* represents the weight of a piece of platinum kept in the National Bureau of Standards in Washington, D.C. Unfortunately, the weight of this piece is not a constant, but varies according to where it is. Thus it would have different weights if measured at the Equator and at the North Pole. While its weight would not vary greatly if measured at these two points, it would weigh only about 1/7 pound on the moon. For this reason the standard pound is defined as the weight of this standard piece at 45° latitude at sea level. Because of the dependence of the force unit on the earth's gravitational pull, this system of units is known as a *gravitational* system of units.

In mechanics the concept of a force distributed over a given area is called *pressure* or *stress*. In the S.I. system, stress is expressed in newtons per square meter, N/m^2, a quantity called the pascal, Pa. In the Customary American system, stress is most often described in terms of pounds per square inch, denoted by psi. A force of 1,000 pounds is frequently called a *kip*. From this comes the stress unit of kips per square inch, ksi, often used in structural design. External loads on structures are frequently given on a basis of square feet, rather than square inches. Hence we may find pressures defined in terms of pounds per square foot, shortened as lb/ft^2, or kips per square foot, $kips/ft^2$.

Table 1.2 Numerical Prefixes

Number	Power-of-Ten Notation	Prefix	Symbol
1,000,000,000	10^9	giga	G
1,000,000	10^6	mega	M
1,000	10^3	kilo	k
0.001	10^{-3}	milli	m
0.000001	10^{-6}	micro	μ
0.000000001	10^{-9}	nano	n

Example: 2,000 N (newtons) $= 2 \times 10^3$ N $= 2$ kilonewtons $= 2$ kN

Whatever system of units is chosen for the solution of problems, it is strongly recommended that the calculations be conducted exclusively in those units. However, since it is likely that both systems of units will be in extensive use for the foreseeable future, a table of factors that allow conversion between the systems of units is given on the inside back cover.

1.4 Numerical Prefixes

A prime feature of the S.I. system of units (although not limited to this system) is the use of prefixes that modify unit symbols to indicate that the quantity is multiplied by some power of 10. The common prefixes and their meanings are given in Table 1.2. Thus, for example 1.0 mm $= 1.0 \times 10^{-3}$ m, and 1.0 MN $= 1.0 \times 10^6$ N. The prefixes shown in Table 1.2 vary by factors of 10^3; this factor is the preferred S.I. multiple of units. Thus the S.I. system tends to measure distance in meters and millimeters, but prefers not to use the unit of centimeter in technical work.

1.5 Numerical Accuracy

Because of the ready availability of hand-held calculators, it is often possible to compute answers to eight, or even more, significant figures. In some applications in physics, this accuracy is necessary. However, in technological situations it is very rare that more than four significant figures of accuracy are required. This is because it is very rare for us to know the forces that are applied to an article (whether it is a machine, an airplane, or a submarine) with such great accuracy. In fact, the true forces that are applied are usually unknown: forces are assumed that, it is hoped, will be similar to the forces that the structure will experience in service. As an example, it may be assumed that the maximum wind that will hit a building during its lifetime will be, say, 100 miles per hour (based on available meteorological data), and yet no one can guarantee that a tornado with a maximum wind speed of, say, 160 miles per hour will not hit the building during its lifetime.

For this reason designers must use all possible care in analyzing a structure or mechanical component but need to be very careful not to claim that their calculations are more accurate than they really are. Part of this care is exercised by specifying answers only to the accuracy that is reasonable.

1.6 Design Codes

In the discussion of practical mechanics the principles are outlined whereby the designer may take a structure or mechanical component with known forces acting on it and proceed to find the various effects those forces will have on the system. In other words, the study of mechanics might start with a beam acted upon by given forces and use established methods to predict the deflection of the beam and the force distribution within the beam due to those applied forces. However, there are several other questions to be asked before construction is completely understood. These questions include: How does the designer know what forces to apply to the beam in the first place? How strong does the material need to be in order safely to sustain the forces? How strong is the material available from which the beam may be built? How much deflection in the beam is tolerable? How can the parts of a structure (such as the parts of a machine) be joined together in a safe yet efficient way?

Some of the answers to these types of questions may be found in *design codes*. There are a large number of such codes covering a wide variety of engineering materials and their applications, and the designer must be familiar with the appropriate design codes covering the intended use before attempting any design or checking of a design. For example, the American Iron and Steel Institute (AISI) issues a specification prescribing the minimum properties of steel suitable for building construction. If something is to be designed using timber, the appropriate code might be the National Design Specification for Stress-Grade Lumber issued by the National Forest Products Association (NFPA), but if the material to be used is concrete, the American Concrete Institute (ACI) would be a more suitable body to seek assistance from.

Not only are there different specifications for different materials, but there are also specific codes for specific applications. Thus a mechanical engineer wanting to design a steam boiler might turn to the Boiler and Pressure Vessel Code issued by the American Society of Mechanical Engineers (ASME). On the other hand, a civil engineer who needs to design a steel building might refer to the Manual of Steel Construction produced by the American Institute of Steel Construction (AISC). If now these designers needed to design a bolted connection between two parts of their design, they might need to make a reference also to a standard issued by the American National Standards Institute (ANSI) to find the necessary properties of the bolts to use. For example, ANSI Standard B1.1-1974 gives the accepted dimensions of Unified and American screw threads. If, however, a designer decided to use welded joints to join two parts together, it would be better for him to seek out the appropriate specification issued by the American Welding Society (AWS).

Many of the standards and specifications have been established by trade or technical groups to help engineers, architects, and technologists in their work. The severity of design requirements, however, is often more dependent on the location, substantial differences occurring between states and even some municipalities. Another circumstance designers must recognize is that the design codes and requirements are updated regularly by the agencies responsible for them. The changes usually reflect recent research results and changes in applications of particular components. The designer must be certain to use the latest available version of a code or specification or it is likely that some design requirement will not be met.

Despite the many design codes available, the principles of mechanics apply equally to a great variety of situations. Reference to particular design codes has been kept to a minimum in this book partly for this reason and also because the requirements of particular design codes change when new versions of the codes are issued.

Forces and Moments   2

1. *To understand that definition of a force requires knowledge of its magnitude, direction, and line of action.*

2. *To be able to determine the resultant of several given forces by using the force-polygon method and the analytical method.*

3. *To be able to derive the moment of a force about a point in a two-dimensional system or about a set of axes in a three-dimensional system.*

4. *To learn to recognize couples and evaluate their moments.*

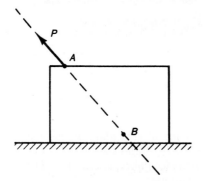

Figure 2.1
A force acting on a block.

2.1 Defining a Force

A force is an action that tends to cause movement of a body. The force may be applied directly to a body, as in the case of pulling an automobile along a road by means of a towrope, or indirectly, as in the case of the earth exerting its gravitational pull on a satellite. For the complete definition of a force, it is necessary to specify the *magnitude* of the force, its *direction,* and its *line of action.* In Figure 2.1 a force *P* is shown acting on a block. The force *P* has a magnitude, or size, which may be represented by the length of the arrow. The numerical magnitude of the force depends on the units in which it is given, such as pounds (American Customary) or Newtons (S.I.). The dashed line in Figure 2.1 is the line of action of the force, and the arrow indicates the direction in which the force is acting.

The point *A* is known as the *point of application* of the force *P.* There is, however, an important principle of statics known as the *principle of transmissibility of forces,* which states that a force may be applied anywhere along its line of action without altering its external effect on a body. The force *P,* therefore, shows the same tendency to move the block whether it is applied at the point *A* or at the the point *B.*

2.2 Resultant of Concurrent Forces and Components

Consider the block of Figure 2.2 acted on by two forces P_1 and P_2, which are applied at points *A* and *B,* respectively, as shown in Figure 2.2a.

These forces are said to be *concurrent* because their lines of action intersect; the intersection is shown as the point *C.* By the principle of transmissibility of forces, the forces P_1 and P_2 may be moved so that they both are applied at the point *C,* as shown in Figure 2.2b. It can be shown by experiment that the two forces acting at *C* may be replaced by a single force *R,* which also acts at *C,* as shown in Figure 2.2c. The force *R* is known as the *resultant* of the two forces P_1 and P_2. The magnitude, direction, and line of action of the resultant force *R* are shown in Figure 2.2c by the line *CE,* which is a diagonal of the parallelogram *CDEF,* of which the sides *CF* and *CD* represent the forces P_1 and P_2, respectively. A parallelogram such as *CDEF* is known as a *parallelogram of forces* and is often a convenient way of depicting the resultant of concurrent forces.

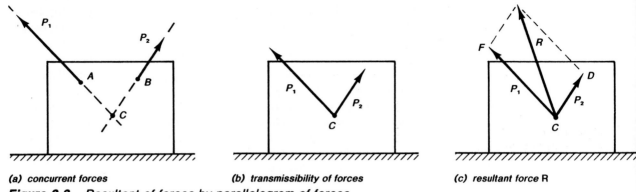

(a) concurrent forces **(b)** transmissibility of forces **(c)** resultant force R

Figure 2.2 *Resultant of forces by parallelogram of forces.*

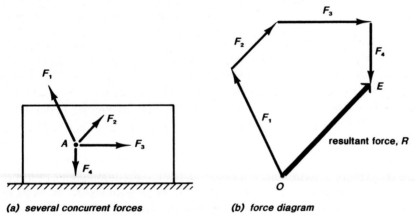

(a) several concurrent forces **(b)** force diagram

Figure 2.3 *The resultant of more than two concurrent forces.*

When more than two concurrent forces are acting on a body it is still possible to draw a *force diagram* in the same way as the parallelogram of forces was drawn in Figure 2.2c. Consider the point A on the block of Figure 2.3a, at which the forces F_1, F_2, F_3, and F_4 are acting as shown. It is desired to find the force R that is the resultant of all the forces acting at the point A. The magnitude and direction of the resultant R are found by constructing a *polygon of forces* as shown in Figure 2.3b. Begin the construction of this polygon of forces at point O and plot each force acting at point A in its correct direction, making the length of each line proportional to the magnitude of the force. When all the forces have been plotted and linked "head-to-tail" as shown, the diagram ends at point E. The line OE represents the resultant force R plotted to the same scale as the original forces at A. R acts in the direction indicated by the arrow.

Just as it was shown in Figure 2.1 to be possible to combine two forces to form a single resultant force, it is also possible to divide one force into two (or more) forces that, if recombined, would have the same value as the original force. Consider the force P of Figure 2.4a: it is desired to divide this force into two forces, one directed along the x-axis and the other along the y-axis. These directions are shown by the dashed lines of Figure 2.4a. The line of action of force P forms the angle θ with the x-axis, as shown. The parallelogram of forces that meets the criteria is shown in Figure 2.4b; the forces P_x and P_y are called *components* of the force P. The components P_x and P_y act along the x- and y-axes, respectively; have the directions indicated by their arrows; and have magnitudes proportional to the length of their lines. When θ is measured from the x-axis, the values of P_x and P_y are given by

(2.1) $$P_x = P \cos \theta$$

and

(2.2) $$P_y = P \sin \theta$$

This approach of finding the components of a force is also a useful analytical means of finding the resultant force of a number of concurrent forces. The x-direction component of the resultant force R of a number of concurrent forces is the sum of the x-direction components of each of these forces. If the x-direction component of the resultant is R_x and the x-direction component of a typical force, F_i, one of a number acting at a point, is F_{xi}, then the value of R_x is given by

(2.3) $$R_x = \Sigma F_{xi}$$

where the symbol Σ represents a summation of all the x-components of forces like F_i. In a similar manner the y-component of the resultant force, R_y, is given by the summation of the y-components of all the forces like F_i, i.e.,

(2.4) $$R_y = \Sigma F_{yi}$$

There are thus at least two methods by which the resultant of a number of concurrent forces may be found. The method of drawing the polygon of forces is a graphical technique, but the method of addition of components is more easily adapted to analytical solution of problems.

It is also clear from Figure 2.4b that it is possible to find the resultant force P if one is given the components P_x and P_y. Using the properties of a right triangle, the following relationships may be verified:

(2.5) $$P = \sqrt{P_x^2 + P_y^2} \quad \text{MAGNITUDE}$$

and

(2.6) $$\tan \theta = \frac{P_y}{P_x} \quad \text{DIRECTION}$$

Thus the magnitude and the direction of the resultant force may be found.

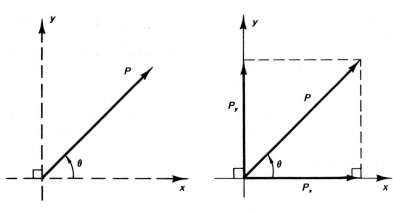

(a) force and its relationship to the x- and y-axes

(b) parallelogram of forces

(c) components in three dimensions

Figure 2.4 *Components of a force.*

When forces act in directions that must be specified in space, it is necessary to specify their values in terms of three components, in the x-, y-, and z-directions. These components are illustrated in Figure 2.4c as P_x, P_y, and P_z. In a similar manner to Equation 2.5 it can be shown that, for three-dimensional forces,

(2.7)
$$P = \sqrt{P_x^2 + P_y^2 + P_z^2}$$

Also

(2.8)
$$P_x = P \cos \theta_x$$
$$P_y = P \cos \theta_y$$
$$P_z = P \cos \theta_z$$

where θ_x, θ_y, and θ_z are angles shown in Figure 2.4c. Each is the angle between the force and the x-, y-, or z-axis.

Example 2.1

The two forces shown are acting on the bolt of Figure 2.5a. Find the resultant force acting on the bolt.

Solutions:
Since the forces are concurrent (i.e., both are acting at the point A), the resultant may be found either graphically or analytically.

Graphical Solution:
The force polygon of Figure 2.5b is constructed, with the length of the sides of the polygon proportional to the sizes of the forces S and T. The resultant force and the angle θ are then measured from the figure. Thus,

$$R = 135 \text{ N}$$

and

$$\theta = 37.2°$$

This may also be written

$$R = 135 \text{ N} \quad \text{acting at} \quad \theta = 37.2°$$

Analytical Solution:
Since both the applied forces and the resultant force may be divided into components as shown in Figure 2.5c, the components of the resultant force may be written as

$$R_x = S_x + T_x$$
$$R_x = (60 \text{ N}) \cos 20° + (80 \text{ N}) \cos 50°$$
$$= 107.8 \text{ N}$$

Similarly

$$R_y = S_y + T_y$$
$$R_y = (60\text{N}) \sin 20° + (80\text{N}) \sin 50° = 81.8 \text{ N}$$

Combining the components of the resultant force gives

$$R = \sqrt{R_x^2 + R_y^2} = \sqrt{107.8^2 + 81.8^2} = 135.3\text{N}$$

and

$$\theta = \tan^{-1}\left(\frac{R_y}{R_x}\right) = \tan^{-1}\left(\frac{81.8}{107.8}\right) = 37.2°$$

$P_x = P\cos\theta$

$P_y = P\sin\theta$

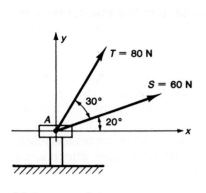

(a) forces on a bolt

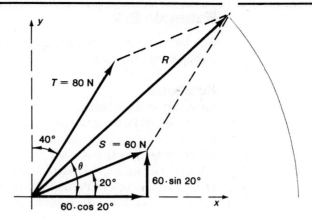

(b) parallelogram of forces

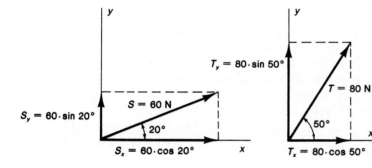

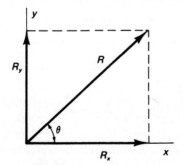

(c) components of forces

Figure 2.5

Example 2.2

The forces acting on a gusset plate at the intersection of a truss and a column are shown in Figure 2.6a. Find the resultant force acting on the gusset plate.

Solutions:
Graphical Solution:

A polygon of forces is constructed as shown in Figure 2.6b. The forces may be taken in any order, but, for convenience, the polygon started with the 70-kip vertical load, and then progressed in a clockwise manner around the gusset plate. Again, the length of each line is proportional to the magnitude of the force acting in that direction. Since the force polygon started at the point O and ended at the point E, the resultant force is represented in magnitude and direction by the line OE as shown. Thus,

$$R = 86 \text{ kips acting at } \theta = 54.4°$$

Analytical Solution:

If the components of the resultant force R are R_x and R_y in the x- and y-directions, respectively, then the component equations may be written as shown below. It is recommended that great care be taken to define which direction is being taken as positive in each force summation. A way to do this is by means of small directional arrows placed before each summation as shown below

$$\Sigma F_x \xrightarrow{+} = R_x = 10 - 50 \cos 30° - 40 \sin 15° + 20 \cos 45° + 80$$
$$= 50.49 \text{ kips}$$

$$\Sigma F_y \uparrow + = R_y = 70 + 50 \sin 30° - 40 \cos 15° + 20 \sin 45°$$
$$= 70.5 \text{ kips}$$

2.3 Moment of a Force

In addition to its tendency to move a body, a force also tends to produce rotation about any axis that does not intersect its line of action. An example of this is shown in Figure 2.7a, where the force F tends to rotate the plate around the x-axis. The force F is thus said to have a *moment* about the x-axis. The magnitude of the moment is given by the product of the force F and the perpendicular distance, d, from the line of action of the force to the x-axis. This is written

(2.9) $$M = F \cdot d \qquad \text{FORCE} \times \text{DISTANCE}$$

In three dimensions the direction of the moment is indicated by the "right hand rule," as shown in Figure 2.7b. If the fingers of the right hand are curled around the x-axis in the same direction as the force causing rotation, then the thumb of the right hand points in the direction the moment is considered to be acting. This is known as the "right-hand rule" sign convention. Screwing in a light bulb, for example, follows this rule. The resulting moment is usually indicated by a double-headed arrow as shown in Figure 2.7c.

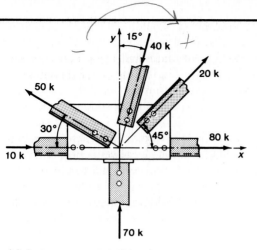

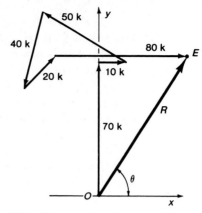

(a) forces on a gusset plate

Figure 2.6

(b) polygon of forces

Thus,

$$R = \sqrt{R_x^2 + R_y^2} = \sqrt{50.49^2 + 70.51^2} = 86.7 \text{ kips}$$

and

$$\theta = \tan^{-1}\left(\frac{R_y}{R_x}\right) = \tan^{-1}\left(\frac{70.51}{50.49}\right) = 54.4°$$

Hence,

$$R = 86.7 \quad \text{kips acting at} \quad \theta = 54.4°$$

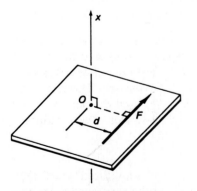

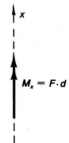

(a) a force F that tends to rotate
the plate about the x-axis

(b) right-hand rule

(c) moment representation in three dimensions

Figure 2.7 Moment of a force.

When representing moments of forces in two-dimensional systems (such as Figure 2.8*a*) it is considered that moments are being taken about the point *O*. It is as if an axis were coming out of the page at that point. For example, if one looked straight down along the *x*-axis of Figure 2.7*a*, then one would see the flat plate of Figure 2.8*a* where *the x-axis meets the plate at the point O.* Thus taking the moment of the force *F* about the point *O* is the same as taking the moment of the force about the *x*-axis. For two-dimensional systems it is sufficient to indicate whether the moment is causing clockwise or counterclockwise rotation of the plate about the point *O*. Depending on the sign convention used, either the clockwise or counterclockwise moment of a force about a point will be positive.

It is possible to sum the moments of different forces about a point in the same manner as we sum the components of forces to find a resultant force. An important theorem of statics is Varignon's theorem, which states that *the moment of a force is equal to the sum of the moments of the components of the force.* Thus if the force *F* of Figure 2.8*a* is resolved into its components, F_x and F_y, in the *x*- and *y*-directions, respectively, as indicated in Figure 2.8*b*, then the following relationship holds:

(2.10)
$$F \cdot d = F_x \cdot d_x + F_y \cdot d_y$$

where d_x and d_y are the perpendicular distances from the lines of action of the forces F_x and F_y to the point *O* respectively. This principle is valid for moments of forces in both two and three dimensions.

When moments of forces are required in three dimensions, it is necessary to recognize that a force has no moment about an axis if its line of action intersects or is parallel to that axis. Thus the force P_x in Figure 2.9*a*, which is acting in the negative *x*-direction through the point *A*(*O, b,* −*a*), has no moment about the *x*-axis since its line of action is parallel to that axis. As indicated by Figure 2.9*b*, however, it does produce a moment about both the *y*- and *z*-axes. Since the directions of moments M_y and M_z are in the positive *y*- and *z*-directions, respectively (using the right-hand rule), both moments are positive. If either arrow had been pointing in the opposite direction, that moment would have been negative.

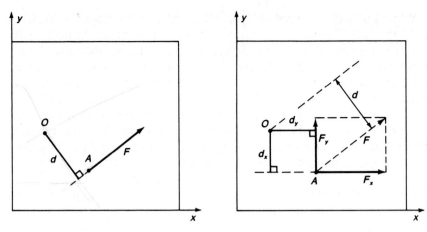

(a) moment about a point in a two-dimensional system

(b) moment of components about a point

Figure 2.8   *Varignon's theorem.*

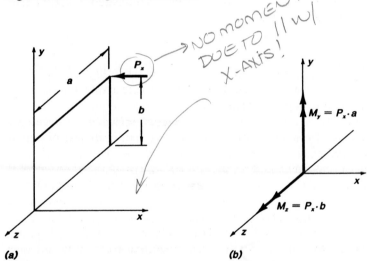

NO MOMENT DUE TO || W/ X-AXIS!

$M_y = P_x \cdot a$

$M_z = P_x \cdot b$

(a)

(b)

Figure 2.9   *Moments about three-dimensional axes.*

Example 2.3

A force F of 800 newtons is acting at A on a bracket as shown in Figure 2.10a. Find the moment of this force about the point B.

Solutions:

I. Using moment definition. The moment of the force F about the point B may be found using Equation 2.9. Before evaluating the moment it is necessary to find the distance d. Consider the triangle CAD in Figure 2.10b, in which

$$CD = CA \cdot \tan 30° = (50 \text{ mm}) \tan 30° = 28.86 \text{ mm}$$

thus

$$BD = BC - CD = 120 - 28.86 = 91.14 \text{ mm}$$

Now, using triangle BDE,

$$d = BE = BD \cdot \sin 60° = (91.14 \text{ mm}) \sin 60° = 78.93 \text{ mm}$$

The moment of the force F is now given by

$$M = F \cdot d = 800 \text{ N} \cdot 78.93 \text{ mm} = 63,140 \text{ N} \cdot \text{mm} = 63.14 \text{ N} \cdot \text{m} \downdownarrows$$

As indicated by the arrow, the force F causes a clockwise moment about the point B.

II. Using Varignon's theorem. The moment of the force may be evaluated using Equation 2.10, provided the values of the components of the force F are found first. The horizontal (or x-direction) and vertical (or y-direction) components of the force are given by

$$F_x = F \cdot \sin 30° = (800 \text{ N}) \sin 30° = 400 \text{ N}$$

and

$$F_y = F \cdot \cos 30° = (800 \text{ N}) \cos 30° = 692.8 \text{ N}$$

In evaluating the sum of the components of the moment about B, care must be taken concerning the sign of the components. If it is assumed that clockwise moments are positive, as indicated by the arrow direction below, the summation of the moments of the components of F gives

$$\Sigma M_B \; \circlearrowright + = M_B = F_y \cdot d_y - F_x \cdot d_x$$
$$= 692.8 \text{ N} \cdot 120 \text{ mm} - 400 \text{ N} \cdot 50 \text{ mm}$$
$$= 63,140 \text{ N} \cdot \text{mm} = 63.14 \text{ N} \cdot \text{m}$$

Since the force F_x would cause a counterclockwise moment about the point B, it is given a negative sign in the summation. Further, the positive sign of the answer indicates that the resultant moment about the point B is clockwise, i.e.,

$$M_B = 63.14 \text{ N} \cdot \text{m} \downdownarrows$$

$M = F \cdot D$

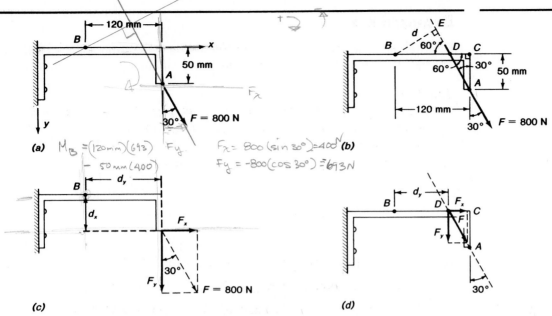

(a)

$M_B = (120mm)(693)\ F_y$

$-50\,mm\,(400)$

$F_x = 800\,(\sin 30°) = 400^N$ **(b)**

$F_y = -800\,(\cos 30°) = 693\,N$

(c)   **(d)**

Figure 2.10 *Moment of a force about a point in two dimensions.*

The fact that the same answer was obtained for Solution I and for Solution II is an indication of the fact that Varignon's theorem is true.

III. Using Varignon's theorem and the principle of transmissibility of forces. Using the principle of transmissibility of forces, the point of application of the force F may be moved from the point A to the point D, which is still on the line of action of F. With the force now considered to be acting at D, the force F may once again be divided into the horizontal and vertical components, F_x and F_y, respectively. Thus

$$F_x = F \cdot \sin 30° = 400\ N$$

and

$$F_y = F \cdot \cos 30° = 692.8\ N$$

Now, however, it is noted that the distance d_x to be substituted into Equation 2.10 is zero, since the line of action of the force component F_x goes through the point B. The distance d_y is given by

$$d_y = BD = 91.14\ mm \quad \text{(see Solution I)}$$

Equation 2.8 then becomes

$$\Sigma M_B \circlearrowright + = M_B = F_x \cdot d_x + F_y \cdot d_y$$
$$= 400 \cdot 0 + 692.8 \cdot 91.14 = 63,140\ N \cdot mm$$
$$M_B = 63.14\ N \cdot m \circlearrowright$$

Example 2.4

Find the moments about the x-, y-, and z-axes caused by the forces of 20 pounds and 30 pounds acting on the cantilever beam shown in Figure 2.11.

Solutions:

Let M_x, M_y, and M_z be the moments of the forces about the x-, y-, and z-axes respectively. Each of these must be evaluated separately.

a. Consider moments about the x-axis:

The 30-pound force has a line of action parallel to the x-axis. Therefore, its moment about that axis is zero. The 20-pound force has a line of action that intersects the x-axis (at the point C). Therefore, its moment about the x-axis also is zero. Therefore,

$$M_x = 0$$

b. Consider moments about the y-axis:

Both the 20-pound force and the 30-pound force will contribute to moment about the y-axis. Thus,

$$M_y = (20 \text{ lb}) \, OC - (30 \text{ lb}) \, EG$$
$$= 20 \cdot 10 - 30 \cdot 1 = 170 \text{ lb} \cdot \text{in}$$

The distances OC and EG are the perpendicular distances from the lines of action of the 20-pound and the 30-pound forces, respectively, to the y-axis. The negative sign on the moment of the 30-pound force indicates that it causes a moment about the y-axis such that the thumb

2.4 Couples and Resultant of a Force at a Distance

When two forces are of equal magnitude but act in opposite directions along parallel lines of action, they are said to form a *couple*. A couple tends to produce rotation only. It has no tendency to move a body in a straight line. Examples of couples abound in technology. One of the most common is shown in Figure 2.12*a*, where a pair of hands is shown turning a steering wheel. In order to turn the wheel to the left the left hand must pull downwards, while the right hand pushes upwards simultaneously. This is represented by the forces shown alongside. If the right hand is pushing upwards with the identical force F that the left hand is exerting downwards, then the hands are exerting a couple on the wheel. The *magnitude of the couple* is expressed in terms of the moment of the couple. Thus the couple of Figure 2.12*a* has a magnitude C given by

(2.11) $$C = F \cdot d$$

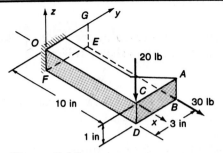

Figure 2.11

points in the negative y-direction, using the right-hand rule for moments. The 20-pound force causes a positive moment about the y-axis, because it requires the thumb to be pointing in the positive y-direction, using the right-hand rule. Since the answer is positive, it too will be such that the thumb is pointing in the positive y-direction.

c. Consider the moments about the z-axis:
The 20-pound force has a line of action parallel to the z-axis: its moment about the z-axis is zero. The 30-pound force produces a moment about the z-axis given by

$$M_z = (-30 \text{ lb}) \, EF = -30 \cdot 3 = -90 \text{ lb} \cdot \text{in}$$

The negative sign of M_z is again derived using the right-hand rule.

Other examples of couples shown in Figure 2.12 are a planetary gear assembly (where two couples are acting) and a specific loading on a beam.

If a force tends to move a point in a straight line and a couple tends to produce rotation of a body, then what is the action of the force F acting on the plate of Figure 2.13, at some distance, d, from the point O, as shown in Figure 2.13a? Since two equal and opposite forces acting along the same line of action will have no net effect on a body's tendency to move in a straight line or to rotate, two equal and opposite forces, each of magnitude F, may be introduced at O, as shown in Figure 2.13b, without in any way affecting the plate. The original force F may now be considered to form a couple when acting in concert with a force of magnitude F acting in the opposite direction at O. Thus, as shown in Figure 2.13c, the net effect of the force acting at a distance d from O is the same as a force F acting at O, plus a couple of magnitude C. The force F acting at a distance therefore tends to cause the plate to move in the direction of F, and also to rotate around the point O in a counterclockwise direction.

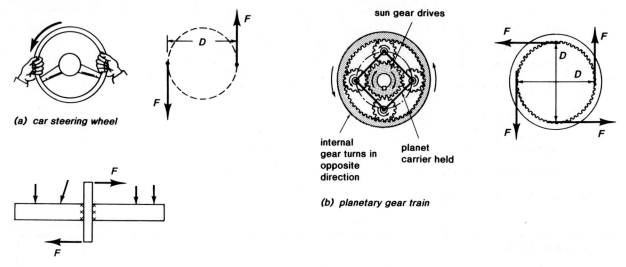

(a) car steering wheel

sun gear drives

internal
gear turns in
opposite
direction

planet
carrier held

(b) planetary gear train

(c) forces on a beam

Figure 2.12 *Examples of couples.*

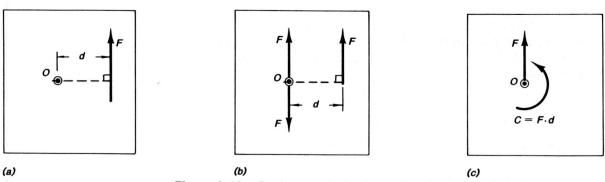

(a)

(b)

$C = F \cdot d$

(c)

Figure 2.13 *Replacement of a force at a distance by a force and a couple.*

Since a couple is specified by the magnitude of its moment, there are an infinite number of arrangements of forces that produce *equivalent couples.* This fact is illustrated by the forces on the beam of Figure 2.14. The force systems of Figures 2.14*a, b,* and *c* may all be replaced by an equivalent counterclockwise couple of size 50 N · m, as shown in Figure 2.14*d.* Notice that the force systems illustrated show forces acting in different directions and even have different points of application on the beam. In each case, however, the forces are in opposite directions, acting along parallel lines of action, and thus they form couples. The couple *C* of Figure 2.14*d,* unlike a force, has no fixed point of application and may be taken to act anywhere on the beam.

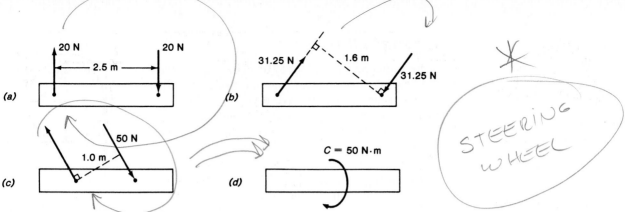

Figure 2.14 *Equivalent couples.*

Example 2.5

The variable speed pulley with center O, shown in Figure 2.15a, is rotated by a motor on the shaft A. If the V-belt is loose along the top, and a force of 1,000 pounds acts in it as it returns underneath from the pulley to the motor, what is the effect of the V-belt force at the center of the pulley at point O?

Solution:
A simplified drawing of the pulley is shown, complete with forces acting on it, in Figure 2.15b, and these forces have been replaced by a force and a couple at O in Figure 2.15c. With the pulley having a radius of nine inches, the values of F and C are

$$F = 1,000 \text{ lb} \leftarrow$$

and

$$C = F \cdot r = (1,000 \text{ lb})(9 \text{ in}) = 9,000 \text{ lb} \cdot \text{in} \mathbf{\searrow}$$

It is noted that the pulley has a tendency both to rotate and to be pulled toward the motor.

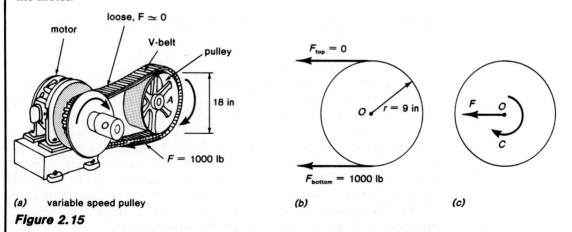

(a) variable speed pulley

(b)

(c)

Figure 2.15

Example 2.6

Four round pegs are attached to a board as shown in Figure 2.16a. Two strings are threaded around the pegs and pulled as indicated with forces F_1 (of 50 newtons) and F_2 (of 60 newtons).

a. Determine the resultant couple acting on the board.
b. If this resultant is to be provided by two forces R acting at A and B, as shown in Figure 2.16b, find the value of R.

Solution:

a. The equivalent couple, C, is given by

$$C = F_1 \cdot 0.2 + F_2 \cdot 0.3$$
$$= 50 \cdot 0.2 + 60 \cdot 03 = 28 \text{ N} \cdot \text{m}$$

As indicated, both the forces F_1 and F_2 produce counterclockwise couples, and the resultant couple will therefore also be in a counterclockwise direction.

b. It is first determined that the forces R indicated in Figure 2.16b would form a couple of magnitude $R \cdot d$, and that this couple would act in a counterclockwise direction with the forces in the direction given. Since the couple C indeed acts in a counterclockwise direction, the forces R are given in the correct direction. In order to compute the magnitude of the couple formed by the forces R, it is necessary to find the distance d, given by

$$d = AB \cdot \sin 60° = (0.7 \text{ m}) \sin 60° = 0.606 \text{ m}$$

Then, in order that the couple C be provided by the forces R,

$$C = R \cdot d$$

thus,

$$R = \frac{C}{d} = \frac{28 \text{ N} \cdot \text{m}}{0.606 \text{ m}} = 46.19 \text{ N}$$

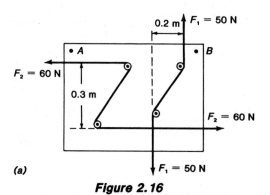

(a)

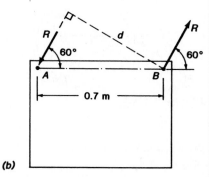

(b)

Figure 2.16

Problems

2.1   Find the rectangular components of the force shown in Figure P 2.1.

2.2   Find the rectangular components of the force shown in Figure P 2.2.

2.3   For the concurrent force system shown in Figure P 2.3, find the resultant force, its magnitude, and its rectangular components.

2.4   Repeat Problem 2.3 for the system shown in Figure P 2.4.

2.5   The forces acting on a plate are shown in Figure P 2.5. Find the moment of the forces about points *A* and *D*.

2.6   The forces acting on the bar are shown in Figure P 2.6. Calculate
     a.   The rectangular components of the moment at *A*.
     b.   The rectangular components of the moment at *O*.

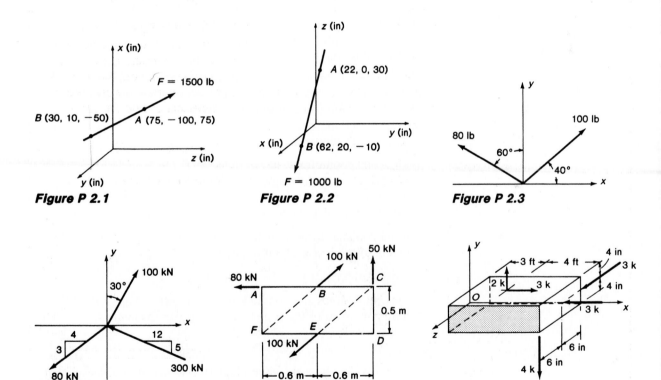

Figure P 2.1   Figure P 2.2   Figure P 2.3

Figure P 2.4   Figure P 2.5   Figure P 2.6

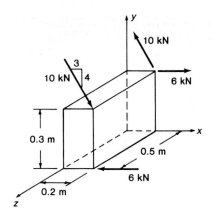

Figure P 2.7

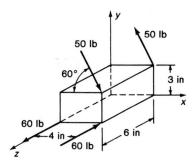

Figure P 2.8

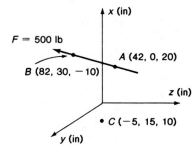

Figure P 2.9

2.7   The forces as shown in Figure P 2.7 are acting in the xy-plane and in a plane parallel to the xy-plane. Find the rectangular components of the equivalent couple and magnitude by
     a. The direct approach.
     b. Varignon's theorem.

2.8   Two couples acting on a rectangular box are shown in Figure P 2.8. The 50-pound forces are in the xy-plane and in a plane parallel to the xy-plane. Find an equivalent couple using
     a. The direct approach.
     b. Varignon's theorem.

2.9   For the 500-pound force shown in Figure P 2.9 , calculate the moment with respect to (a) the origin and (b) the point C (−5, 15, 10). Resolve the force into its rectangular coordinates at point A, then at point B, and give the magnitude. Units are in inches.

2.10   For the coplanar force systems shown in Figure P 2.10, find the moment of the forces about point A.

2.11   Repeat Problem 2.10 for the system of Figure P 2.11.

2.12   A system of forces and couple vectors is shown in Figure P 2.12. Find the moment of the system about the axes through
     a. The origin.
     b. Point A.

2.13   The wooden bracket shown in Figure P 2.13 has three forces acting on it. The 30-pound force acts parallel to the z-axis, the 20-pound force acts parallel to the x-axis, and the 25-pound force acts in a plane parallel to the yz-plane. Find the moment about the axes passing through point A, giving the magnitude and rectangular components.

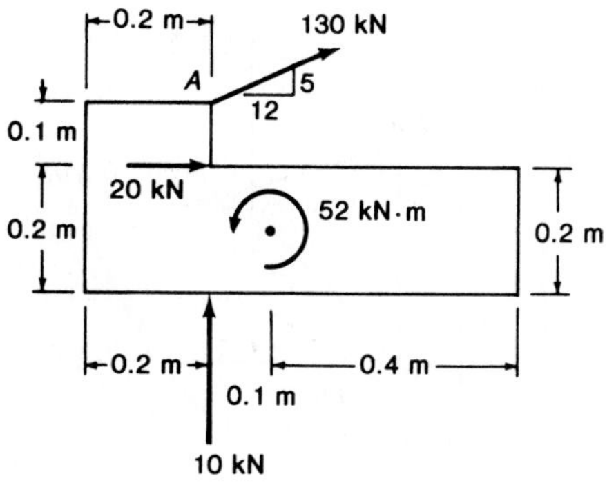

Figure P 2.10

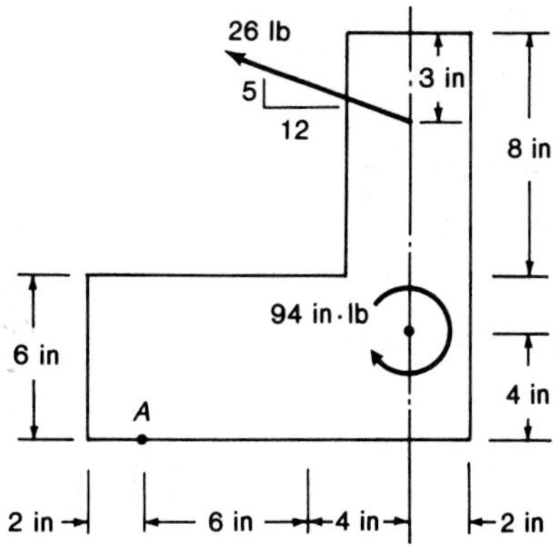

Figure P 2.11

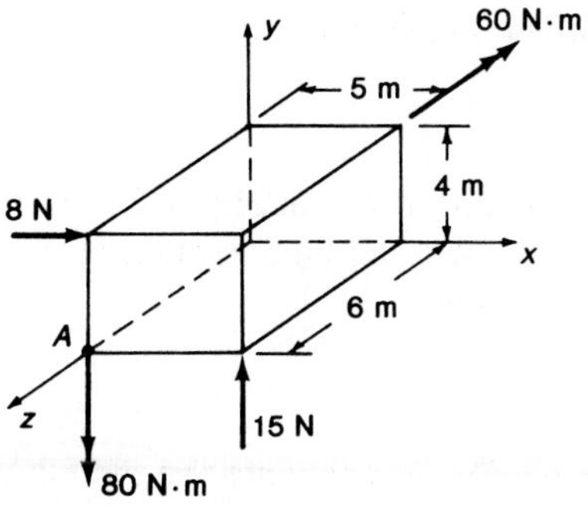

Figure P 2.12

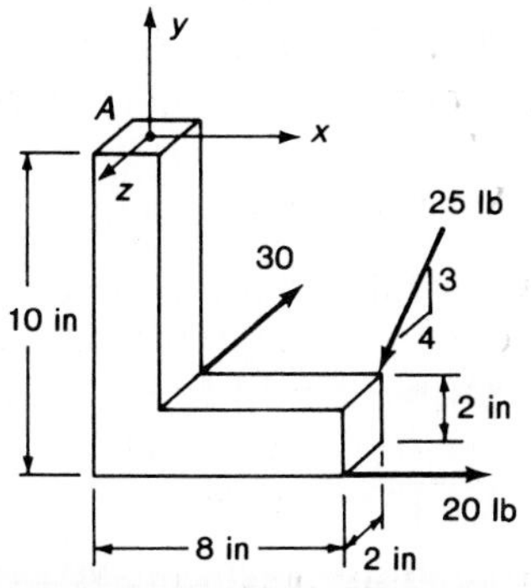

Figure P 2.13

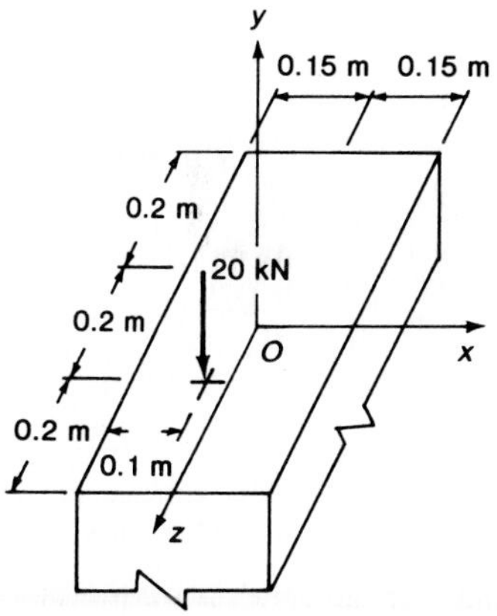

Figure P 2.14

2.14 A 20-kilonewton force acting upon a 0.3- by 0.6-meter column is shown in Figure P 2.14. Calculate the moment about the axes passing through the center of the column (i.e., about the point *O*).

2.15 A system of forces and couple acting on a cantilever beam is shown in Figure P 2.15. Find the moment about the axes passing through point *A*.

2.16 For the force system shown in Figure P 2.16, find an equivalent force couple system at point *A*. The 6-kilonewton force acts in a plane parallel to the *yz*-plane.

2.17 For the coplanar system of forces acting on the flat, rectangular plate *ABCDEF* shown in Figure P 2.17, find an equivalent force-couple system at points *A* and *D*, specifying all necessary magnitudes and directions.

find M_A

cantilever

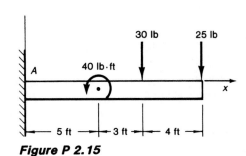

Figure P 2.15

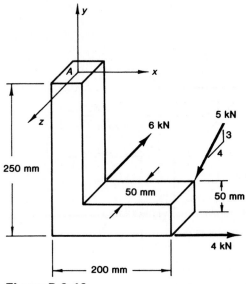

Figure P 2.16

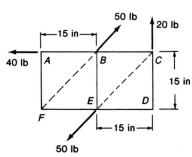

Figure P 2.17

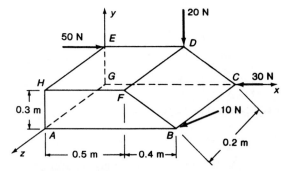

Figure P 2.18

2.18 Determine the resultant moment about point *A* on the car ramp shown in Figure P 2.18, with the 10-newton force acting perpendicular to surface *BCDF* and acting in the *xy*-plane.

2.19 Determine the resultant moment about the origin of the forces shown in Figure P 2.19.

2.20 For the coplanar system of forces shown in Figure P 2.20, determine the resultant moment about point *B*.

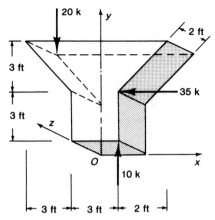

Figure P 2.19

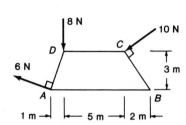

Figure P 2.20

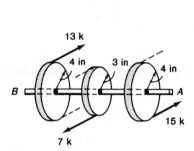

Figure P 2.21

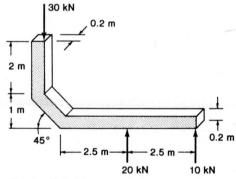

Figure P 2.22

2.21   For the wheels shown on the shaft in Figure P 2.21, determine the moment about the shaft axis AB caused by the forces acting on the wheels.

2.22   Figure P 2.22 represents half of an internal rib for a barge. The forces acting on it are resultant forces from barge load (30 kilonewtons), water waves (15 kilonewtons), and water buoyancy (20 and 10 kilonewtons). Find the resultant moment of these forces about the point O.

Reactions and Equilibrium

3

Objectives

1. *To be able to isolate parts of an object or structure and draw a free body diagram of each part.*

2. *To recognize the type of force interaction between structural or mechanical elements.*

3. *To find the necessary conditions that ensure static equilibrium of the various parts of structures or mechanical devices.*

4. *To understand the role of friction in static equilibrium.*

5. *To be able to compute the equivalent concentrated forces having the same effect as distributed forces.*

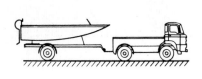

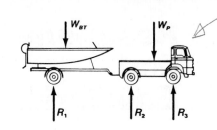

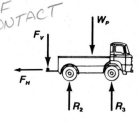

3 POINTS OF CONTACT

| (a) pickup towing a boat on a trailer | (b) free body diagram of boat and trailer | (c) free body diagram of pickup |

Figure 3.1   *Illustrations of a free body diagram.*

3.1 Introduction

A structure or machine is analyzed and designed in a step-by-step process in which the parts of the entire article are considered separately. The parts of the structure or machine all are assumed to be at rest with respect to each other: a condition known as *equilibrium*. Newton's first law defines the mathematical conditions necessary to ensure equilibrium and enables the designer to find the forces acting on each part inside the whole structure.

Before the mathematical conditions of equilibrium can be applied, however, both the whole structure or machine and the parts of it must be carefully isolated. If the entire article is being isolated from its supports, then the designer must be sure to take account of all of the forces that the supports allow to act on it. The isolated structure or machine is known as a *free body* because it is now free of external restraint. Instead, the external restraints on it have been replaced by equivalent forces.

The same principle may be applied to the *parts* of a structure or machine. The first step in analyzing and designing a crane, for example, is to separate the crane as a whole from its ground supports as indicated above. However, it is also possible to isolate any part of the crane, maybe a main support beam, from the rest of the crane in the same manner. A free body diagram of the beam must depict all the effects of the rest of the crane on the individual beam. The purpose in drawing a free body diagram of part of a system is to find the forces acting within the system, but external to the isolated part.

3.2 Free Body Diagram

A free body diagram is a diagram of an object or part of an object on which all external forces acting are indicated. The drawing of a correct free body diagram is the most important step in the process of analyzing and designing a structural or machine element. The free body diagram shows all relevant data, particularly all the forces applied.

Consider, for example, the system shown in Figure 3.1: a pickup that is connected to a boat on a trailer. A free body diagram of the entire pickup-and-trailer system is shown in Figure 3.1b. The forces W_P and W_{BT} denote the weights of the pickup and boat-with-trailer, respectively. The forces R_1, R_2, and R_3 represent the reaction forces exerted by the ground on the pickup and trailer tires. A free body diagram of the pickup alone is shown in Figure 3.1c. Once again the reaction forces on the pickup tires and the pickup weight are shown on the diagram. In this case, however, the symbols F_V and F_H are included to

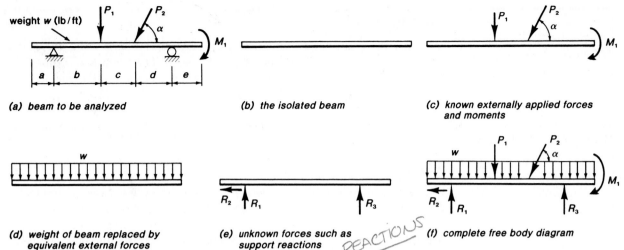

(a) beam to be analyzed

(b) the isolated beam

(c) known externally applied forces and moments

(d) weight of beam replaced by equivalent external forces

(e) unknown forces such as support reactions *REACTIONS*

(f) complete free body diagram

Figure 3.2 *Steps in drawing a free body diagram.*

represent the vertical and horizontal components of the force exerted on the pickup by the trailer (through the towbar).

The steps in drawing a free body diagram are illustrated by means of the beam example shown in Figure 3.2a, where it is assumed that the forces P_1 and P_2 and the applied moment M_1 are known. In addition, the dimensions a, b, c, d, and e are given, as are the angle α and the spread weight of the beam, w pounds per foot. The beam is taken for an example only; the same procedure should be followed in drawing any free body diagram.

Steps in Drawing a Free Body Diagram:

Step 1.   Decide which free body is desired and isolate it from the rest of the structure. If the beam of Figure 3.2a is to be considered, then it is isolated as shown in Figure 3.2b.

Step 2.   Show *all* external forces that act on the free body. These would include:

   a.   All known externally applied forces and moments. For the example considered, these are the forces P_1 and P_2 and the moment M_1 as shown in Figure 3.2c.

   b.   The weight of the body, if applicable. The neglect of the weight forces is a very common error in the drawing of free body diagrams. The equivalent spread (or distributed) load replacing the beam weight is shown in Figure 3.2d.

   c.   Any forces of unknown magnitude acting on the body. The most common unknown forces are the reaction forces, shown in Figure 3.2e as R_1, R_2, and R_3, which represent the forces exerted on the beam by the supports.

Step 3.   Review all stages of Steps 1 and 2 to ensure that the complete free body diagram shows all the external forces acting on the free body, as shown in Figure 3.2f.

Example 3.1

Consider the crane shown in Figure 3.3*a*. The crane is supporting a weight of two tons hanging from the midspan of the beam. The beam supporting the lift gear weighs 100 pounds per foot, and all other parts of the crane are of negligible weight.

Part a. Draw a free body diagram of the entire crane.
Part b. Draw a free body diagram of the beam supporting the weight.

Solution:

Part a. A free body diagram of the entire crane is shown in Figure 3.3*b*. This diagram can be used to determine the reaction forces exerted by the rails in order to stop the crane from sinking into the ground. As shown by the forces marked R_1 to R_{10}, the reaction forces may act in several directions. At the outset the sizes of these reaction forces are unknown, but they must still be shown on the free body diagram. In Figure 3.3*b* the hanging load of two tons is represented by a force of two tons acting downward at the center of the beam. This force is an external force acting on the crane. Notice also in Figure 3.3*b* that the weight of the support beam is replaced by an equivalent spread load of 100 pounds per foot acting along the beam. The effect on the

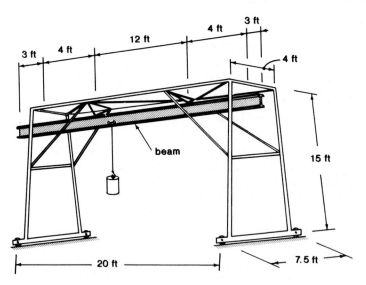

(a) *tramrail gantry crane*
Figure 3.3   *Free body diagram of a crane.*

crane (including the beam) of the weight of the beam is the same as that of a weightless beam with an external spread force of 100 pounds per foot acting on it.

Part b. A free body diagram of the support beam is shown in Figure 3.3c. Once again the hanging weight of two tons has been replaced by an equivalent external force of two tons, and the weight of the beam has been replaced by an equivalent spread force of 100 pounds per foot. The unknown quantities N_1, N_2, and N_3 represent the forces of the crane support on the beam required to keep the beam from falling. Note that in Figures 3.3b and 3.3c all relevant dimensions are shown, since it was established in Chapter 2 that not only the magnitude of a force is important, but also where it acts.

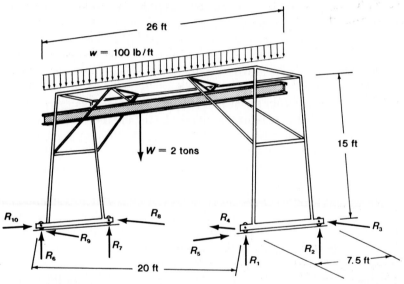

(b) free body diagram of crane

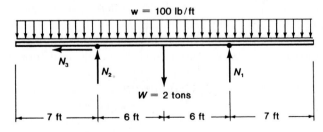

(c) free body diagram of beam

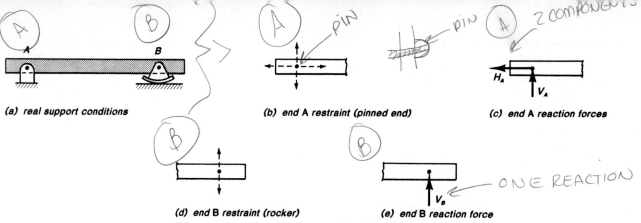

(a) *real support conditions*

(b) *end A restraint (pinned end)*

(c) *end A reaction forces*

(d) *end B restraint (rocker)*

(e) *end B reaction force*

Figure 3.4 *Replacing reactions by equivalent forces.*

3.3 Support Conditions, Real and Assumed

From the explanation of free body diagrams we learned that an important part of drawing one is to replace the support conditions by the appropriate forces. At the start of the analysis, these forces are usually unknown in magnitude. However, the type of support indicates the types of forces to be considered, and often also their line of action.

As an example, consider the beam of Figure 3.4a, which is supported at A on a frictionless pin that goes right through the beam and the support brackets that are rigidly fixed to the base at their lower edge. At the end B, the beam is supported by another frictionless pin, which in this case goes through the beam and a rocker that rests on the base. Such a beam—free to rotate on supports at both ends—is called a simply supported beam. The lower edge of the support is not fixed to the base, but free to rock. (This type of support is very common at the ends of highway bridges.) In order to consider the equivalent forces offered by these supports, consider each end separately. As shown in Figure 3.4b the pinned support at end A offers end restraint in that it prevents end A from moving up or down or in either direction horizontally. Since the end A cannot move vertically, there may be a component of reaction force in the vertical direction. Similarly, since the end A cannot move horizontally, there may be some horizontal reaction force at A to prevent this movement. As shown in Figure 3.4c, the support A is thus replaced by two forces, V_A and H_A, in the vertical and horizontal directions respectively, since movement of the end A is prevented in both horizontal and vertical directions. At present we do not know the directions of these forces (i.e., whether H_A acts to the left or right), and, in fact, one or more of these forces may be zero. However, it is usual to assume a direction, as was done in Figure 3.4c, for later convenience. Since the beam is completely free to rotate at end A, there can be no moment exerted at A by the support.

At the end B of the beam the rocker support prevents movement of the end only in the vertical direction, as indicated by Figure 3.4c. The end is thus free to move horizontally and/or rotate about the pin. Since only vertical movement is restrained, the reaction at end B must have a vertical line of action. Although

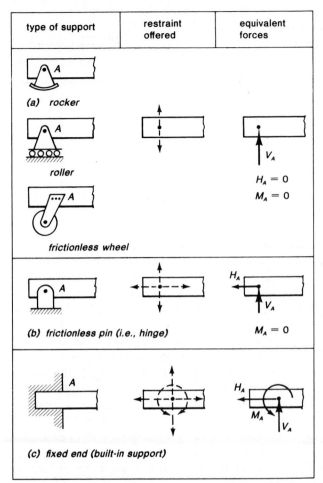

type of support	restraint offered	equivalent forces

(a) rocker

roller

frictionless wheel

$H_A = 0$

$M_A = 0$

(b) frictionless pin (i.e., hinge)

$M_A = 0$

(c) fixed end (built-in support)

Figure 3.5 *Support conditions for planar systems.*

at present it is not known whether the reaction at B acts vertically upward or vertically downward, a reaction force V_B has been shown acting upward on the beam at B on Figure 3.4d. Subsequent calculations will show whether or not the assumed direction is correct.

The two support conditions considered above are commonly encountered. These and one other two-dimensional system are summarized in Figure 3.5. The third condition, shown in Figure 3.5c, is that of the fixed end. In addition to preventing movement in the horizontal and vertical directions, the support at a fixed end (a "built-in" support) also prevents any rotation of the end. As shown in the figure, there are thus three possible restraining actions at the support— the horizontal and vertical reaction forces (H_A and V_A, respectively) and the moment M_A, which must act to prevent rotation of the end.

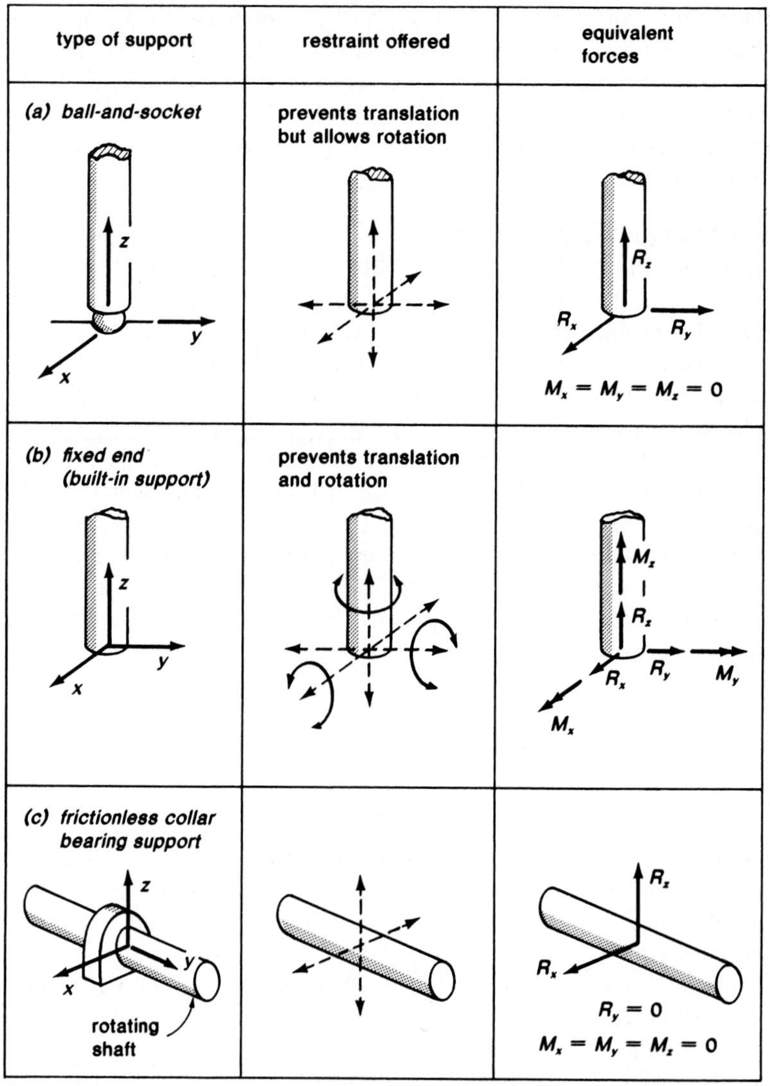

type of support
restraint offered
equivalent forces
(a) ball-and-socket
prevents translation but allows rotation
z
y
x
R_z
R_x
R_y
$M_x = M_y = M_z = 0$
(b) fixed end (built-in support)
prevents translation and rotation
z
y
x
M_z
R_z
R_x
R_y
M_y
M_x
(c) frictionless collar bearing support
z
y
x
rotating shaft
R_z
R_x
$R_y = 0$
$M_x = M_y = M_z = 0$

Figure 3.6 *Common three-dimensional support conditions.*

A similar approach is adopted for considering three-dimensional support conditions, which are encountered when three-dimensional systems are analyzed, as in the case of the crane of Figure 3.3. Common three-dimensional supports are shown in Figure 3.6. As an illustration of the derivation of the support reactions, consider the fixed end support of Figure 3.6*b,* which may be welded to the base or rigidly embedded in the base. It is first noted that the support is restrained from moving horizontally in either the *x*- or *y*- directions, and that it cannot be pulled off the base in the *z*- direction. Since movement in all three of these directions is prevented, there must be restraining forces offered in all three of these directions.

Furthermore, the support cannot be rotated about the *x*- or *y*-axes, or twisted about the *z*-axis, since it is rigid. Since all three of these rotational possibilities are prevented, there must be restraint moments offered by the support about all three axes. These are shown as M_x, M_y, and M_z, acting about the *x*-, *y*-, and *z*-axes, respectively, using the right-hand rule sign convention introduced in Chapter 2.

3.4 Equilibrium Conditions

A rigid body is said to be in *equilibrium* when the resultant of all forces and moments acting on it is zero. Under such conditions, Newton's first law states that a body will continue in its state of rest or uniform motion in a straight line. However, in statics one is concerned only with the mechanics of bodies at rest or traveling at a constant velocity. Hence only the case of bodies in stationary equilibrium will be considered. If a body is in *static equilibrium,* it shows no tendency to move in any direction nor to rotate (or twist) about any axis. For two-dimensional structures or machines this condition may be ensured by two force equations and one moment equation. If the body of Figure 3.7 is acted upon by forces and moments as shown, then the static equilibrium can be guaranteed only if there is no resultant force in any two perpendicular directions and no resultant moment about any point. These three conditions may be summarized as

(3.1) $$\Sigma F_H = 0$$

(3.2) $$\Sigma F_V = 0$$

(3.3) $$\Sigma M_A = 0$$

Although given in terms of the horizontal and vertical directions, the forces may be totaled in any two perpendicular directions. Often, however, the horizontal and vertical directions are the most convenient. Likewise the resultant moment may be shown to be zero about any convenient point.

Throughout this text, the positive sense of the summed forces or moments will be indicated by small arrows.

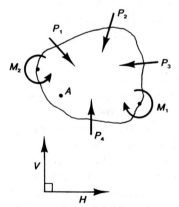

Figure 3.7 *General two-dimensional rigid body under the action of forces and moments.*

Example 3.2

Find the support forces acting on the beam of Figure 3.8a. The crane rail weighs 100 pounds per foot and supports a hanging load of two tons at mid-span.

Solution:

In Figure 3.8a, the loading and support conditions are shown. The support A is a hinged support and the support B a roller support.

If the weight of the beam is 100 pounds per foot and the total length of the beam is 26 feet, then the total weight of the beam is given by

$$W = 100 \times 26 = 2,600 \text{ lb}$$

For the purposes of finding the support reactions, the weight of the beam may be represented by a single force of 2,600 pounds acting at the center of the beam. Replacing the support conditions by equivalent forces (see Figure 3.5), the free body diagram of Figure 3.8b is derived, showing all external forces acting on the beam.

There are now three unknown forces on the free body: V_A, H_A, and V_B. All three have been given assumed directions, although really only the lines of action are known. However, there are also available three equations of equilibrium (Equations 3.1, 3.2, and 3.3), since the beam is in static equilibrium, i.e., it is not moving under the action of the applied forces.

In order to satisfy these equations, the components of the forces are first totaled in the horizontal and vertical directions. For convenience, it is suggested that small arrows be placed over the summations to indicate the positive direction of force summation. The first force equilibrium equation then becomes

$$\Sigma F_H \overset{+}{\rightarrow} = 0$$
$$-H_A = 0$$

Thus

$$H_A = 0 \text{ (i.e., no horizontal reaction)}$$

The second equation becomes

$$\Sigma F_V \uparrow + = 0$$
$$V_A + V_B - 2,600 - 4,000 = 0$$

and

$$V_A + V_B = 6,600 \text{ lb}$$

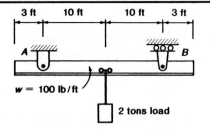

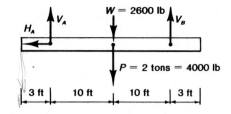

(a) load and support conditions

(b) free body diagram

Figure 3.8 *Beam of Example 3.2.*

We also have a moment equilibrium equation (Equation 3.3), which may be applied about any point. For convenience this equation is applied about the point A, and the positive direction of moment summation is shown (taken here as clockwise). If the moments are taken about the point A, then the forces V_A and H_A have lines of action through the point A, and thus have no moment about the point A. (When taking moments about A, one is really taking moments about an axis perpendicular to the page through point A).

$$\Sigma\, M_A = 0 \;\circlearrowright +$$
$$P \cdot 10 + W \cdot 10 - V_B \cdot 20 = 0$$

i.e.,

$$4{,}000(10) + 2{,}600(10) - V_B \cdot 20 = 0$$

Thus

$$V_B = 3{,}300 \text{ lb}$$

Substituting in the force equilibrium equation above produces also

$$V_A = 3{,}300 \text{ lb}$$

Thus the equilibrium equations have been used to find the unknown support reactions at A and B, which are given by

$$H_A = 0$$
$$V_A = 3{,}300 \text{ lb} \uparrow$$
$$V_B = 3{,}300 \text{ lb} \uparrow$$

Since the equations produced both V_A and V_B as positive numbers, the directions assumed for the forces in the free body diagram of Figure 3.8b must have been correct. If either V_A or V_B had acted in the opposite direction, the answer would have been negative.

Example 3.3

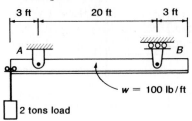

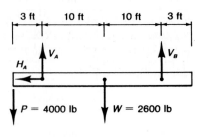

(a) load and support conditions

(b) free body diagram

Figure 3.9   Beam of Example 3.3.

Consider the same beam as in Example 3.2. As before, the beam weighs 100 pounds per foot and is supported at A by a pinned support and at B by a roller support. Now, however, the hanging load of two tons has been moved to the very end of the beam, as shown in Figure 3.9a. Find the unknown support forces at A and B.

Solution:
As in Example 3.2 the weight of the beam is given by

$$W = 100 \text{ lb/ft} \times 26 \text{ ft} = 2{,}600 \text{ lb}$$

Example 3.4

The frame of Figure 3.10a represents a typical cross section of a warehouse. The horizontal forces on the frame are wind loads. The weight of the frame is taken to be 1.5 kilonewtons per meter, acting at the center of each member. If the crane member BF weighs 12 kilonewtons, and it is lifting its rated capacity of 50 kilonewtons at the point of H, find the support reactions at A and G.

Solution:
The first requirement is to draw a free body diagram of the frame. In order to do this, it is necessary to find the weight of each portion of the frame. The weight of each of the vertical legs AC and EG, each having a length of four meters, is given by

$$W_{AC} = W_{EG} = 4 \times 1.5 = 6.0 \text{ kN}$$

The weight of each of these members is represented on the free body diagram of Figure 3.10b by an equivalent force acting at midheight of the member. The weight of each of the sloping roof members of the frame, CD and DE, requires that the length of each member be found first. These lengths are given by the length of the hypotenuse of a right triangle as

$$L_{CD} = L_{DE} = \sqrt{2.5^2 + 6^2} = 6.5\text{m}$$

which may be taken to be acting at the center of the beam. Replacing the supports by equivalent forces, we obtain the free body diagram of Figure 3.9*b*. Applying the equilibrium equations to the forces on the free body diagram yields the equations

$$\Sigma F_H = 0 \xrightarrow{+}$$

$$-H_A = 0$$

$$\Sigma F_V = 0 \uparrow +$$

$$V_A + V_B - 2,600 - 4,000 = 0$$

$$\Sigma M_A = 0 \; \circlearrowright +$$

$$-4,000(3) + 2,600(10) - V_B \cdot 20 = 0$$

Solving these equations gives the support reaction forces as

$$H_A = 0$$

$$V_A = 5,900 \text{ lb}$$

$$V_B = 700 \text{ lb}$$

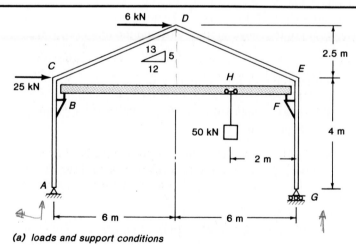

(a) *loads and support conditions*

Figure 3.10 *Portal frame of Example 3.4.*

Thus, the weight of each member is given by

$$W_{CD} = W_{DE} = 6.5 \times 1.5 = 9.75 \text{ kN}$$

Example 3.4 (Continued)

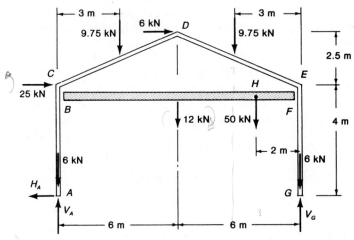

(b) free body diagram

Figure 3.10 (Continued)

Again, the weight of each of these members is replaced by the equivalent force acting at the center of each member. The weight of the crane member BF is also replaced by an equivalent force of 12 kilonewtons acting downward at midspan, and the hanging load by an equivalent force of 50 kilonewtons acting downward on the member BF at H. Inclusion of the horizontal wind forces at C and D, and replacement of the hinge support at A and the roller support at G by their equivalent forces (from Figure 3.5) completes the free body diagram of Figure 3.10b.

Since there are three unknown forces on the frame— V_A, H_A, and V_G— and there are three available equations of equilibrium, it is possible to find these unknown support forces. Application of the equilibrium equations to the free body diagram of Figure 3.10b yields

$$\Sigma F_H = 0 \xrightarrow{+}$$

$$-H_A + 25 + 6 = 0$$

$$\Sigma F_V = 0 + \uparrow$$

$$V_A + V_G - 6.0 - 9.75 - 9.75 - 6.0 - 12.0 - 50.0 = 0$$

Thus

$$H_A = 31.0 \text{ kN} \leftarrow$$

and

$$V_A + V_G = 93.5 \text{ kN}$$

Selection of a suitable point about which moments can be taken will simplify the moment calculation greatly. If moments are taken about the point A, the forces V_A and H_A will have no moment about A, since their lines of action both go through A. If moments are taken about G, the forces H_A and V_G will have no moment about G, since their lines of action both go through G. There is, therefore, no preference in deciding which of these two points to take moments about. If, however, the moment equilibrium equation is applied about point D, all three of the unknown forces H_A, V_A, and V_G will enter the equation. For convenience, the moment equilibrium equation is applied at point A. The force terms are given first, followed by perpendicular distances from A in parentheses.

$$\Sigma \, M_A = 0 \text{⤴}+$$
$$6(0)+25(4)+9.75(3)+6(4+2.5)+12(6)+9.75(6+6-3)$$
$$+50(6+6-2)+6(6+6)-V_G(6+6) = 0$$

Solving this equation gives

$$V_G = 75 \text{ kN↑}$$

Substitution into the equation derived above,

$$V_A + V_G = 93.5 \text{ kN}$$

produces

$$V_A = 93.5 - 75 = 18.5 \text{ kN↑}$$

Since the signs of all three forces H_A, V_A, and V_G are positive, the directions assumed for these forces in the free body diagram of Figure 3.10b are correct.

As a check, the moment equilibrium equation is now applied about the point G:

$$\Sigma \, M_G = 0 \text{⤴}+$$
$$V_A(6+6)-6(6+6)+25(4)-9.75(6+6-3)+6(4+2.5)-12(6)$$
$$-9.75(3)-50(2)+6(0) = 0$$

Solving this equation yields

$$V_A = 18.5 \text{ kN↑}$$

which is exactly the value derived before.

Example 3.5

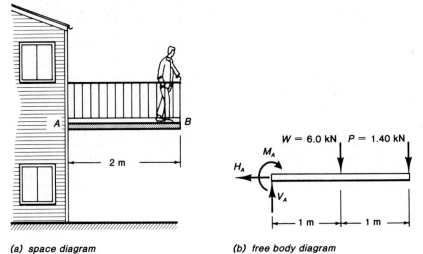

(a) space diagram   (b) free body diagram

Figure 3.11   *Cantilevered balcony of Example 3.5.*

An apartment balcony AB is cantilevered out from the side of an apartment building as shown in Figure 3.11a. The weight of the balcony and railing is 3.0 kilonewtons per meter. For design purposes, two people, each weighing 700 newtons, are to be considered leaning against the outside railing. Find the forces and moment exerted on the wall of the building by the balcony.

Solution:

It is first noted that the balcony must be rigidly fixed to the apartment wall at A. Not only is it impossible for the end A to move either horizontally or vertically, but it is also impossible for the end A to rotate. The end A is thus a fixed end (or built-in support), considered in Figure 3.5c. The end B, however, is able to move either horizontally or vertically, or even to rotate. The end B is thus a "free end," with no restraining forces or moments. Such a beam, rigidly fixed to a support at one end (e.g., end A), and free at the other end (e.g., end B) is known as a cantilever beam.

Figure 3.11a constitutes the space diagram, for it shows the physical situation. However, before the forces exerted by the wall on the beam can be found, a free body diagram of the beam must be constructed as shown in Figure 3.11b. Note that in Figure 3.11b the two people required to be on the balcony have been replaced by an equivalent force, P, given by

$$P = 2 \times 700 \text{ N} = 1,400 \text{ N} = 1.40 \text{ kN}$$

Further, the weight of the balcony and handrail (30 kN/m) is replaced by a weight force, W, acting at the center of the beam, and given by

$$W = 2\,\text{m} \times 3.0\,\text{kN/m} = 6.0\,\text{kN}$$

Since the end A is completely restrained, it has been replaced by the forces V_A and M_A and the moment M_A. All three of these are at present of unknown magnitude. The chosen directions of the forces and moment are completely arbitrary.

The free body of the beam must be in equilibrium, or else the balcony would collapse. We can thus apply the three equilibrium equations to find the three unknown parts of the reaction at A.

$$\Sigma F_H = 0 \overset{+}{\rightarrow}$$
$$-H_A = 0$$

Thus

$$H_A = 0$$

$$\Sigma F_V = 0 \uparrow +$$
$$V_A - 6.0 - 1.40 = 0$$

and

$$V_A = 7.40\,\text{kN}$$

If the moment equilibrium equation is applied about the point A, the forces H_A and V_A have no effect, since both of them have lines of action through A.

$$\Sigma M_A = 0 \,\overset{+}{\circlearrowleft}$$
$$M_A + W(1) + P(2) = 0$$
$$M_A + 6.0(1) + 1.40(2) = 0$$
$$M_A = -8.80\,\text{kN} \cdot \text{m}$$

The negative sign means that the moment at A was originally assumed (Figure 3.11b) in the wrong direction. The correct moment at A is therefore in a counterclockwise direction, i.e.

$$M_A = 8.80\,\text{kN} \cdot \text{m} \,\circlearrowright$$

The determination of reaction forces and moments as demonstrated in the previous example problems is an essential first step in most computations that will be introduced later in this book. To simplify the work, a table of solutions for many loading situations on cantilevered and two-supported structures has been included in Appendix B. Early familiarity with this table will be found advantageous later. Greatly extending the usefulness of such tabulated solutions is the *principle of superposition.* It means that for a particular machine component or structural element the sum of the reactions of loadings considered separately is the same as the reactions computed for the combined loadings. For instance Example 3.5 could have been solved by using Cases 1 and 3 in the table in Appendix B and adding the resulting reactions together.

3.5 Three-Dimensional Equilibrium

Equations 3.1, 3.2, and 3.3 are valid only for objects in two dimensions such as have been discussed in Section 3.4. They are not, however, sufficient to guarantee equilibrium when three dimensions must be considered. The definition of equilibrium must now be applied more carefully, because, in three dimensions a body might move in any of three directions, as well as rotate about any or all of three axes. Thus, if the three-dimensional body of Figure 3.12, acted upon by forces and moments, is in equilibrium, it exhibits no tendency to move in any direction or rotate about any axis. The criterion that there be no tendency to move in any

Example 3.6

A television company wishes to erect a television reception tower with a total height of 170 meters. This is to be done as shown in Figure 3.13a by using a free-standing tower. If the total weight of the tower is 200 kilonewtons (which can be taken as acting at mid-height), and there are horizontal wind forces acting at B and C, of 40 kilonewtons and 5 kilonewtons, respectively, both with a direction making a 30° angle to the positive x-direction as shown, find the reactions at the base of the mast.

Solution:

The first step in finding the unknown forces is to construct a free body diagram as shown in Figure 3.13b. The external forces due to wind have been applied at B and C in the correct direction, and the weight of the tower has been applied as a concentrated force, W, of 200 kilonewtons, applied at mid-height of the tower. Since the support A is fully fixed, movement in all three directions is prevented, as is rotation about all three axes. Reference to Figure 3.6b shows that there are six parts to the reaction at A: three forces (R_x, R_y, R_z) and three moments (M_x, M_y, M_z). All of these are shown on the free body diagram of Figure 3.13b. Positive directions of the assumed forces is shown and the right-hand rule sign convention for moments has been assumed.

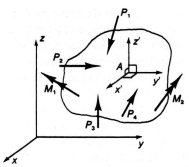

Figure 3.12 *General three-dimensional rigid body under the action of forces and moments.*

direction is satisfied if there is no net force on the body in any direction, i.e., if the sum of the components of all forces in any three mutually perpendicular directions is zero. Thus

(3.4)
$$\Sigma F_x = 0$$

(3.5)
$$\Sigma F_y = 0$$

(3.6)
$$\Sigma F_z = 0$$

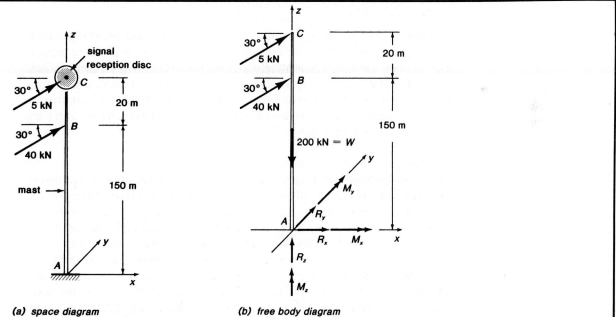

(a) space diagram   (b) free body diagram

Figure 3.13 *Cantilevered television tower for Example 3.6.*

Example 3.6 (Continued)

The six equilibrium equations (Equations 3.4 through 3.9) are now applied. The force equilibrium equations are

$$\Sigma F_x = 0 \overset{\longrightarrow}{+}$$

$$R_x + 40\cos 30° + 5\cos 30° = 0$$

$$\Sigma F_y = 0 \nearrow +$$

$$R_y + 40\sin 30° + 5\sin 30° = 0$$

$$\Sigma F_z = 0 + \uparrow$$

$$-200 + R_z = 0$$

Thus

$$R_x = -38.97 \text{ kN}$$
$$R_y = -22.5 \text{ kN}$$
$$R_z = 200 \text{ kN}$$

Note that the negative signs on the values of R_x and R_y mean that these forces really act in the direction opposite to that which had been assumed. The positive sign on the value of R_z indicates that it had been assumed in the correct direction.

The evaluation of the moment equilibrium equations needs considerable care. Although moments can be taken about axes through any point, it is convenient here to consider moments about the x-, y-, and z-axes through the point A. It is easiest to decide first which forces will not contribute to moment about these axes. It is noted first that all of the forces R_x, R_y, and R_z have lines of action through the point A, which is on all three of these axes. Thus these support forces will have no contribution to the moment equations about any of these axes. It is noted also that the weight W of the mast also has a line of action through A and will thus make no contribution to moments about these axes.

In considering moments it is noted that the wind forces at B and C both have components perpendicular to the x-axis and will thus have moments about the x-axis. The moment equation about the x-axis becomes

$$\Sigma M_{x\text{-axis}} = 0$$

$$M_x - 5 \cdot \sin 30° (170) - 40 \cdot \sin 30° (150) = 0$$

i.e.,

$$M_x = 3{,}425 \text{ kN} \cdot \text{m}$$

The wind forces at B and C also have components perpendicular to the y-axis. The moment equation must therefore include the effects of these forces, i.e.,

$$\Sigma M_{y\text{-axis}} = 0$$
$$M_y + 5 \cdot \cos 30° \, (170) + 40 \cdot \cos 30° \, (150) = 0$$
$$M_y = -5{,}932 \, \text{kN} \cdot \text{m}$$

The negative sign indicates that the direction assumed for the moment M_y was incorrect; in fact it is in the opposite direction.

Taking moments about the longitudinal axis of the tower, the z-axis, is in effect finding the tendency of the tower to twist. For this condition it is noted that the wind forces have lines of action that intersect the z-axis at B and C. These forces will thus exert no moment about the z-axis, and the z-axis moment equation becomes

$$\Sigma M_{z\text{-axis}} = 0$$
$$M_z = 0$$

There is no tendency for the tower to twist.

The support forces and moments at A are thus summarized as

$$R_x = -38.97 \, \text{kN}$$
$$R_y = -22.5 \, \text{kN}$$
$$R_z = 200 \, \text{kN}$$
$$M_x = 3{,}425 \, \text{kN} \cdot \text{m}$$
$$M_y = -5{,}932 \, \text{kN} \cdot \text{m}$$
$$M_z = 0$$

In each of these, the positive direction is defined in Figure 3.13b.

Example 3.7

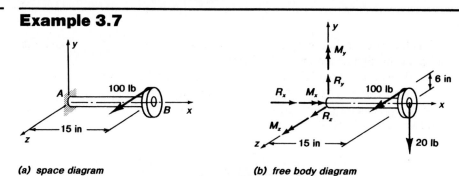

(a) space diagram   **(b) free body diagram**

Figure 3.14   *Loaded shaft of Example 3.7.*

A weightless circular shaft AB is rigidly attached to a vertical wall at its end A, and has a wheel of 12-inch diameter and weighing 20 pounds attached to its free end B. If the shaft is 15 inches long, and a 100-pound force is applied to the rim of the wheel in the positive z-direction as shown in Figure 3.14a, find the support reactions at the end A.

Solution:

A free body diagram of the shaft and wheel is constructed as shown in Figure 3.14b. The weight of the wheel has been replaced by a 20-pound force acting through the point B in the negative y-direction. At the end A there are three possible force reactions and three possible moment reactions, all assumed positive in the directions shown. If the wheel diameter is 12 inches, then its radius is six inches, which is shown on the free body diagram.

Application of the force equilibrium equations results in the following equations:

$$\Sigma F_x = 0 \xrightarrow{+}$$

$$R_x + 0 = 0$$

$$R_x = 0$$

$$\Sigma F_y = 0 \uparrow +$$

$$R_y - 20 = 0$$

$$R_y = 20 \text{ lb}$$

$$\Sigma F_z = 0 \swarrow +$$

$$R_z + 100 = 0$$

$$R_z = -100 \text{ lb}$$

Thus the force R_z in the z-direction was initially assumed in the wrong direction, as the sign of the answer indicates.

The moment equilibrium equations are applied at the point A about the x-, y-, and z-axes for convenience. It is noted that by doing this, none of the forces R_x, R_y, and R_z has any moment about any of the three axes since they all have lines of action through the point A.

$$\Sigma M_{x\text{-axis}} = 0 \overset{+}{\twoheadrightarrow}$$

$$M_x + 100\,(6) = 0$$

$$M_x = -600\,\text{lb}\cdot\text{in} \twoheadrightarrow$$

The 20-pound force representing the weight of the wheel has a line of action that intersects the x-axis at the point B, and thus this force has no moment about the x-axis. The negative sign indicates that the direction assumed and shown on the free body diagram is incorrect. Thus

$$M_x = 600\,\text{lb}\cdot\text{in} \twoheadleftarrow$$

$$\Sigma M_{y\text{-axis}} = 0$$

$$M_y - 100\,(15) = 0$$

$$M_y = 1{,}500\,\text{lb}\cdot\text{in} \uparrow$$

For moments about the y-axis, the 20-pound force is parallel to the axis and thus has no moment about the y-axis.

$$\Sigma M_{z\text{-axis}} = 0$$

$$M_z - 20\,(15) = 0$$

$$M_z = 300\,\text{lb}\cdot\text{in} \swarrow$$

In this case the 100-pound force is parallel to the z-axis and therefore has no moment about the z-axis. The reaction forces at the end A may thus be summarized as

$$R_x = 0$$

$$R_y = 20\,\text{lb} \uparrow$$

$$R_z = 100\,\text{lb} \nearrow$$

$$M_x = 600\,\text{lb}\cdot\text{in} \twoheadrightarrow$$

$$M_y = 1{,}500\,\text{lb}\cdot\text{in} \uparrow$$

$$M_z = 300\,\text{lb}\cdot\text{in} \swarrow$$

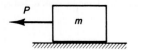

(a) a mass m acted on by a force

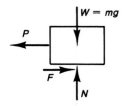

(b) free body diagram of block

Figure 3.15
Illustrations of friction force concept.

are sufficient conditions to ensure that there is no tendency for the body to move. However, the no-rotation condition has not yet been guaranteed. As for two-dimensional equilibrium the rotation condition may be satisfied for rotation about axes through any point. The point A in Figure 3.12 is thus arbitrarily chosen, and three mutually perpendicular axes—x', y', z'—are constructed, using the point A as origin. It is often convenient to adopt the axes x', y', and z' to be parallel to the x-, y-, and z-axes, but it is not essential. The moment equilibrium equations then become

(3.7)
$$\Sigma M_{x'\text{-axis}} = 0$$

(3.8)
$$\Sigma M_{y'\text{-axis}} = 0$$

(3.9)
$$\Sigma M_{z'\text{-axis}} = 0$$

It is assumed that the right-hand rule sign convention is adopted to evaluate these three equations.

There are now six available equilibrium equations (Equations 3.4 through 3.9) for three-dimensional situations, and all six must be obeyed if the structure or machine is in static equilibrium.

3.6 Friction

Of the several different types of friction, the type to be considered here is "dry friction," sometimes called "coulomb friction" in honor of C. A. de Coulomb, who studied this phenomenon in the late eighteenth century. Friction occurs between solid surfaces, and it represents the resistance of surfaces to slipping relative to one another. In technology friction plays an important role in the operation of clutches, belts, and brakes. It is important to recognize that friction forces always oppose motion or impending motion. Friction is dependent on the roughness of the surfaces, as smooth surfaces have little resistance to slip.

If the block of mass m in Figure 3.15a is to be pulled over the surface beneath by the force P, then we can draw a free body diagram of the block, provided it is in equilibrium. This is shown in Figure 3.15b. Since the body has a mass m, its weight, W, is given by Equation 1.1 as

$$W = mg$$

and the force W is placed on the free body diagram along with the force P. There is also a normal force N representing the force of the surface on the block. However, providing P is sufficiently small, the block does not move, and there must therefore be some friction force, shown as F in Figure 3.15b, which prevents such motion. Since the block had tended to move to the left, the friction force, which always opposes the motion, must be a force directed to the right on the block.

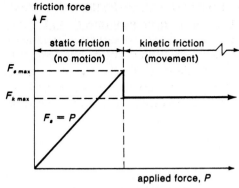

Figure 3.16 *Variation of friction force with applied load.*

If it is considered that the force P is initially zero and it is gradually increased, then it is possible to plot the variation of friction force with applied load P. Such a plot is shown in Figure 3.16. As long as the block does not move, the increase in applied force P from zero must be exactly matched by the friction force F to maintain equilibrium of the free body shown in Figure 3.16. Since there is no motion, the friction force is known as a static friction force F_s. There comes a point, however, beyond which the static friction force cannot be increased. At this point the block is just about to move, and the static friction force, $F_{s\,max}$, has reached a maximum value given by

$$F_{s\,max} = P$$

If the applied force P is increased infinitesimally above the value which produced $F_{s\,max}$, then the block starts to move. Once it is moving then it typically shows a decreased resistance to motion. This state is known as kinetic friction, since the block is moving. That there is a decreased resistance to motion is shown by the fact that the maximum kinetic friction force possible, $F_{k\,max}$, is less than the maximum static friction force possible, $F_{s\,max}$. The diagram of Figure 3.16 shows that the friction force $F_{k\,max}$ remains constant as the block moves. While this may not be exactly true for all surface, it is a good first approximation.

Experiments have shown that the maximum static force possible, $F_{s\,max}$, is proportional to the normal force N between the block and the surface, and thus the equation

(3.10)
$$F_{s\,max} = \mu_s N$$

is postulated, where μ_s is a constant, known as the coefficient of static friction. Thus if the static coefficient of friction is known, and the normal reaction between two surfaces may be computed, then it is possible to compute the maximum resistance force that may be operating to prevent slip between two surfaces.

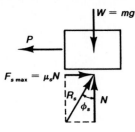

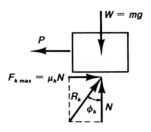

(a) *limiting static condition*

(b) *kinetic condition*

Figure 3.17
**Friction free body
diagrams.**

Table 3.1 Coefficients of Friction

Surfaces in Contact	μ_s	μ_k
Steel on steel (dry)	0.6-0.8	0.4
Steel on steel (greasy)	0.1	0.05
Cast iron on cast iron	1.10	0.15
Brass on steel (dry)	0.5	0.4
Rubber tire on wood (dry)	0.4	0.3
Rubber tire on pavement (dry)	0.9	0.8

Experiments also show that the maximum force resisting slip while the two surfaces are moving relative to one another, $F_{k\ max}$, is given by

(3.11)
$$F_{k\ max} = \mu_k N$$

where μ_k is another constant, known as the coefficient of kinetic friction. Note that the kinetic friction force is also proportional to the normal force between the surfaces. However, since the force $F_{k\ max}$ is generally less than the force $F_{s\ max}$, it is expected that the coefficient μ_k will, in general, be less than the coefficient μ_s.

The values of μ_s and μ_k dependent on the materials that are moving relative to one another, and some typical values are given in Table 3.1. The values may differ from those shown for many reasons (such as surface preparation and the moisture on the surface), but they are typical.

If the block of Figure 3.15a is about to move, then it is possible to construct a free body diagram for the case known as "limiting friction," for it represents the case when the force P is at its maximum possible value without slip occurring. Such a diagram is shown in Figure 3.17a, where it is noted that the maximum friction force is evaluated using Equation 3.10. Since both the normal

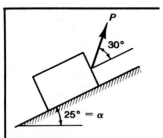

(a) *space diagram*

Figure 3.18
**Block on inclined surface
in Example 3.8.**

Example 3.8

A block weighing 6 kilonewtons is being pulled up a surface inclined at 25 degrees to the horizontal by a force P, as shown in Figure 3.18a. If the coefficient of static friction is 0.4 and the coefficient of kinetic friction is 0.3, find

Part a. The force required to start the block moving up the surface, P_s.
Part b. The force required to keep the block moving up the surface at constant speed, P_k.
Part c. The force required to prevent the block from moving down the surface, P_{min}.

force N and the friction force $F_{s\,max}$ are acting on the bottom of the block, the resultant of these forces, R_s, is given by

$$R_s = \sqrt{F_{s\,max}{}^2 + N^2}$$

but more important, the angle ϕ_s that this resultant makes with the vertical is given by

$$\tan \phi_s = \frac{F_{s\,max}}{N} = \frac{\mu_s N}{N}$$

i.e.,

(3.12)
$$\tan \phi_s = \mu_s$$

The angle ϕ_s is known as the *angle of static friction* or, sometimes, *the angle of repose,* and its value can be found from Equation 3.12.

In a similar manner, *the angle of kinetic friction,* ϕ_k, shown in Figure 3.17b is given by the relationship

(3.13)
$$\tan \phi_k = \mu_k$$

Equations 3.12 and 3.13 may be rewritten as

$$\phi_s = \tan^{-1}\mu_s$$

and

$$\phi_k = \tan^{-1}\mu_k$$

Friction is often thought of as a nuisance, and there is no doubt that it complicates design. However, friction works to our advantage many times every day. If there was no friction, V-belts would slip, brakes and clutches would not work, and the wheels on an automobile would skid rather than propel the vehicle.

Solution:

Part a. If the motion is about to occur up the plane, the maximum possible static friction force is being generated to oppose the motion, i.e., in a downhill direction. Thus the free body of Figure 3.18b is generated, where the weight is represented by the force W and the normal reaction between the surface and the block by the force N. The horizontal and vertical equilibrium equations may be applied since the system is in static equilibrium.

$$\Sigma F_H = 0 \xrightarrow{+}$$
$$P_s \sin 35° - (0.4\ N) \cos 25° - N \sin 25° = 0$$

and

$$\Sigma F_V = 0 +\uparrow$$
$$-6 + P_s \cos 35° + N \cos 25° - (0.4N) \sin 25° = 0$$

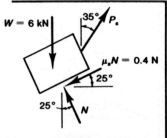

(b) free body diagram—
motion impending uphill

Example 3.8 *(Continued)*

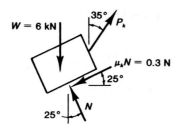

(c) free body diagram—motion continuing uphill

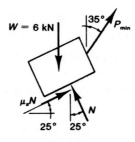

(d) free body diagram—motion impending downhill

Figure P 3.18   *(Continued)*

Solving these equations produces

$$N = 3.23 \text{ kN}$$

and

$$P_s = 4.42 \text{ kN}$$

Part b. If motion is continuing up the plane at constant speed, the block is in a state of kinetic equilibrium, by Newton's first law. Thus we can apply the horizontal and vertical equilibrium equations to the free body diagram of Figure 3.18c to yield the equations

$$\Sigma F_H = 0 \overset{\longrightarrow}{+}$$
$$P_k \sin 35° - (0.3\ N) \cos 25° - N \sin 25° = 0$$

and

$$\Sigma F_V = 0 +\uparrow$$
$$-6 + P_k \cos 35° + N \cos 25° - (0.3N) \sin 25° = 0$$

Solving these equations produces

$$N = 3.39 \text{ kN}$$

and

$$P_k = 4.10 \text{ kN}$$

Note that the value of P_k required to continue motion up the plane is less than the value of P_s required to start the motion.

Part c. It is necessary for us to discover whether, in fact, if there was no applied force P on the block, the block would remain stationary, or would move down the surface. This can be done by computing the angle of static friction, ϕ_s, from Equation 3.12, i.e.

$$\phi_s = \tan^{-1} \mu_s = \tan^{-1} 0.4 = 21.80°$$

Since the angle of inclination of the plane is greater than the angle of static friction, i.e.,

$$a > \phi_s$$

then the block would slide down the plane if there was no force P to prevent it. Thus a force P_{min} is applied to the block in the free body diagram of Figure 3.18d, and it is assumed that this is just sufficient to prevent movement of the block down the plane. Note now that since movement of the block is impending down the surface, the friction force, which always opposes the motion, is directed up the plane. Applying the horizontal and vertical equilibrium conditions gives

$$\Sigma F_H = 0 \overset{\rightarrow}{^+}$$
$$P_{min} \sin 35° + (0.4N) \cos 25° - N \sin 25° = 0$$

and

$$\Sigma F_V = 0 +\uparrow$$
$$P_{min} \cos 35° + (0.4N) \sin 25° + N \cos 25° = 0$$

Solving these equations produces

$$N = 5.17 \text{ kN}$$

and

$$P_{min} = 0.54 \text{ kN}$$

Example 3.9

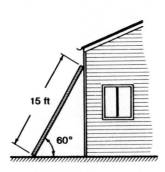

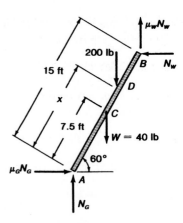

(a) space diagram

(b) free body diagram of ladder

Figure 3.19   *Ladder of Example 3.9.*

A man who weighs 200 pounds has a 40-pound ladder 15 feet long, which he leans against the side of his house in order to go up to clean leaves from the gutters. If he leans the ladder at an angle of 60 degrees to the horizontal (as shown in Figure 3.19*a*) and the coefficient of static friction between the ladder and the house is 0.2 and between the ladder and the ground is 0.4, how far up the ladder can he go before the ladder starts to slip?

Solution:

Let x be the distance from the ground along the ladder to where the man is standing when slip is about to occur. On the free body diagram of the ladder (Figure 3.19*b*) the weight of the ladder has been replaced by a concentrated load, W, of 40 pounds at the center of the ladder, and the weight of the man by a concentrated load at the point D. If the ladder is about to slip, it will slip down the wall and to the left on the ground. Opposing these two actions are the friction forces $\mu_W N_W$ (where μ_W is the coefficient of static friction between ladder and wall, and N_W the normal reaction between ladder and wall) and

3.7 Distributed Forces

In the discussion thus far we have replaced the distributed (or "spread") weight of a beam by an equivalent force acting at the center of the beam. This is sufficient for finding the support reactions of beams if the weight is uniform all along the member. However, if this is not the case, more care need be taken to ensure that the right resultant force is derived, and that it is taken to be acting at the right place.

$\mu_G N_G$ (where μ_G is the coefficient of static friction between ladder and ground, and N_G the normal reaction between ground and ladder).

There are three equilibrium equations that may be applied to this system, Equations 3.1, 3.2, and 3.3. These equations become

$$\Sigma F_H = 0 \xrightarrow{+}$$
$$\mu_G N_G - N_W = 0$$

from which

$$N_W = 0.4 \, N_G$$

Also,

$$\Sigma F_V = 0 \uparrow +$$
$$N_G - 40 - 200 + \mu_W N_W = 0$$
$$N_G - 240 + 0.2 \, N_W = 0$$
$$N_G + 0.2(0.4 \, N_G) = 240$$

Thus

$$N_G = 222.2 \, \text{lb}$$
$$\mu_G N_G = 88.9 \, \text{lb}$$

and

$$N_W = 88.9 \, \text{lb}$$
$$\mu_W N_W = 17.8 \, \text{lb}$$

Taking moments about A.

$$\Sigma M_A = 0 +\circlearrowright$$
$$W(7.5 \cdot \cos 60°) + 200(x \cos 60°) - N_W(15 \cdot \sin 60°) - \mu_W N_W(15 \cdot \cos 60°) = 0$$

Solving this equation produces

$$x = 11.4 \, \text{ft}$$

Figure 3.20 gives three common conditions of distributed loads and their equivalent single forces. Note that, in order to specify correctly the equivalent force, both the resultant force R and the place where it acts (given by x) are required.

In general the more complicated loadings such as the trapezoidal distribution can be derived by the superposition principle. Using Figure 3.21 as an example of this, we see that the trapezoidal distribution (Figure 3.21a) may be

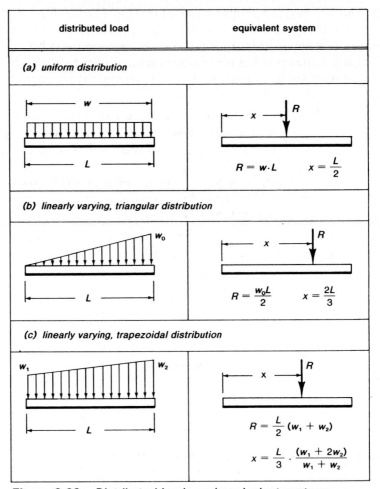

distributed load	equivalent system
(a) uniform distribution	

$R = w \cdot L \qquad x = \dfrac{L}{2}$

(b) linearly varying, triangular distribution

$R = \dfrac{w_0 L}{2} \qquad x = \dfrac{2L}{3}$

(c) linearly varying, trapezoidal distribution

$R = \dfrac{L}{2}(w_1 + w_2)$

$x = \dfrac{L}{3} \cdot \dfrac{(w_1 + 2w_2)}{w_1 + w_2}$

Figure 3.20 Distributed loads and equivalent systems.

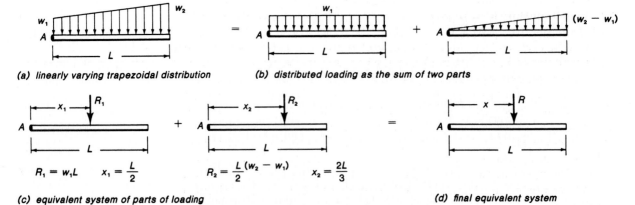

(a) linearly varying trapezoidal distribution

(b) distributed loading as the sum of two parts

$R_1 = w_1 L \qquad x_1 = \dfrac{L}{2}$

$R_2 = \dfrac{L}{2}(w_2 - w_1) \qquad x_2 = \dfrac{2L}{3}$

(c) equivalent system of parts of loading

(d) final equivalent system

Figure 3.21 Distributed loadings by parts.

viewed as the sum of a uniform distribution and a triangular distribution (Figure 3.21b). If each of these two parts is replaced by its equivalent force (Figure 3.21c) then the force equivalent to the original trapezoidal distribution (Figure 3.21d) must also be equal to the sum of the equivalent forces (of Figure 3.21c). Thus for vertical forces to sum to the same value

$$R = R_1 + R_2 = W_1 L + \frac{L}{2}(W_2 - W_1)$$

i.e.,

$$R = \frac{L}{2}(W_1 + W_2)$$

To derive the location x at which this resultant force acts, the equivalent force in Figure 3.21d must have the same moment about the end A as the sum of the moments of the equivalent system forces about the end A in Figure 3.21c. Thus

$$R \cdot x = R_1 \cdot x_1 + R_2 \cdot x_2$$

and

$$\frac{L}{2}(W_1 + W_2) \cdot x = W_1 L \cdot \frac{L}{2} + \frac{L}{2}(W_2 - W_1) \cdot \frac{2L}{3}$$

which yields

$$x = \frac{L}{3}\left(\frac{W_1 + 2W_2}{W_1 + W_2}\right)$$

3.8 Internal Forces within Members

In Section 3.2 the steps that need to be taken to draw the correct free body diagrams of objects or part of objects were outlined. This process may also be used to find internal forces in members. Just as it was possible to find support reactions using free bodies, so it is possible and useful to find internal forces and moments in members due to external forces. Provided that all possible forces that may have been internal to a structure or machine at a point have been replaced by external forces when an article is cut at that point, then the free body must still be in equilibrium.

Before proceeding to find the forces internal to members, it is necessary to understand what types of forces are likely to be acting internal to a body. Four common types of engineering systems are illustrated in Figure 3.22. In each case they have been cut at a section A and the forces that were internal to the member have been replaced by the appropriate external forces (at this point of unknown magnitude) acting at the cut. If the external forces at the cut are identical to those internal forces which had been present before the cut was made, then the two free bodies into which the system has been cut are both in equilibrium.

Figure 3.22
Internal forces in common structural elements.

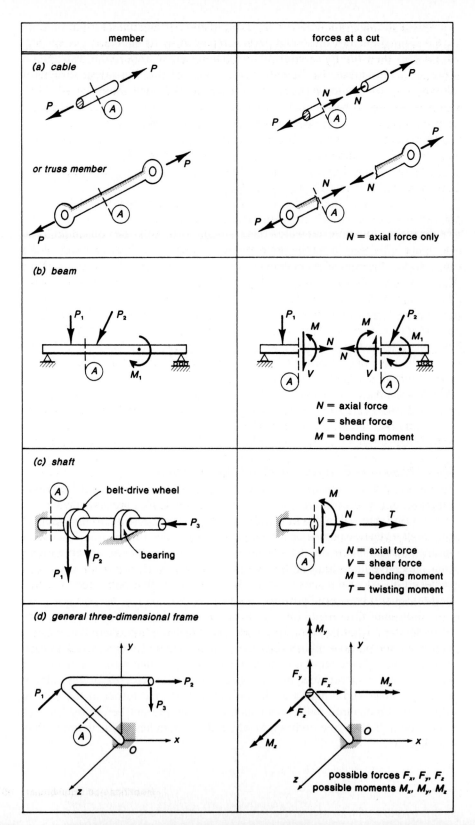

member	forces at a cut
(a) cable or truss member	N = axial force only
(b) beam	N = axial force V = shear force M = bending moment
(c) shaft	N = axial force V = shear force M = bending moment T = twisting moment
(d) general three-dimensional frame	possible forces F_x, F_y, F_z possible moments M_x, M_y, M_z

Consider first the portion of the *cable* and the *truss* member of Figure 3.22a. Such components carry only axial forces so that, when a cut is made internal to the member, there is only one possible direction in which the internal force may act — an *axial* direction. The internal force, *N*, is shown as if it is stretching the member, i.e., the force is a tension, but it may have the opposite direction, in which case it would be a compressive force, tending to compress the member. Such members are often called "two-force members" because there are only two forces acting on the member, one at each end. For equilibrium these two forces must be of identical magnitude, have the same line of action, and be in opposite directions. These conditions are met by the forces *P* on the cable and truss members illustrated in Figure 3.22a.

Another common structural or machine element is the *beam*. A beam is defined as a member to which external loads are applied in such a manner that at least some of these loads have components perpendicular to the longitudinal axis of the member. A beam may have forces and/or moments applied to it externally, and it is distinguished from a truss member in that it tends to bend. An example of a beam is shown in Figure 3.22b. If a cut is made in the beam at any point *A*, then the internal forces must be replaced by an *axial force N*, a *shear force V*, and a *bending moment M* to maintain equilibrium. Notice that the directions of forces and moments at the cut depends on which free body is being studied. In fact, it can be shown that the forces and moments must be of opposite sign in opposite sides of the cut. This follows Newton's law that "to every action there is an equal and opposite reaction."

When a *shaft* is being analyzed, as in Figure 3.22c, an extra moment must be considered. In addition to axial force *N*, shear force *V*, and bending moment *M*, there is also the possibility that a twisting moment, or *torque T*, acts on a shaft. Such actions may be imparted to a shaft by gears and/or belt drive systems. In the equivalent system of Figure 3.22c only the left-hand free body is shown, although an analogous free body diagram of the remainder of the shaft, to the right of the cut, could also be drawn. Once again, all forces and moments at the cut would be in the opposite direction on the other side of the cut.

A general three-dimensional frame is illustrated in Figure 3.22d by a bent rod, rigidly fixed at the origin. A frame may also have moments or couples applied to it anywhere. In such circumstances it is often difficult to tell which forces and moments will be acting at a cut such as at *A*. In such circumstances one may be sure of including all possible forces and moments at the cut by considering forces in three perpendicular directions — say F_x, F_y, and F_z — and moments about three perpendicular axes — say M_x, M_y, and M_z.

The detailed calculation of all necessary internal forces will be demonstrated in later chapters for different types of structural or machine elements.

3.1   For the structures shown in Figure P 3.1, draw the free body diagrams.

3.2   For the structures shown in Figure P 3.2, draw the free body diagrams.

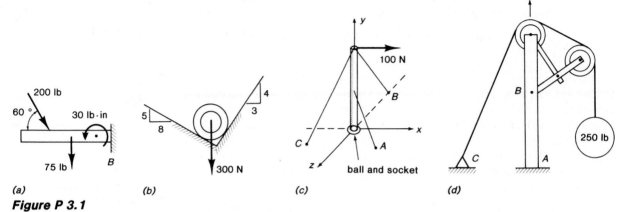

(a)   *(b)*   *(c)*   *(d)*

Figure P 3.1

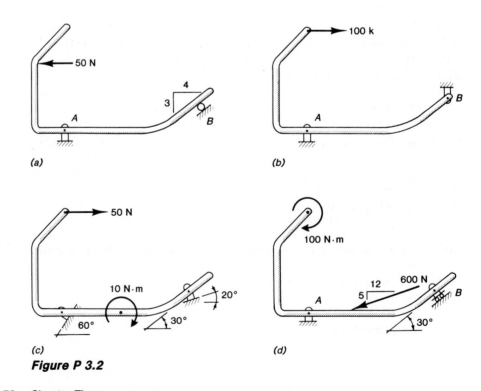

(a)   *(b)*

(c)   *(d)*

Figure P 3.2

3.3 A frame structure is shown in Figure P 3.3; draw the free body diagrams of:
  a. The entire structure.
  b. Member *ABC*.
  c. Member *CDE*.
  d. Member *BD*.
  e. Members *ABC* and *BD* as a unit.
  f. Members *ABC* and *CDE* as a unit.
  g. Members *CDE* and *BD* as a unit.

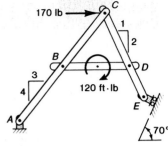

Figure P 3.3

3.4 A frame structure is shown in Figure P 3.4; draw the free body diagrams of:
  a. The entire structure.
  b. Member *BD*.
  c. Member *ABC*.

3.5 A frame structure is shown in Figure P 3.5; draw the free body diagrams of:
  a. The entire structure.
  b. Member *BC*.
  c. Member *ADB*.

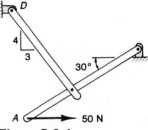

Figure P 3.4

3.6 For the structure shown in Figure P 3.6, draw the free body diagrams of:
  a. The entire structure.
  b. The sphere.
  c. Member *ABC*.

3.7 For the structure shown in Figure P 3.7, all the members except for the cylinder are considered weightless. Draw the free body diagrams for:
  a. The entire structure.
  b. The cylinder.
  c. Member *AEC*.
  d. Member *CD*.

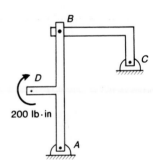

Figure P 3.5

3.8 For the structure shown in Figure P 3.8, draw the free body diagrams for:
  a. The entire structure.
  b. Member *ACD*.
  c. Member *CE*.
  d. Member *DEB*.

3.9 For the structure shown in Figure P 3.9, draw the free body diagrams for:
  a. The entire structure.
  b. Member *AB*.
  c. Member *ECD*.
  d. Member *BC*.

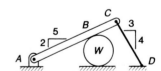

Figure P 3.6

3.10-
3.11 Use independent equations to find the reactions for the structures shown in Figures P 3.10 and P 3.11.

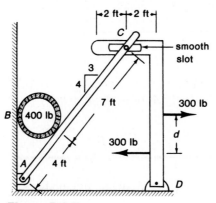

Figure P 3.7

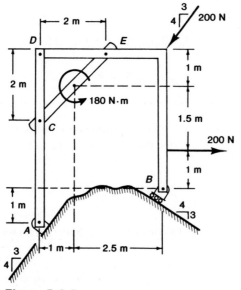

Figure P 3.8

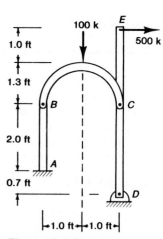

Figure P 3.9

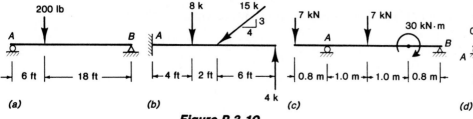

(a)

(b)

(c)

(d)

Figure P 3.10

3.12-
3.14   Find the reactions at *A* and *B* of the structures shown in
       Figures P 3.12 to P 3.14 by using independent equations.

3.15   For the two structures shown in Figure P 3.15*a*, and *b*, find the
       reactions using independent equations of equilibrium.

3.16   Find the reactions at point *A* and *B* for the structures shown in
       Figure P 3.2*a*, *b*, *c*, and *d*.

3.17   Find the reaction at the fixed end for the structure shown in
       Figure P 3.17.

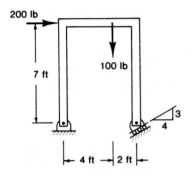

Figure P 3.11

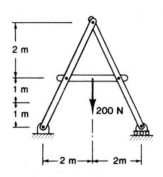

Figure P 3.12

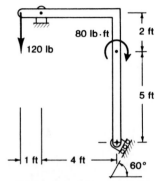

Figure P 3.13

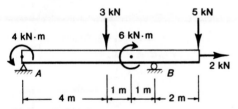

Figure P 3.14

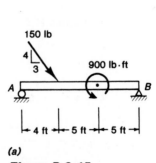

(a)

Figure P 3.15

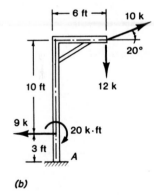

(b)

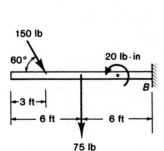

Figure P 3.17

3.18   Find the reactions at *A* and *E* for the frame shown in Figure P 3.18.

3.19   The beam shown in Figure P 3.19 has a pin at *B*. Find the reactions at *A* and *C;* also find the force exerted by the pin at *B*.

3.20   For the A-type frame shown in Figure P 3.20, find the forces on each of the joints.

3.21   For the frame shown in Figure P 3.21, find the forces at all pin connections.

3.22   For the frame shown in Figure P 3.8, find the pin forces at points *C, D,* and *E*.

3.23   Find the components of the reaction at *A* for the beam loaded as shown in Figure P 3.23.

3.24   The mast shown in Figure P 3.24 is supported at *A* by a ball-and-socket joint and two cables *BC* and *ED*. Calculate:
a.   The forces in the cables without involving the unknowns at *A*.
b.   The force components at *A*.

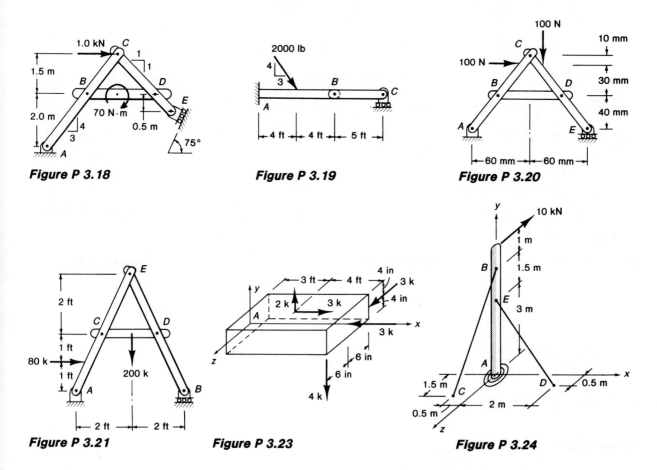

Figure P 3.18   Figure P 3.19   Figure P 3.20

Figure P 3.21   Figure P 3.23   Figure P 3.24

3.25   Calculate the components of the reaction at A for the tubular bar shown in Figure P 3.25. The 65-pound force acts in a plane parallel to the xy-plane.

3.26   A tubular bar is loaded as shown in Figure P 3.26. The 0.2 kilonewton force and the 200 kilonewton meter couple are parallel to the z-axis and the 1.5 kilonewton force acts in a plane parallel to the xy-plane. Find the internal forces and couples at point B.

3.27   The rectangular bar is loaded as shown in Figure P 3.27. The 200-pound force acts in a plane parallel to the yz-plane. Find the components of the reaction at A.

3.28   A uniform ladder weighing 300 newtons resting against a smooth wall is shown in Figure P 3.28. Find the coefficient of static friction between the ladder and the floor if motion is impending.

3.29   A 200-pound block on a 30° inclined plane with the coefficient of static friction $\mu_s = 0.20$ is shown in Figure P 3.29. Find the minimum horizontal force that will cause impending motion of the block up the plane.

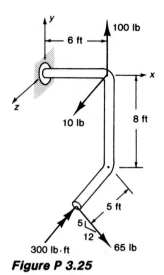

Figure P 3.25

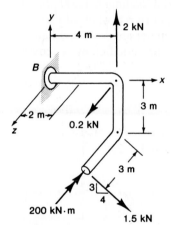

Figure P 3.26

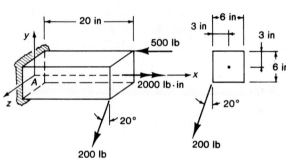

Figure P 3.27

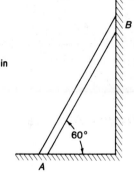

Figure P 3.28

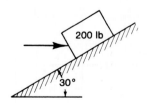

Figure P 3.29

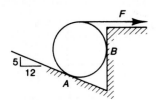

Figure P 3.30

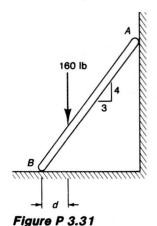

Figure P 3.31

3.30 A cylinder weighing 600 newtons acted upon by a force F is shown in Figure P 3.30. Find the force F to cause motion to be impending. $\mu_s = 0.18$.

3.31 A 25-foot uniform ladder weighing 50 pounds is shown in Figure P 3.31. How far up the ladder can a 160-pound person go before the ladder starts to slide? The coefficients of static friction, μ_s, are 0.2 for the wall and 0.4 for the floor.

3.32 Two blocks A and B are connected by a weightless bar AB as shown in Figure P 3.32. Block B weighs 100 newtons; the static friction coefficient, μ_s, is 0.2 between block B and the inclined surface. Find the minimum weight of block A such that the system does not move.

3.33 A cylinder weighing 400 pounds rests on two inclined planes as shown in Figure P 3.33. Find the force P to produce impending motion. The static friction coefficient μ_s is 0.20 for both surfaces.

3.34 Find the force P to raise the 6 kilonewton block shown in Figure P 3.34. The weight of the wedge is negligible; the static friction coefficient μ_s is 0.2 for all surfaces.

3.35 A triple collar bearing subjected to an axial loading of 1,200 pounds is shown in Figure P 3.35. Each collar supports a portion of the loading P that is directly related to the contact area of the collar. Find the maximum frictional moment M. The coefficient of static friction μ_s is 0.22 for each collar.

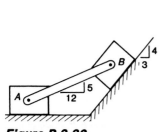

Figure P 3.32

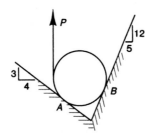

Figure P 3.33

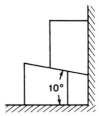

Figure P 3.34

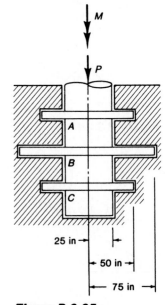

Figure P 3.35

Trusses and Frames   4

1. *To learn to recognize the different types of trusses and to understand what is meant by the terms two-force members, compression forces, and tension forces.*

2. *To be able to analyze a simple truss by the method of joints.*

3. *To learn how to find the forces in any member of a simple or complex truss using the method of sections.*

4. *To be able to determine the interconnecting forces between elements of a frame.*

Objectives

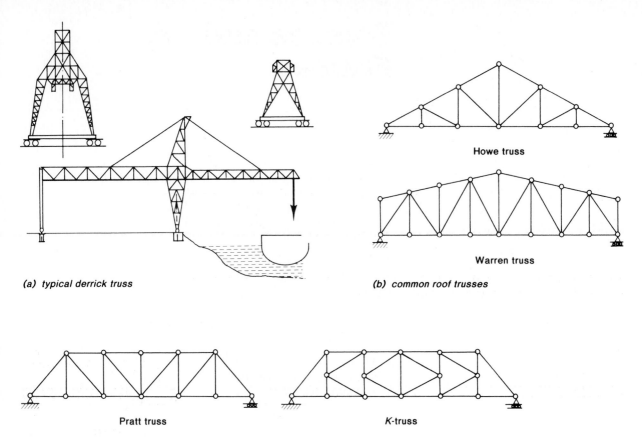

(a) typical derrick truss

(b) common roof trusses

Howe truss

Warren truss

Pratt truss

K-truss

(c) common bridge trusses

Figure 4.1 Types of trusses.

4.1 Types of Trusses

A *truss* may be defined as a structure composed of members joined together with pinned connections such that the members are free to rotate at their ends. Historically, trusses have been used for many purposes, including roof supports, bridges, and derricks. Just as there are many uses of trusses, there are many types of trusses. Some common types of trusses used in roof supports, in bridges, and in derricks are shown in Figure 4.1.

The members of trusses are known as *two-force members* because forces are applied at only two locations on the member—namely, at the pinned connections. Figure 4.2 illustrates two typical truss members—one under the action of a *tensile* (or 'stretching') force and the other under the action of a *compressive* (or 'squeezing') force. In each case it can readily be shown that, for equilibrium, the forces applied at each joint on the member must be of equal magnitude but in opposite directions. It can also be shown that the line of action of these forces must be a straight line joining the pinned connections.

Figure 4.2 also shows that, whether the member is in tension or compression, if a cut is made in the member at any point along its length, then a force

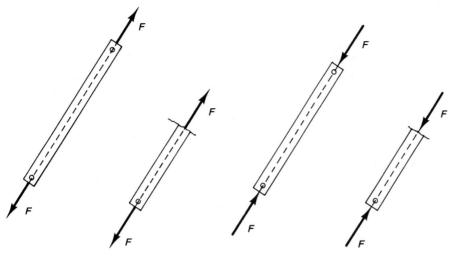

(a) truss member in tension (positive force)   (b) truss member in compression (negative force)

Figure 4.2   *Tension and compression members.*

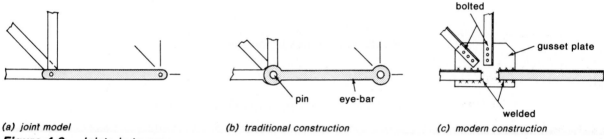

(a) joint model   (b) traditional construction   (c) modern construction

Figure 4.3   *Joints in trusses.*

will need to be inserted at the cut to preserve equilibrium. In this way it is possible to isolate free bodies of parts of members. For subsequent calculations it is also necessary to adopt some convenient sign convention to show whether a particular member is carrying a tensile or compressive force. It is usual to assume that a member in tension (as shown in Figure 4.2a) is carrying a positive force, and that a member in compression (as shown in Figure 4.2b) is carrying a negative force.

The construction of trusses in such a manner that the members are truly two-force members was a concern to engineers for many years. If a particular joint was to be constructed, as shown in Figure 4.3, then the theory of two-force members assumes a joint model as shown in Figure 4.3a. It is, however, impossible to construct such a member, as the joints would be very weak. Traditionally, therefore, engineers constructed trusses from members such as shown in Figure 4.3b. The members in such trusses were known as 'eye-bars' because each member had an 'eye' at each end. The members were held together at the joints

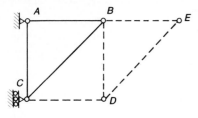

(a) simple truss construction

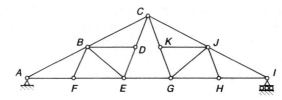

(b) complex truss

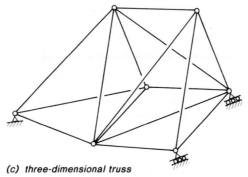

(c) three-dimensional truss

Figure 4.4 *Types of trusses.*

by pins, which theoretically permitted each member to rotate freely at each joint. The pins, particularly in bridge construction, were often very large—up to about 18 inches (0.46 meter) in diameter. Such joints typically did not function as intended because corrosion and other environmental effects quickly 'froze' the joints. It was later discovered that the behavior of the members was not greatly influenced by the type of joint used, and today it is more usual to form truss joints by welding or bolting the members to a gusset plate as shown in Figure 4.3c.

Before attempting any truss analysis, it is necessary to recognize that there are several types of trusses that may readily be analyzed. A *simple* truss is one that may be formed by a combination of triangles, such that when two more members are added they join to form another joint. This process is illustrated in Figure 4.4a. If one starts with the triangular truss *ABC,* then the joint *D* is added to the truss merely by adding the two members *BD* and *CD.* Similarly the joint *E* may be added by adding the members *BE* and *DE.* Sometimes, however, this process is not possible. The truss of Figure 4.4b, for example, is such a truss. In order to construct this truss it is necessary first to construct two simple trusses *ABCDEF* and *CKGHIJ* before joining them at *C* and with the link piece *EG*. Such a truss is known as a *compound truss*. The analytical techniques described in this chapter are intended specifically for simple trusses, although with careful planning they may also be applied to find the forces in the members of compound trusses.

The trusses of Figure 4.4a and 4.4b are both two-dimensional, or planar, trusses. It is possible however, to construct three-dimensional trusses. Such trusses are called *space* trusses, as shown in Figure 4.4c. The trusses analyzed in the examples in this chapter are all planar trusses, although it is a relatively simple matter to extend the analyses to cover space trusses.

Another common assumption made in considering trusses is that the members composing the truss are weightless. This assumption—which makes it unnecessary to consider loads distributed along a member, as treated in Chapter 3—will also be adopted in this chapter.

4.2 Method of Joints

In this chapter two methods of analysis of trusses are considered. The first of these is the method of joints. In Chapter 3 the principle of dissecting a structure into free bodies was introduced, and this same principle is now used to analyze a truss by the method of joints. It is possible to create as many free bodies as desired by considering as many portions of a structure as necessary. For the method of joints the truss is divided so that each joint is formed into a free body. Care must be taken to ensure that all forces acting on each joint are included in the free body diagram of that joint.

It is already known that at each joint the members are joined together by pinned joints. There are therefore no moments exerted by the member forces at that joint. However, since each joint must be in static equilibrium, there are still two equilibrium conditions that must be satisfied at each joint—namely, that the sums of the horizontal and vertical forces at each joint both must be zero. Thus

(4.1)
$$\Sigma F_H = 0$$

and

(4.2)
$$\Sigma F_V = 0$$

The application of these equations to each joint in a truss constitutes the method of joints. One of the real benefits of this method is that there is always at least one joint in a simple truss that has not been used in the determination of the member forces. This joint constitutes a 'check' joint. The two equilibrium equations must be satisfied at the check joint also or else some mistake has been made in the analysis prior to this step.

The procedure in finding the forces in the members of a truss by the method of joints is best shown in a stepwise manner. If each of the following steps is followed in order, then a correct analysis will result.

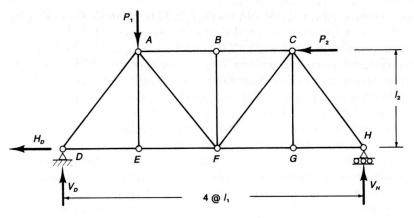

(a) truss diagram

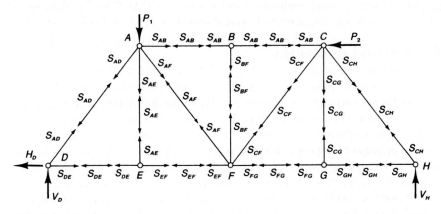

(b) exploded diagram

Figure 4.5 *Analysis by methods of joints.*

Steps in Analyzing a Truss by the Method of Joints

Consider the truss of Figure 4.5a, in which both the applied forces P_1 and P_2 are given, as well as the dimensions l_1 and l_2. It is required to find the forces in all the members of the truss due to the given loads.

Step 1. The unknown reactions at the supports—H_D, V_D, and V_H— must first be found. To do this the whole truss is regarded as a free body, and the three available equations of equilibrium are applied to the truss as described in Chapter 3. Up to three unknown forces may be found in this manner.

Step 2. An '*exploded diagram*' of the truss is drawn as shown in Figure 4.5b. The purpose of this diagram is to assist in making sure all the forces on the free bodies have been included and that they are in the correct relationship to each other. In the exploded diagram the force in each member of the truss—currently unknown—has been designated S with a subscript identifying which bar the force acts in. For example, the force S_{AB} is the unknown force in the member connecting joints A and B. Since it is not yet known whether the truss members are in tension or compression, all members have been assumed to be in tension. Clearly this may not be the case for all members. However, the sign of the values of the forces will indicate which members are in tension and which are in compression. In the exploded diagram, each bar has been shown as a free body with the member assumed to be in tension. By Newton's third law, the forces acting on each joint are equal and opposite. Note also that all the external forces—both applied loads and support reactions—have been included on the exploded diagram.

Step 3. The two available equilibrium equations now are applied to each joint in turn. Since there are only two equations to apply at each joint, only two unknown member forces may be found at each joint. The order in which the joints are considered must therefore be chosen carefully. In the example of Figure 4.5, the analysis may be started at either joint D or joint H. At any other joint there would be too many unknown member forces to find by using the two equilibrium equations. Assume that the analysis was commenced by considering joint D and that the two available equations were applied to find the member forces S_{AD} and S_{DE}. The question of which joint to consider next is answered by searching for a joint at which there are now only two unknown member forces. Joints E and H both meet these requirements since the member force S_{DE} is now known. In this manner each joint of the truss is considered until all of the unknown member forces have been found.

Step 4. In the analysis of simple trusses there will always be one joint at which the equilibrium equations have not been applied when all the member forces have been found. There will further be another joint at which only one of the two available equations has been used. There will thus be three equations remaining and these may be used to check the previous calculations. All three of the unused equations must be valid for the answer to be correct. If any one of these equations has not been satisfied then an error has been made at some stage in the analysis.

Step 5. For the convenience of the reader it is usual to present the results of the completed analysis on a truss diagram similar to that given in Figure 4.5a.

These steps will be followed in the subsequent examples.

Example 4.1

Find the forces in the members of the bridge truss shown in Figure 4.6a.

Solution:
Step 1. To find the support reactions it is first necessary to draw a free body diagram of the whole truss, as shown in Figure 4.6b. Applying the three available equilibrium equations to the truss yields:

$$\Sigma F_V = 0 \uparrow +$$

$$V_A + V_D - 90 + 30 = 0$$

$$V_A + V_D = 60$$

$$\Sigma M_A = 0 \; \circlearrowright +$$

$$V_D \cdot 6 + 30 \cdot 4 - 90 \cdot 2 = 0$$

$$V_D = 10 \text{ kN}$$

and

$$V_A = 50 \text{ kN}$$

$$\Sigma F_H = 0 \; \overset{+}{\leftarrow}$$

$$H_A = 0$$

Step 2. The exploded diagram for this truss is shown in Figure 4.6c. As usual, all unknown member forces have been shown as if all members are in tension. The slopes of inclined members are also shown on this diagram.

Step 3. The analysis may be started at either joint A or joint D, since there are only two unknown forces at each joint. The joint A is arbitrarily chosen as the first joint to be analyzed. A free body diagram of each joint is drawn beside the computations for that joint below. The individual joint free body diagrams are extracted from the exploded diagram.
Joint A:

$$\Sigma F_V = 0 \uparrow +$$

$$S_{AE} \cdot \frac{1}{\sqrt{2}} + 50 = 0$$

$$S_{AE} = -50 \sqrt{2} = -70.7 \text{ kN}$$

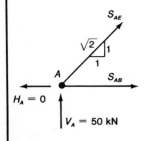

joint A

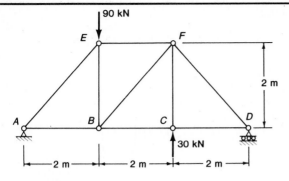

(a) truss diagram

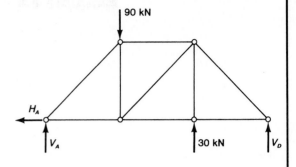

(b) free body diagram of whole truss

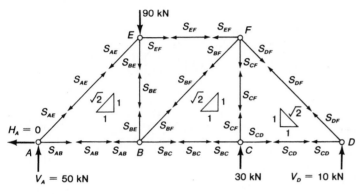

(c) exploded diagram

Figure 4.6

Also, at Joint A:

$$\Sigma F_H = 0 \overset{+}{\rightarrow}$$

$$S_{AE} \cdot \frac{1}{\sqrt{2}} + S_{AB} = 0$$

$$S_{AB} = -(-50\sqrt{2}) \frac{1}{\sqrt{2}} = 50 \text{ kN}$$

The negative sign on the derived value of S_{AE} indicates that the force in the member AE is a compressive force. On the other hand, the positive value of the derived force S_{AB} indicates that the force in the member AB is a tensile force. Once the values of S_{AB} and S_{AE} are known it is necessary to find another joint at which there are only two unknown forces. Both joints E and D meet this requirement. Joint E is arbitrarily chosen, and the selection of the order of subsequent joints considered is made also on the basis of the necessity of including no more than two unknown forces at any joint.

Example 4.1 (Continued)

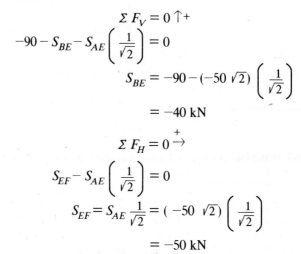

Joint E:

$$\Sigma F_V = 0 \uparrow +$$

$$-90 - S_{BE} - S_{AE}\left(\frac{1}{\sqrt{2}}\right) = 0$$

$$S_{BE} = -90 - (-50\sqrt{2})\left(\frac{1}{\sqrt{2}}\right)$$

$$= -40 \text{ kN}$$

$$\Sigma F_H = 0 \xrightarrow{+}$$

$$S_{EF} - S_{AE}\left(\frac{1}{\sqrt{2}}\right) = 0$$

$$S_{EF} = S_{AE}\frac{1}{\sqrt{2}} = (-50\sqrt{2})\left(\frac{1}{\sqrt{2}}\right)$$

$$= -50 \text{ kN}$$

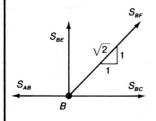

joint E

Joint B:

$$\Sigma F_V = 0 \uparrow +$$

$$S_{BE} + S_{BF}\left(\frac{1}{\sqrt{2}}\right) = 0$$

$$S_{BF} = -\sqrt{2} S_{BE} = -\sqrt{2}(-40)$$

$$= 40\sqrt{2} = 56.6 \text{ kN}$$

$$\Sigma F_H = 0 \xrightarrow{+}$$

$$S_{BC} = -S_{BF}\left(\frac{1}{\sqrt{2}}\right) + S_{AB}$$

$$= -(40\sqrt{2})\left(\frac{1}{\sqrt{2}}\right) + 50 = 10 \text{ kN}$$

joint B

Joint C:

$$\Sigma F_V = 0 \uparrow +$$

$$S_{CF} + 30 = 0$$

$$S_{CF} = -30 \text{ kN}$$

$$\Sigma F_H = 0 \xrightarrow{+}$$

$$S_{CD} - S_{BC} = 0$$

$$S_{CD} = S_{BC} = 10 \text{ kN}$$

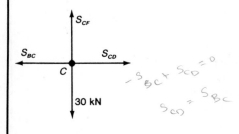

joint C

Joint F:

$$\Sigma F_V = 0 \uparrow +$$

$$-S_{BF}\left(\frac{1}{\sqrt{2}}\right) - S_{CF} - S_{DF}\left(\frac{1}{\sqrt{2}}\right) = 0$$

$$S_{DF} = -S_{BF} - S_{CF} \cdot \sqrt{2}$$

$$= -(40\sqrt{2}) - (-30)\sqrt{2}$$

$$= -10\sqrt{2} = -14.1 \text{ kN}$$

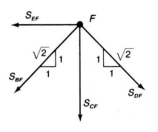

joint F

It is noted that at this stage all the unknown member forces have been found.

Step 4. There are still a number of unused equilibrium equations available for checking the answers. These equations are the horizontal force summation at joint F and both horizontal and vertical force summations at joint D. These equations are now used as a check.

Joint F:

$$\Sigma F_H \xrightarrow{+} = S_{DF}\left(\frac{1}{\sqrt{2}}\right) - S_{EF} - S_{BF}\left(\frac{1}{\sqrt{2}}\right)$$

$$= (-10\sqrt{2})\left(\frac{1}{\sqrt{2}}\right) - (-50) - (40\sqrt{2})\left(\frac{1}{\sqrt{2}}\right)$$

$$= 0 \text{ (check OK)}$$

Joint D:

$$\Sigma F_V \uparrow + = S_{DF}\left(\frac{1}{\sqrt{2}}\right) + 10$$

$$= (-10\sqrt{2})\left(\frac{1}{\sqrt{2}}\right) + 10$$

$$= 0 \text{ (check OK)}$$

$$\Sigma F_H \xrightarrow{+} = -S_{CD} - S_{DF}\left(\frac{1}{\sqrt{2}}\right)$$

$$= -(10) - (-10\sqrt{2})\left(\frac{1}{\sqrt{2}}\right)$$

$$= 0 \text{ (check OK)}$$

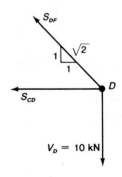

joint D

Example 4.1 (Continued)

Step 5. For convenience, the results are presented on a truss diagram in Figure 4.6d. It is advisable to indicate whether a number is subject to a tensile force (T) or a compressive force (C), as has been done in this figure.

Example 4.2

Find the forces in the members of the crane truss of Figure 4.7a when the crane is lifting a load of 10 kips at the point E.

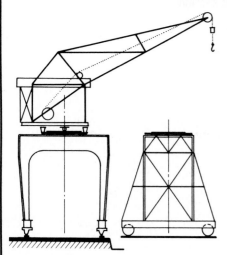

(a) travelling crane

Figure 4.7

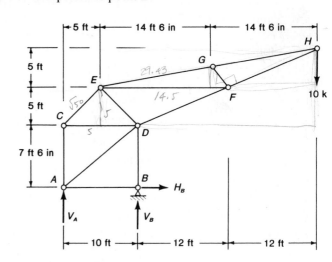

(b) free body diagram of whole truss

Solution:

Step 1. A free body diagram of the truss is drawn as shown in Figure 4.7b in order to find the support reactions. It is assumed for this example that the truss is on a roller support at A and a pinned support at B. The three equilibrium equations are now applied to the truss:

$$\Sigma M_A = 0 \; \circlearrowleft+$$
$$V_B \cdot 10 - 10 \cdot 34 = 0$$
$$V_B = 34 \text{ k}$$

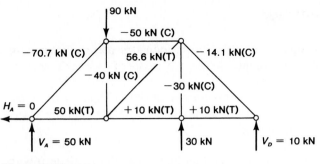

(d) results diagram

Figure 4.6 (Continued)

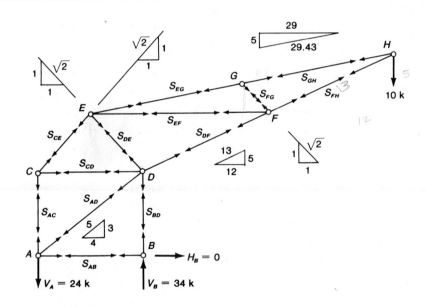

(c) exploded diagram

$$\Sigma F_V = 0 \uparrow+$$

$$V_A + V_B - 10 = 0$$

$$V_A = 10 - V_B = 10 - 34 = -24 \text{ k}$$

$$\Sigma F_H = 0 \xleftarrow{+}$$

$$H_B = 0$$

Example 4.2 (Continued)

The negative sign of the value of V_A indicates that the truss must be held down at that point.

Step 2. The exploded diagram for this truss is shown in Figure 4.7c. Again all member forces have been assumed to be tensile, and the slopes of the members are also given.

Step 3. The analysis may be started at either joint B or joint H, since at both of these joints there are only two unknown member forces acting. The joint H is arbitrarily chosen to begin the analysis, and the subsequent joints are chosen working towards the supports.

Joint H:

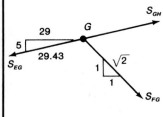

joint H

$$\Sigma F_H = 0 \xrightarrow{+}$$

$$-S_{GH}\left(\frac{29}{29.43}\right) - S_{FH}\left(\frac{12}{13}\right) = 0$$

$$\Sigma F_V = 0 \uparrow+$$

$$-10 - S_{GH}\left(\frac{5}{29.43}\right) - S_{FH}\left(\frac{5}{13}\right) = 0$$

Thus, solving these equations

$$S_{FH} = -44.35\text{k}$$
$$S_{GH} = 41.54\text{k}$$

Joint G:

joint G

$$\Sigma F_H = 0 \xrightarrow{+}$$

$$-S_{EG}\left(\frac{29}{29.43}\right) + S_{GH}\left(\frac{29}{29.43}\right) + S_{FG}\left(\frac{1}{\sqrt{2}}\right) = 0$$

$$\Sigma F_V = 0 \uparrow+$$

$$S_{GH}\left(\frac{5}{29.43}\right) - S_{EG}\left(\frac{5}{29.43}\right) - S_{FG}\left(\frac{1}{\sqrt{2}}\right) = 0$$

Thus, solving these equations

$$S_{FG} = 0$$
$$S_{EG} = S_{GH} = 41.54\text{k}$$

Joint F:

$$\Sigma F_H = 0 \xrightarrow{+}$$

$$S_{FH}\left(\frac{12}{13}\right) - S_{DF}\left(\frac{12}{13}\right) - S_{EF} - S_{FG}\left(\frac{1}{\sqrt{2}}\right) = 0$$

$$\Sigma F_V = 0 \uparrow+$$

$$S_{FH}\left(\frac{5}{13}\right) - S_{DF}\left(\frac{5}{13}\right) + S_{EF}\left(\frac{1}{\sqrt{2}}\right) = 0$$

Thus, solving these equations

$$S_{EF} = 0$$

$$S_{DF} = S_{FH} = -44.35k$$

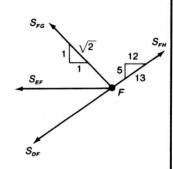

joint F

Joint E:

$$\Sigma F_H = 0 \xrightarrow{+}$$

$$S_{EF} + S_{EG}\left(\frac{29}{29.43}\right) + S_{ED}\left(\frac{1}{\sqrt{2}}\right) - S_{EC}\left(\frac{1}{\sqrt{2}}\right) = 0$$

$$\Sigma F_V = 0 \uparrow+$$

$$S_{EG}\left(\frac{5}{29.43}\right) - S_{EC}\left(\frac{1}{\sqrt{2}}\right) - S_{ED}\left(\frac{1}{\sqrt{2}}\right) = 0$$

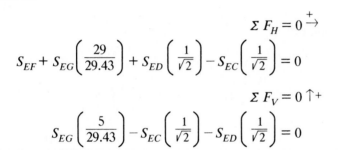

Thus, solving these equations

$$S_{DE} = -23.95k$$

$$S_{CE} = 33.94k$$

joint E

Joint C:

$$\Sigma F_V = 0 \uparrow+$$

$$S_{CE}\left(\frac{1}{\sqrt{2}}\right) - S_{AC} = 0$$

$$S_{AC} = 24.0k$$

$$\Sigma F_H = 0 \xrightarrow{+}$$

$$S_{CE}\left(\frac{1}{\sqrt{2}}\right) + S_{CD} = 0$$

$$S_{CD} = -24.0k$$

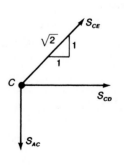

joint C

Example 4.2 *(Continued)*

Joint D:

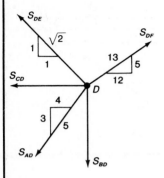

joint D

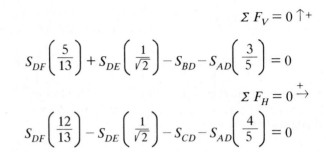

$$\Sigma F_V = 0 \uparrow +$$

$$S_{DF}\left(\frac{5}{13}\right) + S_{DE}\left(\frac{1}{\sqrt{2}}\right) - S_{BD} - S_{AD}\left(\frac{3}{5}\right) = 0$$

$$\Sigma F_H = 0 \xrightarrow{+}$$

$$S_{DF}\left(\frac{12}{13}\right) - S_{DE}\left(\frac{1}{\sqrt{2}}\right) - S_{CD} - S_{AD}\left(\frac{4}{5}\right) = 0$$

Thus, solving these equations

$$S_{AD} = 0$$
$$S_{BD} = -34.0k$$

Joint A:

$$\Sigma F_H = 0 \xrightarrow{+}$$

$$S_{AD}\left(\frac{4}{5}\right) + S_{AB} = 0$$

$$S_{AB} = 0$$

At this stage all the unknown member forces have been found.

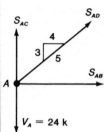

joint A

Step 4. The three equilibrium equations currently available for checking purposes are the vertical force summation at joint A and both horizontal and vertical force summations at joint B.

Joint A:

$$\Sigma F_V \uparrow + = S_{AC} + S_{AD}\left(\frac{3}{5}\right) - 24$$

$$= 24.0 + 0 - 24.0 \ 01 = 0 \text{ (check OK)}$$

Joint B:

$$\Sigma F_H \overset{+}{\rightarrow} = -H_B - S_{AB} = 0 - 0 = 0 \text{ (check OK)}$$
$$\Sigma F_V \uparrow + = S_{AB} + 34.0$$
$$= -34.0 + 34.0 = 0 \text{ (check OK)}$$

joint B

Step 5. The results of the analysis are presented on the results diagram of Figure 4.7d. On this diagram it is noted that in many members the member force is very much larger than the 10-kip applied load. This fact is typical of cranes of this sort, which therefore must be analyzed very carefully.

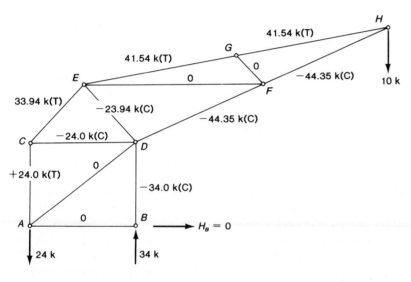

(d) results diagram

Figure 4.7 *(Continued)*

4.3 Method of Sections

The method of joints for analyzing trusses is particularly suited to situations where the forces in all members of the truss are required. Sometimes, however, the forces in only a few members of a truss are required. Under such circumstances the method of joints may become very long and tedious. For this reason the method of sections is used when the member forces are required in only a few of the truss members. Once again the method is based on the concept of isolating free bodies within the truss. However, the free bodies used in the method are usually much larger than the free bodies of joints, which were considered in the method of joints. In principle, a free body of a part of a truss may be isolated from anywhere within a truss provided that all external forces acting on the free body are shown. Thus, if the free body is formed by cutting one or more members of the truss, then the unknown forces in those members appear as external forces on the free body. The free body then isolated must be in equilibrium, and so the equilibrium equations may be applied to find any unknown member forces:

(4.1)
$$\Sigma F_H = 0$$

(4.2)
$$\Sigma F_V = 0$$

(4.3)
$$\Sigma M_{\text{any point}} = 0$$

As an example consider the truss of Figure 4.8a, in which only the forces in members BD and DE are required to be found. If the truss is cut along the line 1–1, then the truss is divided into two free bodies as shown in Figures 4.8b and 4.8c. The unknown forces in the members AB, BD, and DE have been shown as external forces acting on each free body, i.e., as forces S_{AB}, S_{BD}, and S_{DE}, respectively. In general the cut made in the truss should cut at least one of the members in which the force is required to be found. The cut 1–1 happens to cut both the members BD and DE and so the forces in both these members should be found with this single cut. Either of two free bodies may now be used to find the unknown forces. Furthermore, it is not always necessary to use all three equilibrium equations.

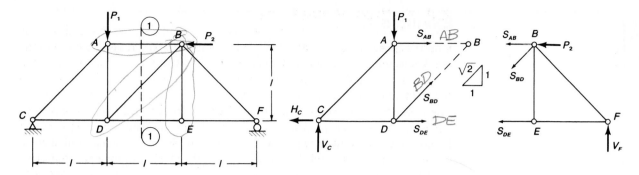

(a) complete truss

(b) left-hand free body

(c) right-hand free body

Figure 4.8 Method of sections.

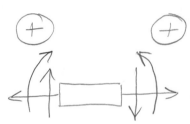

In both free bodies, for example, the force S_{BD} is the only one of the three unknown member forces that has a component in the vertical direction. Summation of forces on either free body in the vertical direction will therefore yield the value of S_{BD} directly. In addition it is noted that both of the unknown member forces S_{AB} and S_{BD} have lines of action through the point B. Thus the summation of moments about the point B will permit an equation to be derived in which the only unknown force is the force S_{BE}, which is the other force required.

It is somewhat difficult to list a general procedure for the determination of member forces by the method of sections, because the procedure varies with the truss and the members in which the value of the force is required. In many instances, however, it is good practice to start by finding the support reactions in the same manner in which the method of joints started. After this, it is usually a good idea to make cuts that include no more than three unknown forces. Examples 4.3 to 4.5 illustrate the fact that none of these rules is necessary all of the time. It is also noted that the cut made in a truss does not have to follow a straight line, as Example 4.4 shows.

Unlike the method of joints, there is <u>no automatic check</u> of the results when the method of sections is used. Great care must therefore be taken in the derivation of the equilibrium equations.

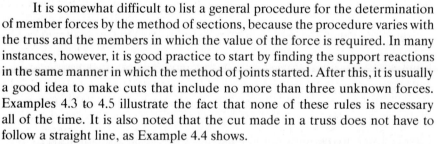

Example 4.3

Find the forces in the members *GH*, *BC*, and *BI* of the truss shown in Figure 4.9*a* when it is loaded as shown.

Solution:

The first step in finding the unknown member forces is to find the support reactions, and for this purpose the free body of the whole truss shown in Figure 4.9*b* has been drawn. On this free body the support reactions have been shown as V_F, H_F, and V_L. Application of the three equilibrium equations gives the equations:

REACTIONS:

$$\Sigma F_H = 0 \xrightarrow{+}$$

$$H_F - 12 = 0$$

$$H_F = 12 \text{ kips}$$

$$\Sigma F_V = 0 \uparrow +$$

$$V_F + V_L - 20 - 20 - 5 - 5 - 5 = 0$$

i.e.,

$$V_F + V_L = 55 \text{ kips}$$

$$\Sigma M_F = 0 \; \lrcorner +$$

$$20(20) + 20(40) + 5(60) + 5(80) + 5(100) - V_L(120) - 12(20) = 0$$

$$V_L = 18 \text{ kips}$$

and thus

$$V_F = 37 \text{ kips}$$

Clearly it will not be sufficient to make just one cut through the truss to find all the unknown member forces. The cut 1–1 cuts the member *GH* while the cut 2–2 cuts the members *BC* and *BI*.

The cut 1–1 produces the free bodies of Figure 4.9*c*. Either of these free bodies may be used to find the force S_{GH}. For convenience the left-hand free body is chosen, since there are fewer forces acting on this free body. If moments are taken about the point *A* on the left-hand free body, the unknown forces in members *AB* and *AH* will not enter into the equation, since both have lines of action through the point *A*. Thus

$$\Sigma M_A = 0 \; \lrcorner +$$

$$37(20) - 12(20) - S_{GH}(20) = 0$$

$$S_{GH} = 25 \text{ kips}$$

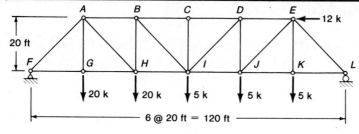

(a)

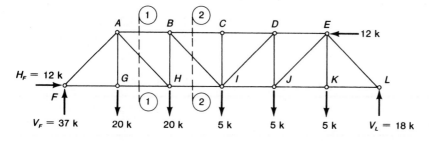

(b) free body of whole truss

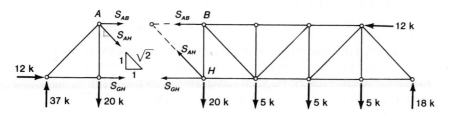

(c) free bodies at cut 1-1

NO MOMENT AT I
DUE TO LINE OF ACTION

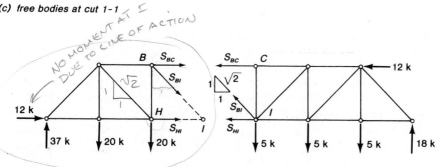

(d) free bodies at cut 2-2

Figure 4.9

Example 4.3 (Continued)

The positive sign of the answer indicates that the member is in tension. The cut 2–2 produces the free bodies of Figure 4.9c. Once again either free body may be chosen, but the left-hand free body is chosen because it has fewer forces acting on it. Applying the equilibrium equations yields:

$$\Sigma F_V = 0 \uparrow +$$

$$37 - 20 - 20 - S_{BI}\left(\frac{1}{\sqrt{2}}\right) = 0$$

$$S_{BI} = -3\sqrt{2} = -4.24 \text{ kips}$$

The negative sign indicates that the member BI is in compression.

Example 4.4

Find the forces in the members HI and IM of the K-truss shown in Figure 4.10a when it is subjected to the loads shown.

Solution:
It is noted that this is a complex truss, according to the definitions of Section 4.1. As usual, the first step is to derive a free body of the whole truss and use it to find the support reactions. Such a diagram is shown in Figure 4.10b. Applying the equilibrium equations to this diagram gives:

$$\Sigma F_H = 0 \xrightarrow{+}$$

$$H_G - 54 = 0$$

$$H_G = 54 \text{ kN}$$

$$\Sigma F_V = 0 \uparrow +$$

$$V_A + V_G - 80 - 15 = 0$$

$$V_A + V_G = 95 \text{ kN}$$

$$\Sigma M_A = 0 \; \curvearrowright +$$

$$15(6) + 80(18) - V_G(36) - 54(5) = 0$$

$$V_G = 35 \text{ kN}$$

and thus

$$V_A = 60 \text{ kN}$$

$$\Sigma M_I = 0 \ \text{\reflectbox{$\circlearrowleft$}}+$$

$$37(60) - 20(40) - 20(20) + S_{BC}(20) = 0$$

$$S_{BC} = -51 \text{ kips}$$

The member BC is also subjected to a compressive force.

As a demonstration that either free body may be taken in order to find the unknown member forces, the force in the member GH is now found using the right-hand free body of Figure 4.9c. Once again moments are taken about the point A, yielding the equation

$$\Sigma M_A = 0 \ \text{\reflectbox{$\circlearrowleft$}}+$$

$$S_{GH}(20) + 20(20) + 5(40) + 5(60) + 5(80) - 18(100) = 0$$

$$S_{GH} = 25 \text{ kips, as before}$$

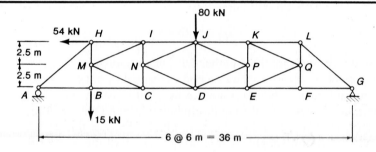

(a) K-truss diagram

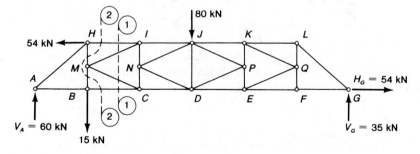

(b) free body diagram of whole truss

Figure 4.10

Example 4.4 *(Continued)*

It is now necessary to decide which cut or cuts to make in the truss. It may be thought that the cut 1–1 would be appropriate, and so the free bodies formed by the cut 1–1 are shown in Figure 4.10c. Examination of either free body, however, shows that this cut leaves too many unknown forces to be found. An alternative cut 2–2 is examined, and the free bodies formed by this cut are shown in Figure 4.10d. (Note that the cut made need not form a straight line.) Only one of the required unknown forces is included in these free bodies—the force S_{HI}. Once again it appears that there are too many unknown forces (four) on the free bodies for them to be found by using the equilibrium equations. However, closer examination reveals that three of these forces (S_{BC}, S_{BM}, and S_{HM}) all have lines of action that pass through the point B. Taking moments about B will therefore produce an equation in which the only unknown force is the force in the member HI. For convenience this equation is applied to the left-hand free body of Figure 4.10d:

$$\Sigma M_B = 0 \; \lambda +$$

$$S_{HI}(5) + 60(6) - 54(5) = 0$$

$$S_{HI} = -18 \text{ kN}$$

The negative sign indicates that the member HI is in compression.

It is now possible to return to the free bodies formed by cut 1–1. Since one of the unknown forces at the cut is now known, it is possible to find the three other unknown forces using the equilibrium conditions. Only the force in the member IM is required, and so the most efficient way of finding the force in this member is now sought. A free body diagram of the section to the left of cut 1–1 is redrawn in Figure 4.10c. It is clear that both the unknown forces S_{BC} and S_{CM} have lines of action through the point C. Taking moments about this point would then include only one unknown force—the required force in the member IM. According to the principle of transmissibility of forces, the force S_{IM} may be considered to act at the point I, as shown. At this point only the horizontal component of S_{IM} has any moment about the point C, since the line of action of the vertical component also goes through the point C. Thus:

$$\Sigma M_C = 0 \; \lambda +$$

$$60(12) - 15(6) - 54(5) + S_{HI}(5) + S_{IM}\left(\frac{12}{13}\right)(5) = 0$$

$$S_{IM} = -58.5 \text{ kN}$$

The member IM is thus also in compression.

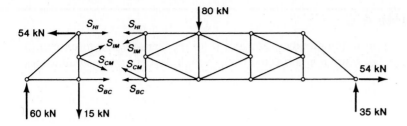

(c) free bodies formed by cut 1-1

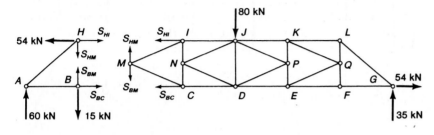

(d) free bodies formed by cut 2-2

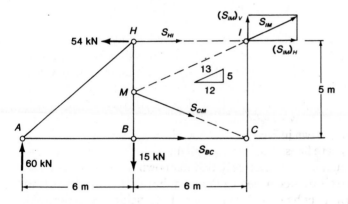

(e) free body to left of cut 1-1

Figure 4.10 *(Continued)*

Example 4.5

Find the forces in the members *DH* and *JL* of the transmission tower shown in Figure 4.11*a* when it is subjected to the weight forces and line-pull forces shown.

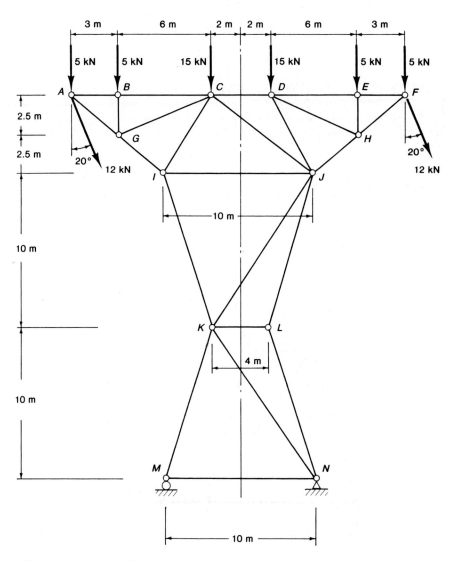

(a) transmission tower diagram

Figure 4.11

Solution:

A free body diagram of the whole truss is drawn, as in Figure 4.11b, in order to find the support reactions. Applying the equilibrium equations to this free body gives

$$\Sigma F_H = 0 \xrightarrow{+}$$

$$12 \cdot \sin 20° + 12 \cdot \sin 20° + H_N = 0$$

$$H_N = -8.21 \text{ kN}$$

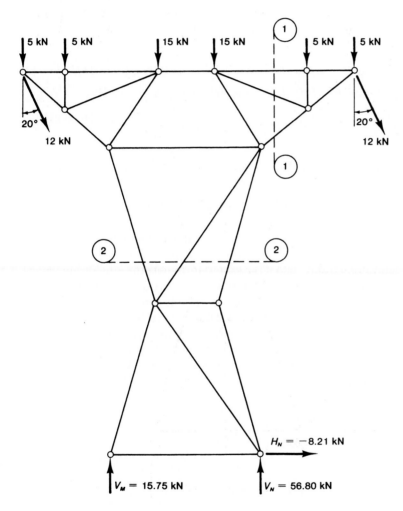

(b) free body diagram of whole tower

Example 4.5 (Continued)

The negative sign of the value of H_N indicates that the reaction force was initially assumed to be directed in the wrong direction: it is really directed to the left.

$$\Sigma F_V = 0 \uparrow +$$

$$V_M + V_N - 5 - 5 - 15 - 15 - 15 - 5 - 5 - 12 \cdot \cos 20°$$

$$-12 \cdot \cos 20° = 0$$

$$V_M + V_N = 72.55 \text{ kN}$$

$$\Sigma M_M = 0 \downarrow +$$

$$-12 \cdot \cos 20° \cdot (6) + 12 \cdot \sin 20° \cdot (25) - 5(6) - 5(3) + 15(3) + 15(7) + 5(13)$$

$$+ 5(16) + 12 \cdot \sin 20° \cdot (25) + 12 \cdot \cos 20° \cdot (16) - V_N \cdot (10) = 0$$

$$V_N = 56.80 \text{ kN}$$

and thus

$$V_M = 15.75 \text{ kN}$$

The cut 1–1 is now made in the truss in order to find the force in the member *DH*, and a free body diagram of the portion of the truss to the right of this cut is shown in Figure 4.11c. (Of course, a diagram of the portion of the truss to the left of the cut 1–1 could also be drawn. This portion, however, is acted upon by many more forces, and it is easier to formulate the equilibrium equations on the portion shown.) Note that the cut reveals three unknown forces— in the members *ED, DH,* and *HJ*—whereas only the force in the member *DH* is required. Note also that the forces in the members *HJ* and *ED* both have lines of action that pass through the point *F*. Taking moments about the point *F* will therefore eliminate both of these forces. As an additional attraction, taking moments about *F* also eliminates both forces acting at *F*. The moment equilibrium equation then becomes

$$\Sigma M_F = 0 \downarrow +$$

$$-5(3) + S_{DH} \cdot \left(\frac{5}{13} \right) (9) = 0$$

$$S_{DH} = 4.33 \text{ kN}$$

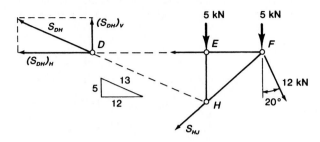

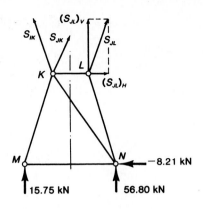

(c) free body diagram to the right of cut 1-1

(d) free body diagram below cut 2-2

Figure 4.11 *(Continued)*

In order to write this equation, the principle of transmissibility of forces was used to transfer the point of application of the force S_{DH} to the point D. At this point only the vertical component of the force has any moment about the point F. Note that in order to find the force in the member DH, it was not strictly necessary to find the support reactions, since they were not used in the equilibrium equation.

Consider now the cut 2–2, which cuts through the member JL and should enable the force in this member to be found. In Figure 4.11c a free body diagram of the portion of the truss below this cut has been drawn. The equilibrium of this portion will be considered. If moments are taken about the point K, the only unknown force to be included is the required force in the member JL:

$$\Sigma M_K = 0 \ \zeta+$$

$$15.75(3) - S_{JL} \cdot \left(\frac{10}{10.44}\right)(4) - 56.80(7) + 8.21(10) = 0$$

$$S_{JL} = -70.0 \text{ kN}$$

The member JL is thus also in compression. In order to formulate the above equation it is necessary to recognize that only the vertical component of the force S_{JL} has any moment about the point K.

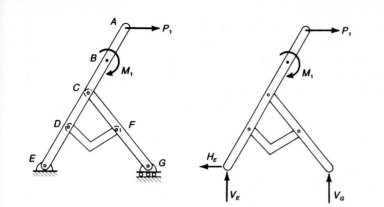

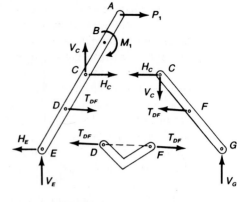

(a) frame diagram   (b) free body diagram of whole frame   (c) exploded diagram

Figure 4.12   *Free body diagrams for frames.*

4.4 Frames

As does a truss, a frame consists of a number of members connected together. However, unlike a truss, a frame includes at least one member in which loads are applied at points other than at its ends. In other words, whereas a truss is composed of two-force members, a frame includes at least one multi-force member. A frame is essentially a rigid structure intended to support a given loading.

      In the same way as trusses were analyzed by dividing the truss into a series of free bodies, so the interconnecting forces between the elements of a frame may be found by considering the frame broken into the necessary free bodies. Although there are numerous examples of three-dimensional frames, this section will be restricted to the consideration of two-dimensional frames.

      A procedure for finding the forces between the elements of a frame resembles that describing the method of sections.

Steps in Finding the Interconnecting Forces Between Elements of a Frame

The steps outlined here will be illustrated with reference to the frame of Figure 4.12a.

Step 1.   Start solving a frame problem by finding the support reactions. To this end a free body diagram of the entire frame is drawn as shown in Figure 4.12b. In this diagram the unknown support reactions have been replaced by the equivalent forces in the appropriate directions. By applying the equilibrium equations the support reactions may be found, provided there are no more than three unknown support forces or moments.

Step 2.   To find the forces between the elements of a frame an exploded diagram is drawn as shown in Figure 4.12c. The known forces and/or moments (P_1, M_1) acting on the isolated elements are shown, as are the support reactions, which are now known in magnitude and direction.

Considerable care must be exercised to ensure that the correct forces are included at the connections between elements of the frame. For example, at the pinned-joint connection at C in the example frame there may be both horizontal and vertical force components of the connection force, but, since it is a pinned joint, there will be no moment. Thus the two forces V_C and H_C must be included. Since the sense of these forces is as yet unknown, the directions of the forces are merely guesses. It is now necessary to recall Newton's law that "to every action there is an equal and opposite reaction." Thus if the forces V_C and H_C are included in the directions shown on the element $ABCDE$, then these forces must be of equal magnitude but in opposite directions on the element CFG. This requirement is satisfied in Figure 4.12c, and it is a general requirement for all joints in a frame.

In many instances the problem may be considerably simplified by recognizing two-force members where they occur. In the frame of Figure 4.12 the element DF is indeed a two-force member, since loads are applied to it at its pinned ends only. As discussed in Section 4.1, a two-force member has forces at its ends that have lines of action that pass through the pinned connections. In addition it is known that the forces at each end are of equal magnitude but of opposite direction (as was the case in truss members). Thus the force T_{DF}, the unknown force in the member DF, is acting along a known line of action, even though the member is not straight. Newton's law is again applied to find the consistent directions of the forces on the various elements of the frame, even though the correct directions of this force are not known as yet.

Once the elements of the frame have been isolated, the elements of the frame may be considered in turn to find the interconnecting forces. For example, there are three unknown forces acting on the element $ABCDE-V_C$, H_C, and T_{DF}. The line of action of each of these forces is known, although not the direction. Since three equilibrium equations are applicable to the element $ABCDE$, the values of the unknown forces may be found. In the example shown, there are only three unknown forces in the whole frame, and so no further elements need be considered to find other unknown forces. In general, however, it may be necessary to consider more than one multiforce element in a frame in order to find all the unknown forces.

Step 3. In the frame of Figure 4.12 it is noted that the equilibrium equations did not have to be applied to the element CFG of the frame in order to find the required forces. This element CFG then serves as a check element. If the previous computations are all correct, it must be possible to prove that the element CFG is also in equilibrium by applying the three available equilibrium equations. There should always be such a multiforce check element available in a frame when all the forces have been found.

Example 4.6

Find the forces between the elements of the frame shown in Figure 4.13a.

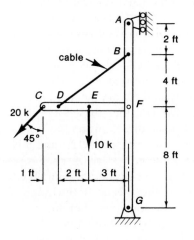

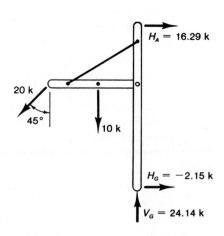

(a) frame diagram

Figure 4.13

(b) free body diagram of whole frame

Solution:

Step 1. A free body diagram of the complete frame is drawn in Figure 4.13b in order to find the support reactions. For this free body the following equations are derived:

$$\Sigma F_V = 0 \uparrow +$$

$$V_G - 10 - 20 \cdot \cos 45° = 0$$

$$V_G = 24.14 \text{ kips}$$

$$\Sigma F_H = 0 \xrightarrow{+}$$

$$H_G + H_A - 20 \cdot \sin 45° = 0$$

$$H_G + H_A = 14.14 \text{ kips}$$

$$\Sigma M_F = 0 \; \text{↻} +$$

$$H_A(14) - 10(3) - 20 \cdot \cos 45° \cdot (8) - 20 \cdot \sin 45° \cdot (6) = 0$$

$$H_A = 16.29 \text{ kips}$$

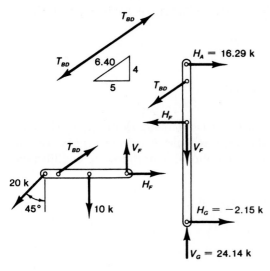

$H_A = 16.29$ k

$H_G = -2.15$ k

$V_G = 24.14$ k

(c) exploded diagram

and thus

$$H_G = -2.15 \text{ kips}$$

Step 2. The exploded diagram of Figure 4.13c is drawn. Since the cable *BD* is a two-force member it is replaced by a single force between its ends. It is now possible to use either element *CDEF* or element *ABFG* to find the unknown forces at the frame connections. If the element *CDEF* is chosen, then the following equations are found:

$$\Sigma F_V = 0 \uparrow +$$

$$V_F + T_{BD} \cdot \left(\frac{4}{6.40}\right) - 10 - 20 \cdot \cos 45° = 0$$

$$V_F + 0.625 \, T_{BD} = 24.14 \text{ kips}$$

$$\Sigma F_H = 0 \xrightarrow{+}$$

$$H_F + T_{BD} \cdot \left(\frac{5}{6.40}\right) - 20 \cdot \sin 45° = 0$$

$$H_F + 0.781 \, T_{BD} = 14.14 \text{ kips}$$

$$\Sigma M_F = 0 \; \text{2+}$$

$$T_{BD} \cdot \left(\frac{4}{6.40}\right) (5) - 10(3) - 20 \cdot \cos 45° \cdot (6) = 0$$

$$T_{BD} = 36.78 \text{ kips}$$

Example 4.6 (Continued)

Thus,

$$V_F = 1.15 \text{ kips}$$

and

$$H_F = -14.59 \text{ kips}$$

Step 3. All the unknown forces have now been found. However, the element
ABFG has not been used in any way. Thus this element is now used in
order to check previous calculations. The following equilibrium
checks are valid:

$$\Sigma F_V \uparrow + = V_G - V_F - T_{BD} \cdot \left(\frac{4}{6.40} \right)$$

$$= 24.14 - 1.15 - (36.78)(0.625)$$

$$= 0 \text{ (check OK)}$$

Example 4.7

Find the forces at the connections between the elements of the frame of Figure 4.14a, when it is subject to the 15 kilonewton force at B in a horizontal
direction and a mass of 500 kilograms is hung from the small frictionless
pulley at C by a wire that is connected to a wall at H, on the same horizontal
level as C.

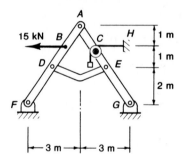

(a) frame example

Figure 4.14

(b) free body diagram of whole truss

Solution:
Step 1. In order to find the support reactions at F and G, it is necessary
to isolate the frame by cutting the wire CH and put it in equilibrium.
The weight of the 500 kilogram mass is given by

$$\Sigma F_H \overset{+}{\rightarrow} = H_A + H_G - H_F - T_{BD} \cdot \left(\frac{5}{6.40} \right)$$

$$= 16.29 + (-2.15) - (-14.59) - (36.78)(0.781)$$
$$= 0 \text{ (check OK)}$$

$$\Sigma M_G \, \rotatebox{90}{$\curvearrowright$}^+ = H_A(14) - T_{BD} \cdot \left(\frac{5}{6.40} \right) \cdot (12) - H_F \cdot (8)$$

$$= 16.29(14) - (36.78)(0.781)(12) - (-14.59)(8)$$
$$= 0 \text{ (check OK)}$$

$$W = mg = (500 \text{ kg}) \cdot (9.81 \text{ m/s}^2) = 4.9 \text{ kN}$$

Since the pulley is frictionless, the tension in the wire on both sides of the pulley is equal to the weight of the mass hanging on the wire. The free body diagram of the whole frame is shown in Figure 4.14b. The support reactions are now given by

$$\Sigma F_H = 0 \overset{+}{\leftarrow}$$
$$15 + H_F - 4.9 = 0$$
$$H_F = -10.1 \text{ kN}$$

$$\Sigma F_V = 0 \uparrow +$$
$$V_F + V_G - 4.9 = 0$$
$$V_F + V_G = 4.9 \text{ kN}$$

$$\Sigma M_F = 0 \, \rotatebox{90}{$\curvearrowright$} +$$
$$-15(3) + 4.9(3.75) + 4.9(3) - V_G(6) = 0$$
$$V_G = -2.0 \text{ kN}$$

and thus

$$V_F = 6.9 \text{ kN}$$

Example 4.7 *(Continued)*

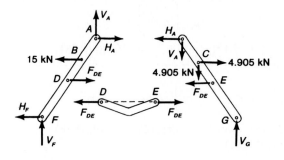

(c) exploded diagram

Figure 4.14   *(Continued)*

Step 2. In the exploded diagram of Figure 4.14c, the member DE is treated as a two-force member since the only forces on it are applied at its pinned connections to the other members. Using the element $ACEG$ the appropriate equilibrium equations are

$$\Sigma F_V = 0 \uparrow +$$

$$-V_A - 4.905 + V_G = 0$$

Since

$$V_G = -1.981 \text{ kN}$$

$$V_A = -6.887 \text{ kN}$$

$$\Sigma F_H = 0 \xrightarrow{+}$$

$$-H_A + 4.905 - F_{DE} = 0$$

4.5 Different Types of Frames

The designer should be aware that there are two different types of frames. The type of frame discussed above is characterized by having nonrigid joints connecting the pieces of the frame: the joints have all been either pinned joints or sliding joints. This enables the forces at the joints to be readily calculated using the principles of statics applied to free bodies of the individual parts of a frame. Such a frame is shown in Figure 4.15a.

Some frames, however, have rigidly connected members. The members might be welded together so that no relative movement or rotation may take place between the parts of a frame. Such a frame is shown in Figure 4.15b. The principles presented above are not sufficient to analyze such frames, and the reader should seek further help with such structures.

i.e.,

$$F_{DE} + H_A = 4.905 \text{ kN}$$

$$\Sigma M_A = 0 \circlearrowleft+$$

$$4.905 \, (0.75) - 4.905 \, (1) + F_{DE} \cdot (2) - V_G \cdot (3) = 0$$

i.e.,

$$F_{DE} = -2.360 \text{ kN}$$

and thus

$$H_A = 7.265 \text{ kN}$$

Step 3. The element *ABDF* has not been used in the determination of the interconnection forces and thus is now used as the multiforce checking element. Checking equilibrium of this element gives

$$\Sigma F_H \overset{+}{\rightarrow} = H_A - 15 + F_{DE} - H_F$$

$$= 7.265 - 15 - 2.360 + 10.095$$

$$= 0 \text{ (check OK)}$$

$$\Sigma F_V \uparrow+ = V_A + V_F = -6.887 + 6.887 = 0 \text{ (check OK)}$$

$$\Sigma M_A \circlearrowleft+ = H_F \cdot (4) + V_F \cdot (3) - F_{DE} \cdot (2) + 15 \, (1)$$

$$= (-10.095) \, (4) + (6.887) \, (3) - (-2.360) \, (2) + 15 \, (1)$$

$$= 0 \text{ (check OK)}$$

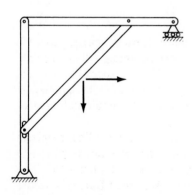

(a) *a typical frame with nonrigid joints*

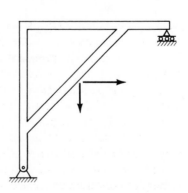

(b) *a typical frame with rigid joints*

Figure 4.15   *Types of frames.*

Problems

4.1   Determine the bar force in each lettered member of the truss shown in Figure P 4.1.

4.2   Find all zero-force members in the structure shown in Figure P 4.2.

4.3–
4.8   By the joint method compute the bar forces in the lettered bars of the trusses shown in Figures P 4.3. to P 4.8.

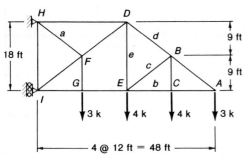

Figure P 4.1

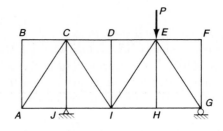

Figure P 4.2

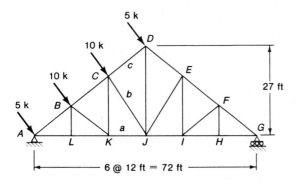

Figure P 4.3

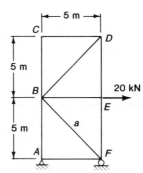

Figure P 4.4

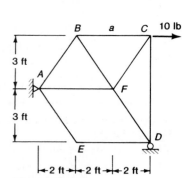

Figure P 4.5

4.9   Determine the forces in each member of the Howe truss loaded as shown in Figure P 4.9. Use the method of joints to solve.

4.10   For the truss shown in Figure P 4.10, all the forces applied are perpendicular to the chord. Find the forces in all members by method of joints.

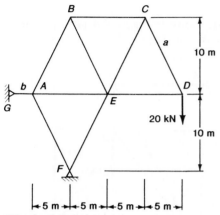

Figure P 4.6

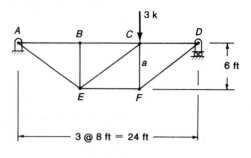

Figure P 4.7

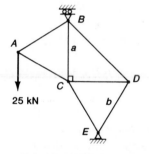

all members are 5 m in length except "BD"

Figure P 4.8

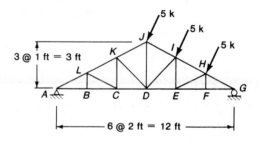

Figure P 4.9

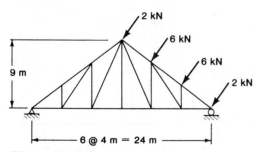

Figure P 4.10

4.11 Find the forces in all the members and reactions for the truss loaded as shown in Figure P 4.11.

4.12 Find the reactions at A and E, and the axial forces in members BE and CD for the truss shown in Figure P 4.12.

4.13 Determine the axial forces in each of the members of the truss shown in Figure P 4.13.

4.14 Repeat Problem 4.13 for the truss shown in Figure P 4.14.

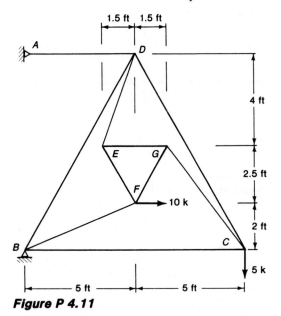

Figure P 4.11

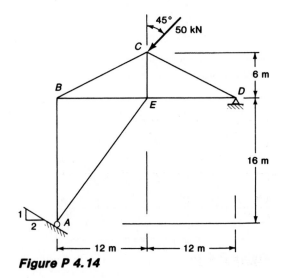

Figure P 4.12

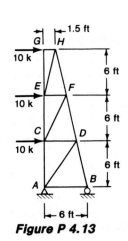

Figure P 4.13

Figure P 4.14

4.15   The signboard truss is designed to support a horizontal wind load
        of 1,200 pounds. A separate analysis shows that 5/8 of the wind load
        is transmitted to joint *B,* while the remainder is equally divided
        between joints *A* and *D.* Determine the forces in members *BD*
        and *AC.*

4.16–
4.17   Calculate the forces in members *a* and *b* for the structures shown in
        Figures P 4.16 and P 4.17.

4.18   Make a complete analysis of the truss shown in Figure P 4.18.

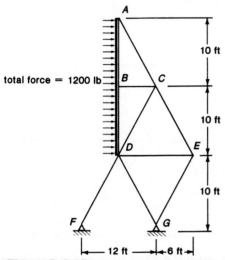

Figure P 4.15

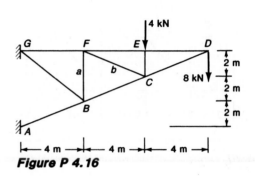

Figure P 4.16

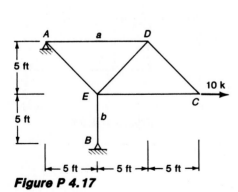

Figure P 4.17

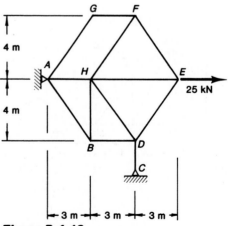

Figure P 4.18

4.19–
4.24 Compute the forces in the lettered bars of the trusses shown in
Figures P 4.19 to P 4.24.

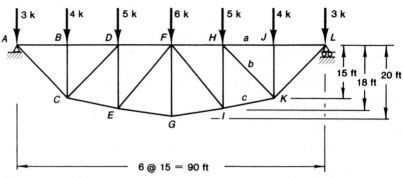

Figure P 4.19

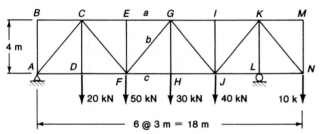

Figure P 4.20

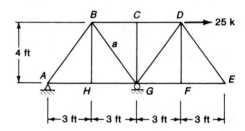

Figure P 4.21

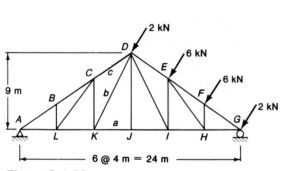

Figure P 4.22

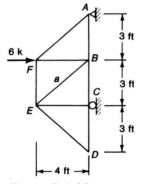

Figure P 4.23

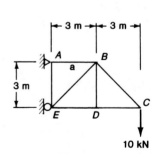

Figure P 4.24

4.25–

4.29 Find the forces at the connections between the elements of the frames in Figures P 4.25 to P 4.29.

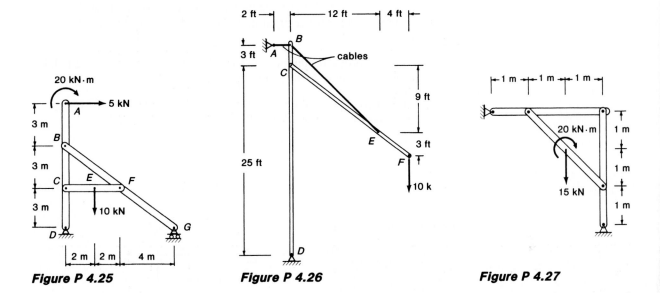

Figure P 4.25

Figure P 4.26

Figure P 4.27

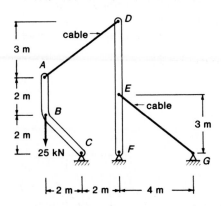

Figure P 4.28

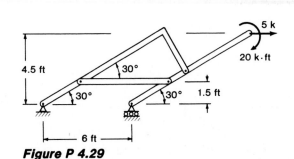

Figure P 4.29

4.30   Find the forces in all members of the truss shown in Figure P 4.30
        by method of joints.

   $L^* = SS + 50$ (length in feet)
   $P^* = (SS + 50)^2$ (force in pounds)

        SS in these formulas represents either the last two digits of the
        student's Social Security number, or two digits assigned by the
        instructor.

4.31   Find the forces in members AB and BE of the truss of Figure P 4.30
        using the method of sections.

4.32   Find the forces in all members of the truss of Figure P 4.32 using
        method of joints. For L^* and P^* see Problem 4.30.

4.33   Find the forces in members BG, CD, and ED in Figure P 4.32 by the
        method of sections.

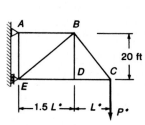

Figure P 4.30

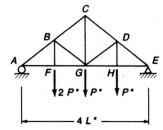

Figure P 4.32

Shear Force Diagrams 5

Objectives

1. *To understand the concept of internal shear force in beams.*

2. *To develop the ability to compute the internal shear force in any cross section of a beam using free body diagrams.*

3. *To understand the application of the technique of superposition to shear force determinations.*

4. *To learn how to draw a shear force diagram by inspection.*

5.1 The Concept of Shear Force

Structural or machine elements that are subjected to external loads transverse to their longitudinal axis are called *beams*. For instance, from the standpoint of the designer, the handle of a wrench, the shaft of an electric motor, or the floor of a building is considered to be a beam. Since the analysis of curved beams is a little more complicated than that of straight beams, our discussion will be limited to the latter. In practical applications beams may be placed vertically, horizontally, or in an inclined manner. For the sake of convenience, we will draw them in a horizontal position when considering the examples that follow.

External forces acting on a beam consist of loadings and their corresponding reactions. The computation of reactions at the points of support for any given external loading on statically determinate beams was presented in Chapter 3. Methods for determining reactions at all supports for statically indeterminate beams will be considered in Chapter 14. In either case, a beam under external loading with all reactions properly determined is in static equilibrium. For the purposes of this chapter no discrimination is needed between active external loads and their corresponding reactions.

External forces acting on a mechanical or structural component are transferred through the material by internal forces. These internal forces, and moments, may be computed at any arbitrarily selected cross section of the beam by applying the method of the *free body diagram* that was introduced in Chapter 3. Let us consider, for example, the case of a wrench, shown in Figure 5.1a. The function of the wrench is to create a torque on the bolt in order to loosen or tighten it. The torque T applied to the bolt is the product of the force P applied by the hand and l, the distance (called the arm) between the hand and the bolt. Figure 5.1b shows the problem in its idealized condition. The forces acting on the beam are found by using the equilibrium conditions: $\Sigma F_H = 0$, $\Sigma F_V = 0$, and $\Sigma M = 0$. If the magnitude of the internal forces is desired, at section a–a of the wrench, for example, then the beam is cut at that section and the equilibrium formula is applied for either part. It does not matter which part of the bisected beam is used for the computation of the internal shear force. The answer should be the same, except that the direction is opposite on the two sides of the cut.

Internal shear forces may be positive or negative. The sign depends on the *sign convention* selected by the designer. In this chapter the positive internal shear will be the one that would result in a clockwise rotation in a small segment of the beam cut out of the whole, in the absence of external forces providing equilibrium. This concept is shown in Figure 5.2. According to this sign convention, the small element of the wrench between cuts a–a and b–b, shown in Figure 5.1d, is subjected to negative internal shear forces.

Depending on the shape of the beam and on the external loading, the magnitude of the internal shear force may vary from point to point along the axis of the beam. The graphical representation of this internal shear force distribution is the *shear force diagram*.

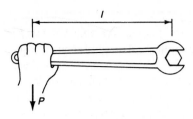

(a) actual conditions

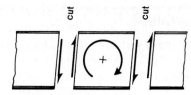

Figure 5.2
Sign convention for internal shear forces: positive shear results in clockwise rotation.

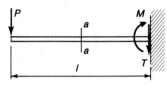

(b) idealized loading diagram

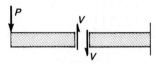

(c)   the shear force in a cut
        (internal moments not here)

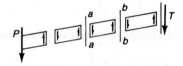

(d) the transfer of shear in the beam

Figure 5.1   *Interpretation of the internal shear force.*

5.2 Shear Force Diagram

A straightforward method of constructing shear force diagrams is to use the free body diagrams. The beam may be cut at any arbitrary point along its axis, and the equilibrium formula for vertical forces is used to determine the internal shear force. By analyzing a number of such cuts at characteristic points corresponding to the external loading distribution, the magnitude of the internal shear forces at these points can be found. Plotting these values along the axis of the beam will result in the shear force diagram. The following examples demonstrate this method.

Example 5.1

Determine the shear force diagram for the beam loaded as shown in Figure 5.3.

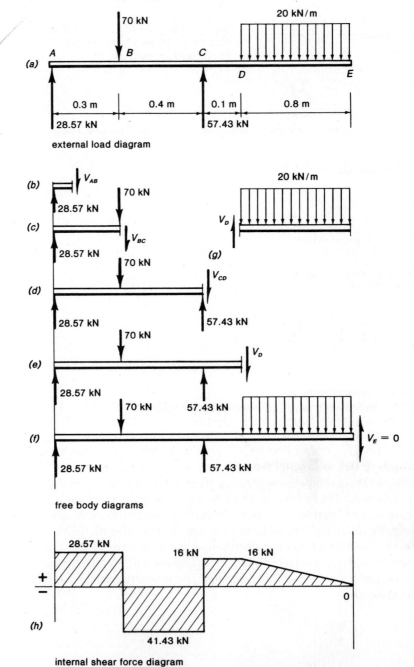

Figure 5.3 *Determination of shear force diagram by free body equilibrium equations.*

Solution:

Before any shear force diagram may be derived, the beam must be shown to be in equilibrium. By writing the three formulas of equilibrium at any point of the beam, one may ascertain that the loading given is indeed in complete equilibrium.

Points A through E may be selected as those characteristic for the external loading. If the beam is cut immediately to the right of point A, the free body is the small piece of beam between A and the cut. The equilibrium formula for transverse forces will contain only one external force, that of 28.57 kilonewtons acting upward at point A. Since

$$\Sigma F_V = 0 \uparrow +$$

$$28.57 - V_{A-B} = 0$$

and the shear force must equal

$$V_{A-B} = 28.57 \text{ kN}$$

Since the free body of Figure 5.3 is the same wherever the cut is made between A and B, the value of the shear between A and B is a constant equal to 28.57 kilonewtons. Continuing the process, the beam is to be cut next immediately to the right of point B. Writing the equilibrium formula we find that

$$\Sigma F_V = 0 \uparrow +$$

$$28.57 - 70 - V_{B-C} = 0$$

i.e.,

$$V_{B-C} = -41.43 \text{ kN}$$

Recalling the sign convention adopted in Figure 5.2, one may notice that while the internal shear force was positive between points A and B, it reverses to the negative sense beyond point B. Once again the shear force is constant between B and C. The next cut may be made to the right of point C. The equilibrium of transverse forces will now involve three known external forces and the internal shear force in the cut; that is,

$$\Sigma F_V = 0 \uparrow +$$

$$28.57 - 70 + 57.43 - V_{C-D} = 0$$

Thus,

$$V_{C-D} = 16 \text{ kN}$$

Example 5.1 (Continued)

We may note at this point that the summation of forces could be simplified by considering the free body diagram to the right-hand side of the cut. On the right-hand free body diagram, the only external loading present is the distributed load of 20 kilonewtons per meter acting over a length of 0.8 meter. Hence the total load of the right side is 20 (0.8), or 16 kilonewtons. This load would produce a downward movement of the beam unless a shear force directed upward maintained the balance. In comparing the free body diagrams shown in Figure 5.3e and g, we may note that while V_D in both cases equals 16 kilonewtons, it acts in opposite directions in the two free bodies and has a positive sign in the two cases.

Example 5.2

The floor joist shown in Figure 5.4 spans 18 feet. It is uniformly loaded with a spread load of 200 pounds per foot. Construct the corresponding internal shear force diagram.

Solution:
First the reactions are to be computed. The net applied force, $R = 200 \times 18$, or 3,600 pounds, acting at the center of the beam. Thus

$$\Sigma M_A = 0 \; \circlearrowright +$$

$$3,600 \cdot \left(\frac{18}{2}\right) - R_R \cdot (18) = 0$$

$$R_R = 1,800 \text{ lb}$$

$$\Sigma F_V = 0 \uparrow +$$

$$R_R + R_L - 3,600 = 0$$

$$R_L = 1,800 \text{ lb}$$

The free body isolated by a cut at an arbitrary distance x feet to the right of the left end of the beam is subject to the equilibrium equation

$$\Sigma F_V = 0 \uparrow +$$

$$R_L - [200 \text{ (lb/ft)} \cdot x \text{ (ft)}] - V = 0$$

from which we get

$$V = 1,800 - 200 \cdot x$$

At point E the beam ends. For lack of material contact at the ends the internal shear force must be zero. One useful check on the correctness of our work is that the shear force must reduce to zero at the ends of the beam. If this is not the case, the most likely source of error is that the beam reactions were incorrectly determined. Between D and E the shear force diagram is a straight sloping line connecting values of 16 kilonewtons at D and zero at E (see Figure 5.9b).

The final step in the construction of the shear force diagram is plotting the results. The diagram is shown in Figure 5.3h. As indicated in the graph, the maximum internal shear exists between points B and C. The internal shear force there is a constant, -41.4 kilonewtons. The slope of the line between points D and E corresponds to the distributed load of 20 kilonewtons per meter.

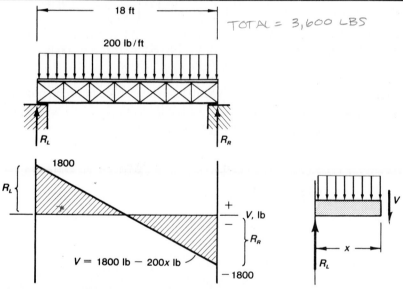

Figure 5.4 *Shear force diagram under a uniformly distributed load.*

at any arbitrary location. The equation represents a line with a negative slope of 200 pounds per foot along the beam's axis. At $x = 0$, just right of the left-side reaction, the internal shear force is 1,800 pounds. At the midpoint of the joist the internal shear force is zero, since for

$$x = 18/2 = 9 \text{ ft}, \ V = 1,800 - 200 \ (9) = 0$$

At the right support, the internal shear force reaches its maximum negative value of $-1,800$ pounds.

Example 5.3

A 12-foot-long simply supported beam carries a concentrated force of 10,000 pounds 3 feet away from the left support. Determine the internal shear forces in the beam using the table in Appendix B.

Solution:

Reviewing the table we find that Case 12 corresponds to the loading situation described. The notations are shown in Figure 5.5. Accordingly,

$$a = 3 \text{ ft}$$
$$b = 9 \text{ ft}$$
$$l = 12 \text{ ft}$$
$$W = 10,000 \text{ lb}$$

Using the equations given in the table, the reactions are

$$R_1 = + W\left(\frac{b}{l}\right) = 10,000\left(\frac{9}{12}\right) = 7,500 \text{ lb}$$

$$R_2 = + W\left(\frac{a}{l}\right) = 10,000\left(\frac{3}{12}\right) = 2,500 \text{ lb}$$

both acting in the upward direction. The internal shear force between points A and B is

$$V_{AB} = + W\left(\frac{b}{l}\right) = + 7,500 \text{ lb}$$

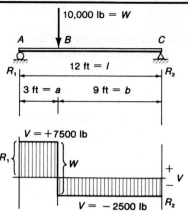

Figure 5.5 Tabular solution of internal shear force distribution in a two-supported beam.

and between points B and C is

$$V_{BC} = - W\left(\frac{a}{l}\right) = - 2,500 \text{ lb}$$

The results, depicted in Figure 5.5, show that between concentrated forces acting on a beam the internal shear force is a constant value. The shear forces at the two ends of the beam are both zero.

Example 5.4

A 12-foot-long simply supported beam is loaded with a uniformly distributed force of 2,000 pounds per foot. Determine the reactions and the distribution of the internal shear force and its magnitude 7 feet away from the left support, using the table in Appendix B.

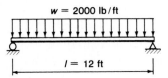

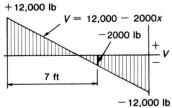

Figure 5.6   *Internal shear force distribution under linearly varying external load.*

Solution:

The corresponding loading condition in our table is Case 15. The notations are shown in Figure 5.6. The reactions at the supports are given by the table as

$$R_1 = R_2 = +\frac{1}{2}(w \cdot l) = +\frac{1}{2}(2{,}000\,\text{lb/ft}) \cdot 12\,\text{ft} = 12{,}000\,\text{lb}$$

Internal shear forces in simply supported or cantilevered beams subjected to common cases of loading may be determined by tabulated solutions. One such table is included in Appendix B. Similar tables are available in design codes and in engineering handbooks.

According to the laws of statics, collinear forces may be added together, or subtracted from each other, according to their signs. This principle, called the *principle of superposition*, allows us to treat combined loading conditions for which direct solutions may not be found in tables. For instance, when several concentrated forces or combinations of concentrated and distributed

The equation giving the internal shear force at any point x along the beam is

$$V = \frac{1}{2}(w \cdot l)\left(1 - \frac{2x}{l}\right) = \frac{1}{2}(2,000)\,12\left(1 - \frac{2x}{12}\right)$$

$$= 12,000 - 2,000\,(x)\ \text{lb}$$

From the form of this equation, one may realize that the internal shear force varies linearly along the beam. Its maximum positive value appears where $x = 0$,

$$V_A = 12,000\ \text{lb}$$

The maximum negative value is at $x = l = 12$ ft, where

$$V_B = 12,000 - 2,000(12) = -12,000\ \text{lb}$$

At midspan, $x = 6$ ft, and the shear force is zero, since

$$V_{midspan} = 12,000 - 2,000(6) = 0\ \text{lb}$$

The equation allows the plotting of the shear force diagram as shown in Figure 5.6. The magnitude of the shear force at $x = 7$ ft is

$$V_{7\,ft} = 12,000 - 2,000(7) = -2,000\ \text{lb}$$

forces appear together on a simple beam, the internal shear force diagram can be constructed by finding solutions for each one of the individual cases of loading separately, and adding together the individual results. Example 5.5 demonstrates this technique.

The complete understanding of the construction of internal shear force diagrams make it possible for the reader to develop any one of the formulas shown in the table of Appendix B or to derive appropriate formulas for shear subject to any type of loading. This requires that all force and length variables be introduced in a symbolic form rather than numerically. To demonstrate the method of derivation for shear force diagrams, Example 5.6 may be considered. It presents a derivation for Case 17 of the table in Appendix B.

Example 5.5

The 12-foot-long simply supported beam considered in the two previous examples is loaded both by the concentrated force of 10,000 pounds and the distributed force of 2,000 pounds per foot. Determine the internal shear force distribution along the beam.

Solution:

The individual solutions for the internal shear force distribution for the concentrated load 3 feet to the right of the left support and for the distributed load are available in Examples 5.3 and 5.4. These are as follows: between the left support and the 10,000-lb force,

$$V = + 7,500 \text{ lb, a constant value;}$$

between the force and the right support,

$$V = - 2,500 \text{ lb, a constant value;}$$

from the distributed force,

$$V = 12,000 - 2,000 \, x.$$

Adding the corresponding results together will give: between the left support and the concentrated force:

$$V = 7,500 + 12,000 - 2,000 \, x = 19,500 - 2,000 \, x \text{ (lb)}$$

between the concentrated force and the right support:

$$V = - 2,500 + 12,000 - 2,000 \, x = 9,500 - 2,000 \, x \text{ (lb)}$$

To plot the combined shear force diagram, the functional values need to be computed at the two supports and at the location of the concentrated force. Substituting $x = 0$ and $x = 12$ ft, respectively, in the first and second equations results in

$$V_0 = 19,500 \text{ lb}$$

$$V_{12 \text{ ft}} = - 14,500 \text{ lb}$$

A careful study of Examples 5.1—5.6 leads one to an important general conclusion: *The internal shear force in any cut of a beam equals the area under the diagram of external forces to either side of the cut.* Based on this general concept, shear force diagrams may be constructed by inspecting the functional nature of the loading. This method is developed in detail in Section 5.3.

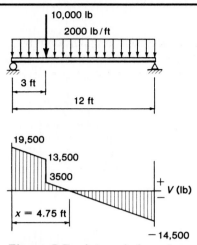

Figure 5.7 *Internal shear force diagram for combined loading.*

At $x = 3$ ft, the first equation gives the shear force immediately to the right of the concentrated force:

$$V = 19,500 - 2,000(3) = + 13,500 \text{ lb}$$

and the second equation gives the shear force immediately to the left as

$$V = 9,500 - 2,000(3) = + 3,500 \text{ lb}$$

To find the location on the beam where the shear force is zero, we may substitute $V = 0$ into the second equation, leading to

$$0 = 9,500 - 2,000\, x$$

which results in $x = 4.75$ ft. The shear force diagram is plotted in Figure 5.7.

5.3 Construction of Shear Force Diagrams by Inspection

The foregoing three examples have involved the most common loading conditions found in simple beam problems. These were concentrated loads, uniformly distributed loads, and linearly varying distributed loads. Careful

Example 5.6

Derive the equation for the shear force diagram for the triangular load distribution shown in Figure 5.8.

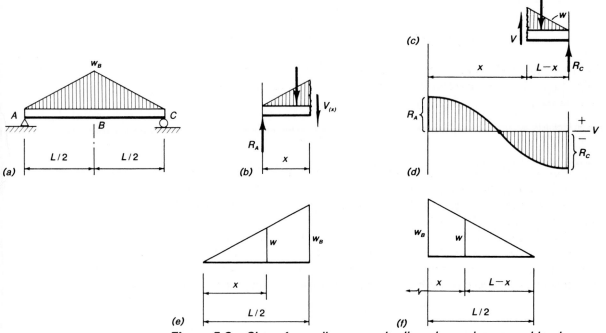

Figure 5.8 *Shear force diagram under linearly varying spread load.*

Solution:

The loading reaches its peak at the middle of the beam and is symmetrical. Accordingly the two reactions are the same in magnitude and direction. Their value is equal to the half of the area under the external loading diagram, that is

$$R_A = R_C = \frac{1}{2}\left(\frac{Lw_B}{2}\right) = Lw_B/4$$

The magnitude of the linearly varying loading at any point at a distance x from the left support between points A and B can be expressed by proportionality, such as

$$\frac{w_B}{L/2} = \frac{w}{x}$$

both sides defining the slope of the external loading diagram, as shown in Figure 5.8e, and from here,

$$w = w_B\,(2x/L)$$

For a free body of the beam cut at length x, there are three forces maintaining equilibrium. These are R_A; the resultant of the triangular spread load, which equals $\frac{1}{2}(wx) = w_B L(x/L)^2$; and the internal shear force V in the cut. Writing the equilibrium formula for these three loads, we have

$$\Sigma\, F_V = 0 \uparrow +$$

$$R_A - w_B L(x/L)^2 - V = w_B L/4 - w_B L(x/L)^2 - V = 0$$

In this equation V is taken with the negative sign due to the sign convention selected by Figure 5.2. Expressing the internal shear force from this equation, one obtains

$$V_{AB} = w_B L \left[\frac{1}{4} - (x/L)^2 \right]$$

Since the spread load was defined for the portion of the beam between points A and B, this equation is valid for that portion of the beam. In the region B to C of the beam, a different relationship will be required, since the spread load varies in a different manner.

The magnitude of the linearly changing spread load at any point at distance x from the left side of the beam (since we maintain the origin of the coordinate at the left support A) can now be expressed by first writing the equation of proportionality for the slope of the loading as shown in Figure 5.8f:

$$\frac{w_B}{L/2} = \frac{w}{L-x}$$

And from this equation the magnitude of w between B and C is

$$w = \frac{(L-x)}{L/2} w_B = 2\left(1 - \frac{x}{L} \right) w_B$$

For the right side of the beam, it is simpler to draw a new free body diagram. By the sign convention shown in Figure 5.2, positive shear is directed upward. Writing the equilibrium formula for our new free body diagram, we obtain

$$\Sigma\, F_V = 0 \uparrow +$$

$$V - \frac{1}{2} w(L-x) + R_c = 0$$

$$V - \frac{1}{2}\left[2\left(1 - \frac{x}{L} \right) w_B \right](L-x) + \frac{L w_B}{4} = 0$$

Simplifying and expressing for the shear force V,

$$V_{BC} = - w_B L \left[\frac{1}{4} - \left(1 - \frac{x}{L} \right)^2 \right]$$

Example 5.6 *(Continued)*

The negative sign of the value of the shear force V indicates that the shear force is really directed downward. The two equations expressing the internal shear force may now be used to calculate and plot the shear force diagram. Using the formula valid for the portion of the beam between A and B, the shear force at the support (where x equals zero) is determined as

$$V_A = + w_B L/4 (\text{at } x = 0)$$

At the middle of the span, point B, both formulas are valid. Both would give a zero value for the shear force at $x = L/2$. At the right end of the beam,

comparison of the loading conditions and the resulting internal shear force diagrams sheds light on the following mathematical relationship between them:

a. Concentrated loads on the structure result in sudden jumps in the shear diagram. The magnitudes of these changes are equal to the magnitudes of the forces. The direction of the changes is the same as the direction of the forces causing them.

b. Between concentrated loads, if there is no loading on the beam, the shear diagram has a constant value; i.e., it is horizontal line segment.

c. The shear force must be zero at a free end of the beam.

d. Under a uniformly distributed load, the shear force diagram increases or decreases at a constant rate. The rate of change, or slope, of the shear diagram corresponds to the magnitude of the uniform loading.

e. Under linearly varying loads, the shear force diagram changes according to a second-order (parabolic) function. The slope of the shear diagram at any arbitrary point of the beam is equal to the magnitude of the spread load at the point in question. Where the spread load happens to be zero (as at points A and C in Example 5.6), the slope of the parabola representing the shear force is horizontal. Horizontal slopes generally represent extreme values, either minima or maxima. Therefore these points are of special importance in design.

The foregoing relationships may be generalized by stating that the internal shear force is described by a function that is always one order higher than the function describing the loading condition. Hence, under a parabolic (second-order function) spread load the shear force diagram will take the shape of a third order ($V = kx^3$) function. Figure 5.9 shows the relationships between loading and internal shear diagrams in a tabulated form.

By use of these relationships, the internal shear force diagram can be constructed for most relatively simple cases by inspection. The following examples will demonstrate this technique.

immediately to the left of the support where R_C acts, the shear force equation would give

$$V_C = - w_B L/4 \text{(at } x = L)$$

At the support, the shear force reduces to zero as the reaction R_C is applied.

Between the points A, B, and C, the shape of the shear force diagram corresponds to a curve of the second degree. Its parabolic shape is described by the $(x/L)^2$ function. At the supports the slopes of these parabolas are horizontal. The steepest slope appears at point B at midspan, where the curve has an inflection point. Figure 5.8d shows the complete shear force diagram for this problem. It corresponds to the equation given for Case 17 of our table.

equation for loading	loading diagram shear force diagram	equation for shear force
(a) *no loading between concentrated forces* **w = 0**	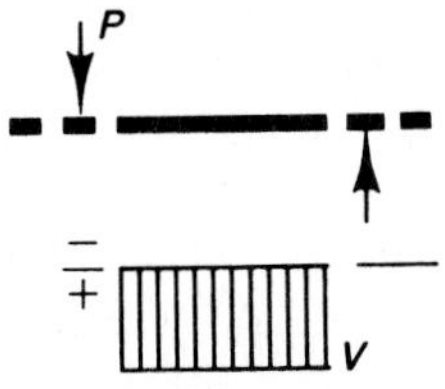	shear force is constant and equal to the sum of the external loads to the left $V = \text{Constant} = C$
(b) *uniformly distributed load* **w = constant**	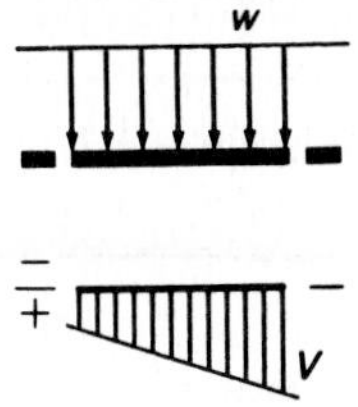	shear force varies linearly $V = k \cdot x + C$
(c) *linearly increasing load* **w = C·x**	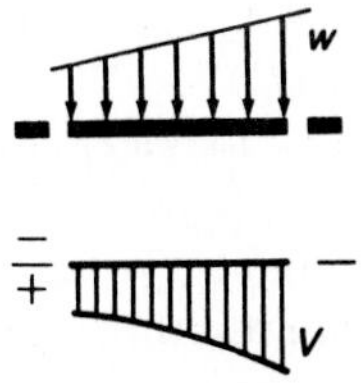	shear force increases parabolically $V = k \cdot x^2 + C$ the constant C depends on the sum of the loading to the left
(d) *parabolically increasing distributed loading* **w = C·x²**	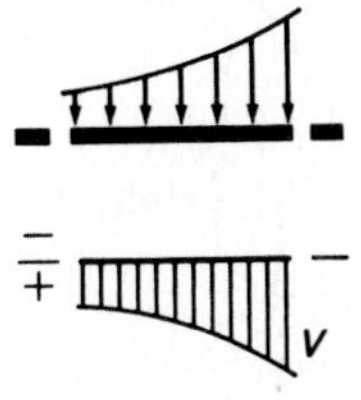	The shear force increases with the cube of the distance x $V = k \cdot x^3 + C$ C again is dependent on the loading to the left

Figure 5.9 *Relationships between the shape of the loading diagram and the resulting shear force diagram.*

Example 5.7

Determine the magnitude of the maximum shear force to be resisted by the wings of the twin-engine airplane shown in Figure 5.10a. In this figure the forces denoted by P_1 represent the weight of the two engines, and P_2 is the gross weight of the fuselage. To maintain equilibrium and to keep the plane in the air, the wings generate an uplift force represented by the uniformly distributed force w.

Solution:

The magnitude of the uniformly distributed uplift force may be determined by denoting the wingspan by L and employing the equilibrium condition as

$$\Sigma\, F_V = 0 \uparrow +$$

$$wL - 2\,P_1 - P_2 = 0$$

$$wL = 2\,P_1 + P_2$$

hence,

$$w = \frac{2\,P_1 + P_2}{L}$$

By observing the loading diagram, construction of the shear force diagram may begin at the tip of the left wing. As this is a free end of the structure, the internal shear force there is zero. From the zero point the shear force increases at a uniform rate of w. As the uplift force is directed upward, the shear force diagram grows in the same direction. As the left engine is reached, the shear force diagram must have a sudden drop downward. The magnitude of the drop equals P_1. Proceeding toward the right, the uplift force will, again, govern. At the base of the left wing the magnitude of the internal shear force will reach its maximum. Its value there equals the sum of all external forces to the left, i.e.,

$$V_{\text{left wing base}} = (wL/2) - P_1$$

At this point one may notice that by virtue of the equilibrium condition, and because of the symmetry of the structure, the maximum shear force also equals

$$V_{\text{left wing base}} = \frac{1}{2}P_2$$

This is value *ef* on the shear force diagram of Figure 5.10b. On the other side of the fuselage, at the base of the right wing, the shear force may be determined as before. In this manner the shear force can be defined as the sum of

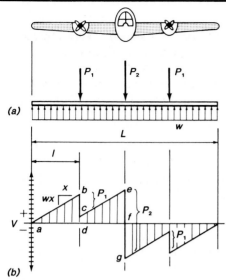

Figure 5.10 *Construction of the shear force diagram by inspection.*

the shear force at the base of the left wing and the weight of the fuselage, that is,

$$V_{\text{right wing base}} = (wL/2) - P_1 - P_2$$

This is value fg on the shear force diagram of Figure 5.10b.

The complete shear force diagram is shown in Figure 5.10b. Let us now review some of the elements of the construction of this diagram. At point a the value of the shear force must be zero because this is a free end of the structure. Between points a and b, the rate of growth of V is given by the distributed uplift load, w. Hence the value of V at point b (given by height bd) must equal the product of l and w, l being the distance of the engine from the wing tip. The distance between b and c in terms of internal shear equals P_1. This means that immediately to the right of the engine, the shear in the wing is reduced to the magnitude of the distance between points d and c. Plotting the shear diagram with a suitably selected scale, the magnitude of this value may be read off easily. From point c to point e the shear again grows at the rate of w. At the fuselage the correct value for the magnitude of the shear force is given by the distance between e and f for the left side, and by the distance between f and g for the right side. A common error made by beginners is to consider the distance between e and g. This represents the magnitude of P_2, but *not* the shear. The right half of the shear force diagram is an inverted mirror image of the left.

Example 5.8

Construct the shear force diagram for the floor joist shown in Figure 5.8a by inspection.

Solution:

As before (in Example 5.6), the reaction at each end of the beam is equal to

$$R_A = R_C = w_B L/4$$

First, we recognize that at the two supports the shear force increases abruptly from zero to R_A and R_C, respectively. On the left support this results in a positive shear value of magnitude R_A, and on the right support its magnitude is the same but it is of opposite sign. These are plotted as the distances ad and ec, respectively, in the shear force diagram of Figure 5.11.

Next, observe that the loading diagram between a and c represents a linearly varying loading distribution. Consulting Figure 5.9, we find that the corresponding shear force diagram must be parabolic.

The slope of the shear force diagram is given by the magnitude of the distributed load at the point where it is sought. Accordingly, one may conclude that the slope of the diagram is zero at the end points of the beam, since the magnitude of the spread load is zero at both ends. At the middle of the beam, at point B, the loading is w_B; this represents the maximum slope encountered. Since the loading is symmetrical around the center of the beam, the shear force distribution will also be symmetric. This determines the magnitude of

Example 5.9

Determine the shear force distribution along the fuselage of the airplane shown in Figure 5.12.

Solution:

Figure 5.12a shows the distribution of the loading on the airframe. On the upper part of this drawing the distribution of the weight of the structure, including the cargo, is shown. On the lower part the distributed uplift force transferred from the wings is shown, along with the concentrated force transferred through the elevator attachments. For the plane to fly in an equilibrium condition, these forces must be in equilibrium.

On Figure 5.12b the loading shown on the drawing above is combined. This is now the actual idealized loading distribution to be considered in the design. The construction of the shear force diagram may proceed from here in the following steps:

Step 1.   Realizing that the shear force increases parabolically between points A and B, we first draw the initial tangent of the line at

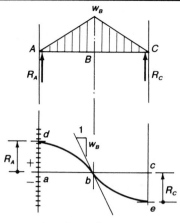

shear force diagram

Figure 5.11 *Construction of the shear force diagram under a triangular spread load by inspection.*

the shear force as zero at point B. All of these slopes are known, hence they may be drawn into the diagram of the shear forces.

Once the tangents to the second-order curve are known at the three characteristic points, the parabolas db and bc may be drawn. This step will complete the construction of the shear force diagram.

point A. The slope of this tangent equals w_A. The line must start from zero, since this is a free end of the airframe.

Step 2. The parabola between A and B will curve downward since the load is increasing from a to b. The spread load increases between A and B, and the magnitude of w_B determines the slope of the line at B. The position of the line there is given by the sum of the loading to the left; that is,

$$V_B = - \left(\frac{w_A + w_B}{2} \right) L_1$$

Knowing the position as well as the slope of the shear force diagram at point B, we may draw the parabola.

Step 3. To the right of point B the loading is constant. Hence the shear force diagram will change linearly. The slope of this line will equal $(w_u - w_B)$. The line must start at V_B as shown in

Example 5.9 *(Continued)*

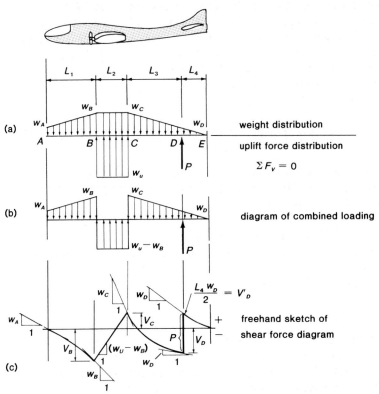

Figure 5.12 *Construction of shear force diagram by inspection.*

Figure 5.12*c*. The magnitude of V_C is given by the sum of all vertical forces to the left, hence

$$V_C = - \left(\frac{w_A + w_B}{2} \right) L_1 + (w_u - w_B) L_2$$

Step 4. Between points C and D the loading, again, is linearly varying. The difference, however, is that while between A and B it increased, between C and D it decreases. This means that the parabola will now have a decreasing negative slope. First we draw the slope of the initial tangent, which equals $(-w_C)$. Before the curve can be drawn, the location of its end point must be known. This we may define as V_D. Summing the forces to the left to obtain V_D, we have

$$V_D = - \left(\frac{w_A + w_B}{2} \right) L_1 + (w_u - w_B) L_2 - \left(\frac{w_C + w_D}{2} \right) L_3$$

The tangent at this point is defined by $(-w_D)$. The location and the slope of the parabola are now known, and it may be plotted in the drawing.

Step 5. At point D the concentrated force P is acting upward. As this is of known magnitude, it may be plotted directly. The magnitude of the negative shear force immediately to the right of D is defined as V_D'. This equals

$$V_D' = V_D + P$$

Step 6. Between points D and E the previous linearly varying load continues. The initial tangent of the parabola here equals $(-w_D)$. At the tail end of the plane the loading reduces to zero. This means that the tangent of the parabola will be horizontal. Static equilibrium requires that the shear force be zero here, since this end also is a free end. To assure that the diagram closes at zero, the shear force at D was recomputed, taking the part of the plane to the right of point D. Summing the loading to the right of D we obtain

$$V_D' = \frac{1}{2} w_D L_4$$

If our analysis was correct, this must equal the earlier result.

Problems

5.1   For the beam loaded as shown in Figure P 5.1, find the axial force and shear force to the left and right of the 2,000-pound load.

5.2   For the beam shown in Figure P 5.2, determine the shear force equation.

5.3–
5.10   For the planar structures shown in Figures P 5.3 to P 5.10, determine the shear force equation, and draw the shear force diagram.

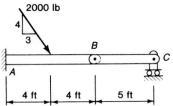

Figure P 5.1

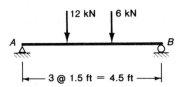

Figure P 5.2

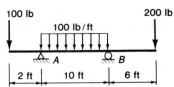

Figure P 5.3

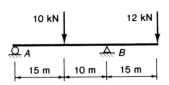

Figure P 5.4

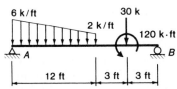

Figure P 5.5

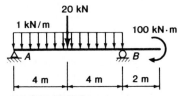

Figure P 5.6

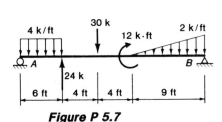

Figure P 5.7

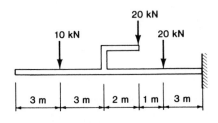

Figure P 5.8

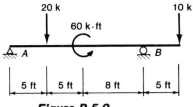

Figure P 5.9

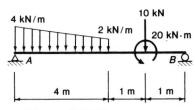

Figure P 5.10

144   Chapter Five

5.11 For the beam shown in Figure P 5.1, find the shear force equation and draw a shear force diagram.

5.12–
5.22 For the structures shown in Figures P 5.7, P 5.10, P 5.12, and P 5.15 to P 5.22, draw the corresponding shear force diagrams by inspection.

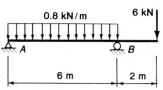

Figure P 5.12

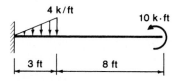

Figure P 5.15

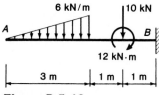

Figure P 5.16

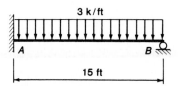

Figure P 5.17

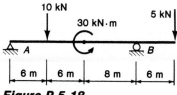

Figure P 5.18

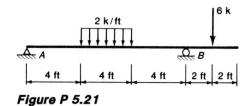

Figure P 5.19

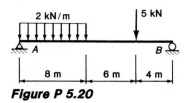

Figure P 5.20

Figure P 5.21

Figure P 5.22

Moment Diagrams

6

1. *To understand the concept of internal bending moments caused by external loadings on beams.*

2. *To learn how bending moments are calculated at arbitrarily selected points along a beam by the use of free body diagrams.*

3. *To understand the technique of drawing moment diagrams from known shear force diagrams.*

Objectives

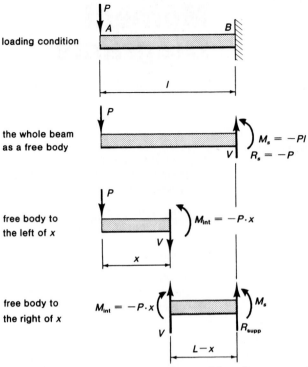

Figure 6.1 *The concept of internal bending moment.*

6.1 Internal Bending Moments

The primary effect of transverse loading on a beam is *bending.* Bending results from *internal bending moments* that develop as the material of the beam resists the external loads placed on it. The magnitude of the internal bending moment may be determined at any point on the beam by considering one part of the beam as a free body, cut at the point in question. Let us consider, for instance, a cantilever beam loaded by a concentrated force at its end, shown in Figure 6.1. By taking the whole structure as a free body, the reactions at the fixed end may be computed. The equilibrium conditions for forces and moments will have to be applied. As long as the external force P at the end is perpendicular to the beam, there will be no axial forces generated and hence the equilibrium equation $\Sigma F_H = 0$ will not be needed. The other two equations, $\Sigma F_V = 0$ and $\Sigma M = 0$, will result in a reaction force R_s, and a moment M_s, acting at the support. Taking moments about the support

$$\Sigma M_{support} = 0 \; \curvearrowright +$$
$$-M_s - Pl = 0$$
$$M_s = -Pl$$

To determine the internal moment at any point along the axis of the beam at a distance x from the force P, we may now consider it cut at that point. Writing the equilibrium formula, the resulting internal bending moment will be found to equal

(a) positive   **(b) negative**

Figure 6.2   *Sign convention for internal bending moment.*

(6.1)
$$\Sigma M_{cut} = 0 \circlearrowleft +$$
$$- M_{int} - Px = 0$$
$$M_{int} = - Px \quad (\text{ for left side})$$

$$\Sigma F_V = 0 \uparrow +$$
$$- P - V = 0$$
$$V = - P$$

These expressions are valid throughout the length of the beam.

If now the equilibrium of the other (right-side) free body is considered, the loading condition will be somewhat different. In addition to a transverse force V, we also have a moment at the right end. Applying moment equilibrium about B:

$$\Sigma M_B = 0 \circlearrowright +$$
$$M_s - V(l - x) - M_{int} = 0$$
$$M_{int} = M_s - V(l - x)$$

Substituting M and P, we obtain

(6.2)
$$M_{int} = - Pl + P(l - x) = - Px \text{ (for right side)}$$

Note that, in general, there will be both shear force and bending moment acting at a cut.

For internal bending moments the *sign convention* may be selected in an arbitrary manner as long as it is consistent with the one selected for the internal shear forces. Whatever sign convention one selects, it must be adhered to throughout the computation. The sign convention to be used in this text will be the one shown in Figure 6.2. As this figure shows, positive internal moments will cause counterclockwise rotation if the cut is on the right side of the element, and clockwise rotation if the cut is on the left. According to this sign convention, Equation 6.1 is to be considered and plotted as a negative internal moment. An easy way to remember this convention is to pretend that "happy beams have positive bending moment" (as shown in Figure 6.2a). Conversely, "unhappy beams have negative bending moment" (as shown in Figure 6.2b).

6.2 Bending Moment Diagram

When designing structural or machine components subjected to bending, the magnitude of the internal bending moments must be known along the entire axis of the beam. These moments vary depending on the distribution of the external loading. To show this variation and to allow the designer to locate the maximum value of the bending moment, as well as its value at critical points of the beam, the internal bending moment is depicted in the form of a diagram.

Example 6.1

Construct a diagram showing internal bending moments calculated at even 2-foot intervals for the two-supported beam loaded with a concentrated force as shown in Figure 6.3.

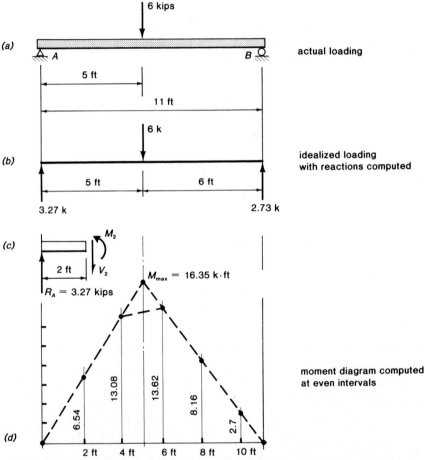

Figure 6.3 *Construction of moment diagram by moments computed at even intervals along the axis of the beam.*

Most commonly the horizontal coordinate of the diagram represents the distance along the axis of the beam from a convenient starting point. The magnitude of the bending moment is plotted on the vertical coordinate, using a suitable scale. Positive internal bending moments are usually plotted upward of the horizontal axis.

A simple but laborious way of constructing bending moment diagrams is to calculate the bending moment at definite intervals along the axis of the beam. One such example follows.

Solution:
First the reactions at the two supports are calculated, using the equations of equilibrium. The results are

$$R_A = \frac{6(6)}{11} = 3.27 \text{ k} \qquad R_B = \frac{5(6)}{11} = 2.73 \text{ k}$$

Computing the moments at even 2-foot distances from the left support is achieved by making a cut at each location. Only the left-hand free body from a cut made two feet from the end A is shown, but the results of successive cuts are:

$$M_0 = 0 \text{ (pinned end)}$$
$$M_2 = 2(3.27) = 6.54 \text{ kip} \cdot \text{ft}$$
$$M_4 = 4(3.27) = 13.08 \text{ kip} \cdot \text{ft}$$
$$M_6 = 6(3.27) - 1(6) = 13.62 \text{ kip} \cdot \text{ft}$$
$$M_8 = 8(3.27) - 3(6) = 8.16 \text{ kip} \cdot \text{ft}$$
$$M_{10} = 10(3.27) - 5(6) = 2.7 \text{ kip} \cdot \text{ft}$$
$$M_{11} = 0 \text{ (pinned end)}$$

Plotting these results on a suitably selected graph paper (say, 1 in $= 2$ kip $\cdot$ ft) and connecting the computed points, a moment diagram may be obtained as shown in Figure 6.3d. Observing the result, the following conclusions may be drawn:

Between concentrated forces the moment diagram varies linearly.

Unless the external concentrated force (the 6-kip load in this case) happens to fall on a plotting point, a possible extreme value in the moment diagram may be missed. Indeed, in this case the maximum moment is at the location of the 6-kip load, and its value is $M_5 = 5 (3.27) = 16.35$ kip $\cdot$ ft. This may be verified by considering moment equilibrium of the left-hand portion of the beam when a cut is made at point C.

6.3 Use of Moment Diagrams for Common Cases

As demonstrated in Example 6.1, the construction of a moment diagram by repeated computation of the bending moment at predetermined intervals along the beam is an inefficient procedure. Furthermore, this process may not locate critical values of the bending moment. (In Example 6.1 the maximum value of the bending moment was not found when the moment was computed at 2-foot intervals.)

Fortunately, for many simple cases the bending moments have already been derived, and general expressions for the variation of bending moment are available. Some such expressions are given in Column 3 of Appendix B. The following examples in this section illustrate the use of this table.

It helps in the construction of moment diagrams to recall that the moment at certain structural locations is always zero. Figure 6.4 shows some of these cases. The ends of cantilever beams, hinges, and pinned or roller-supported ends of beams are all points where the bending moment will be zero.

Example 6.2

The simply supported beam ABC in Figure 6.5a has an applied load of 10,000 pounds applied at B. Draw the bending moment for this beam and give the equations of the moment variation.

Solution:
Examining Appendix B, it is noted that Case 12 (left-hand column of table) corresponds to the beam in this example, as both the end support conditions and the applied loading are the same. To derive the expressions for bending moment (Column 3) the following substitutions are made:

$$a = 3 \text{ ft}$$
$$b = 9 \text{ ft}$$
$$l = 12 \text{ ft}$$
$$W = 10,000 \text{ lb}$$

Thus the bending moment between A and B, M_{AB}, is given by

$$M_{AB} = W\left(\frac{b}{l}\right) \cdot x = 10,000 \cdot \left(\frac{9}{12}\right) \cdot x$$
$$= 7,500x \text{ lb} \cdot \text{ft}$$

Similarly the bending moment between B and C is given by

$$M_{BC} = W\left(\frac{a}{l}\right)(l-x) = 10,000 \cdot \left(\frac{3}{12}\right)(12 - x)$$
$$= 2,500(12-x) \text{ lb} \cdot \text{ft}$$

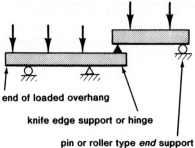

end of loaded overhang

knife edge support or hinge

pin or roller type *end* support

Figure 6.4 *Points where bending moment is definitely zero.*

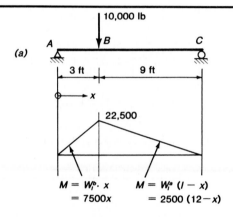

(a)

10,000 lb

A   B   C

3 ft | 9 ft

x

22,500

$M = W_r^b \cdot x$ $M = W_r^a (l - x)$
$= 7500x$ $= 2500 (12-x)$

(b)   *bending moment diagram (lb·ft)*

Figure 6.5 *Moment diagram for a simple beam with a point-applied load.*

In both of these expressions the origin for x is point A, as shown. The maximum value of the bending moment occurs at B and is given by

$$M_B = W\left(\frac{ab}{l}\right) = 10,000 \cdot \left(\frac{3 \cdot 9}{12}\right) = 22,500 \, \text{lb} \cdot \text{ft}$$

The final bending moment diagram is plotted in Figure 6.5*b*.

Example 6.3

Derive the equation and plot the bending moment diagram for the simply supported beam with a variable spread load applied as shown in Figure 6.6a.

Solution:
Examination of Appendix B shows that Case 15 fits the beam, both in support conditions and in loading applied. To find the bending moment expression, the following substitutions are made:

$$l = 12 \text{ ft}$$
$$w = 2{,}000 \text{ lb/ft}$$

Thus

$$W = \frac{1}{2}wl = \frac{1}{2} \cdot 2{,}000 \cdot 12 = 12{,}000 \text{ lb}$$

(This represents the total force acting on the beam.) The bending moment expression is then

$$M = \frac{1}{3}W\left(x - \frac{x^3}{l^2}\right) = \frac{12{,}000}{3}\left(x - \frac{x^3}{12^2}\right)$$

$$= 4{,}000\left(x - \frac{x^3}{144}\right) \text{ lb} \cdot \text{ft}$$

6.4 Principle of Superposition

Equations showing the moment distribution along the axis of beams under commonly encountered loadings are included in the table shown in Appendix B. However, in practice, a combination of various loadings may be placed on a beam. In such cases the analysis may be performed by using the *principle of*

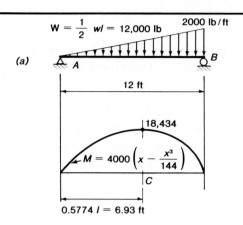

(a)

(b)   bending moment diagram (lb·ft)

Figure 6.6   *Bending moment diagram for a simply supported beam with a variable spread load.*

The maximum bending moment occurs at

$$x = 0.5774 \, l = 6.93 \text{ ft}$$

(i.e., just to the right of the center line).
The value of the maximum bending moment is

$$M_{\text{max}} = 0.128 \, Wl = 0.128 \cdot 12{,}000 \cdot 12$$
$$= 18{,}434 \text{ lb} \cdot \text{ft}$$

The bending moment diagram is plotted in Figure 6.6*b*.

superposition. This means that the various loads are considered separately and, after obtaining the moment diagram for each, the results are added together. The sum of the individual results will be the same as if all the different loads were considered in one step. Two such instances are Examples 6.5 and 6.6.

Example 6.4

Give the equations and plot the diagram showing the way the bending moment varies along the cantilever beam *ABCD* when it is loaded with a uniform spread load between points *B* and *C*, as shown in Figure 6.7a.

Solution:

Examination of Appendix B shows that Case 4 meets both the support conditions (built-in at one end and free at the other) and the loading conditions of the given problem. The appropriate substitutions are:

$$a = 3 \text{ m}$$
$$b = 6 \text{ m}$$
$$l = 9 \text{ m}$$
$$w = 5 \text{ kN/m}$$

Thus,

$$W = w(b-a) = 4(6-3)$$
$$= 12 \text{ kN}$$

(This is the resultant applied load.) Using Column 3 of Appendix B, the expressions for the bending moment are:

$$M_{AB} = 0$$

$$M_{BC} = -\frac{1}{2}\left(\frac{W}{b-a}\right)(x-l+b)^2$$

$$= -\frac{1}{2}\cdot\left(\frac{15}{6-3}\right)(x-9+6)^2$$

$$= -\frac{5}{2}(x-3)^2 \text{ kN}\cdot\text{m}$$

Thus at *C* ($x = 6$ m),

$$M_C = -\frac{5}{2}(6-3)^2 = -22.5 \text{ kN}\cdot\text{m}$$

$$M_{CD} = -\frac{1}{2}W(2x-2l+a+b)$$

$$= -\frac{1}{2}\cdot 15(2x-2.9+3+6)$$

$$= -\frac{15}{2}(2x-9) \text{ kN}\cdot\text{m}$$

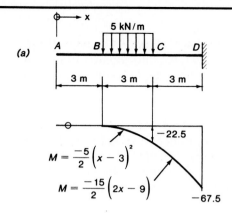

(a)

(b)   *bending moment diagram (kN·m)*

Figure 6.7   *Bending moment diagram for a cantilever beam with a partial uniform load.*

The maximum moment occurs at D and is given by

$$M_{max} = -\frac{1}{2}W(a+b) = -\frac{1}{2} \cdot 15(3+6)$$

$$= -67.5 \text{ kN} \cdot \text{m}$$

The resulting bending moment diagram is plotted in Figure 6.7b.

Example 6.5

A simple supported beam shown in Figure 6.8 is loaded by a concentrated force at the center and a uniformly distributed force throughout. Construct a moment diagram by obtaining the moment diagrams for each loading and adding them.

Solution:

Examination of the required loads on the beam shows that they may be divided into a point load applied at midspan (Figure 6.8b) and a uniform spread load applied over the entire span (Figure 6.8d). Furthermore, both of these load cases are given in Appendix B. Case 11 represents the case of the concentrated load at midspan. In this example,

$$l = 12 \text{ ft}$$
$$W = 900 \text{ lb}$$

Thus the maximum moment, which occurs at midspan, is

$$M = \frac{Wl}{4} = \frac{900 \cdot 12}{4} = 2{,}700 \text{ lb} \cdot \text{ft}$$

Since both equations for the moment given in Column 3 are linear, the bending moment diagram for this case will have the shape given in Figure 6.8c.

In a similar manner, Case 13 represents the uniform spread load of Figure 6.8d with the following substitutions:

$$l = 12 \text{ ft}$$
$$W = wl = 120 \cdot 12 = 1{,}440 \text{ lb}$$

Thus the expression for the bending moment is

$$M = \frac{W}{2}\left(x - \frac{x^2}{l}\right) = \frac{1{,}440}{2}\left(x - \frac{x^2}{12}\right) \text{ lb} \cdot \text{ft}$$

(i.e., a parabolic curve).

Again, the maximum moment occurs at midspan and is given by

$$M_{max} = \frac{Wl}{8} = \frac{1{,}440 \cdot 12}{8}$$
$$= 2{,}160 \text{ lb} \cdot \text{ft}$$

The moment diagram for this case is given in Figure 6.8e.

The principle of superposition says that the total bending moment due to the combined loading is the sum of the bending moments due to the indi-

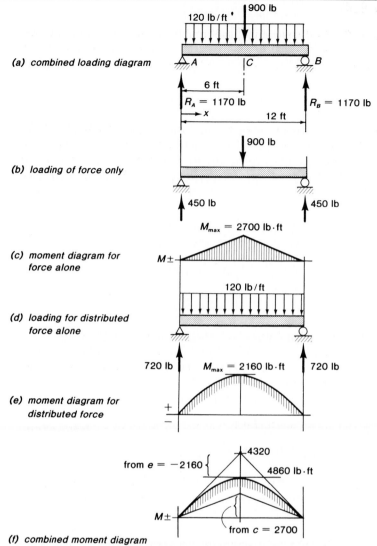

(a) combined loading diagram

(b) loading of force only

(c) moment diagram for force alone

(d) loading for distributed force alone

(e) moment diagram for distributed force

(f) combined moment diagram

Figure 6.8 The principle of superposition for moments.

vidual loads taken separately. Thus the moment diagram due to the combined loading is the sum of Figures 6.8c and e and is given in Figure 6.8f. Since the maximum bending moment occurs at midspan in both cases, the maximum bending moment of the combined loading will also occur at midspan, and its value is

$$M_{max} = 2{,}700 + 2{,}160$$
$$= 4{,}860 \text{ lb} \cdot \text{ft}$$

Example 6.6

Find the bending moment diagram for the combined loading condition shown in Figure 6.9a.

Solution:

Once again, the combined moment diagram may be derived as the sum of moments from two loading conditions shown in Figures 6.9b and 6.9d. Each of these separate loadings has been previously considered. In Example 6.2, the moment diagram for the case shown in Figure 6.9b was found to be that given in Figure 6.9c (using Case 12 in Appendix B). Similarly, in Example 6.3 the moment diagram for the case shown in Figure 6.9d was shown to be that given in Figure 6.9e (using Case 15 in Appendix B). The combined moment diagram is shown in Figure 6.9f.

Between A and B the equation for the combined case is the sum of the moments from the two constituent moments. Thus

$$M_{AB} = 7{,}500\,x + 4{,}000\left(x - \frac{x^3}{144}\right)$$
$$= 11{,}500\,x - 27.78\,x^3 \text{ lb}\cdot\text{ft}$$

At B, the moment is the sum of the two constituent moments, giving

$$M_B = 22{,}500 + 4{,}000\left(3 - \frac{3^3}{144}\right)$$
$$= 33{,}750 \text{ lb}\cdot\text{ft}$$

Between B and C the bending moment equation can likewise be derived as the sum of the two constituent moment equations. Thus

$$M_{BC} = 2{,}500\,(12 - x) + 4{,}000\left(x - \frac{x^3}{144}\right)$$
$$= 30{,}000 + 1{,}500\,x - 27.78\,x^3 \text{ lb}\cdot\text{ft}$$

Since the peak of the moment diagram for the varying spread load was at a distance x of 6.93 feet from end A, it is well to evaluate the combined moment at this point, i.e.,

$$M = 30{,}000 + 1{,}500\,(6.93) - 27.78\cdot(6.93)^3$$
$$= 31{,}109 \text{ lb}\cdot\text{ft}$$

Although it cannot be guaranteed that the maximum bending moment is located at one of these points, it is likely that the maximum moment is either the larger of these (i.e., 33,750 lb·ft) or only slightly higher.

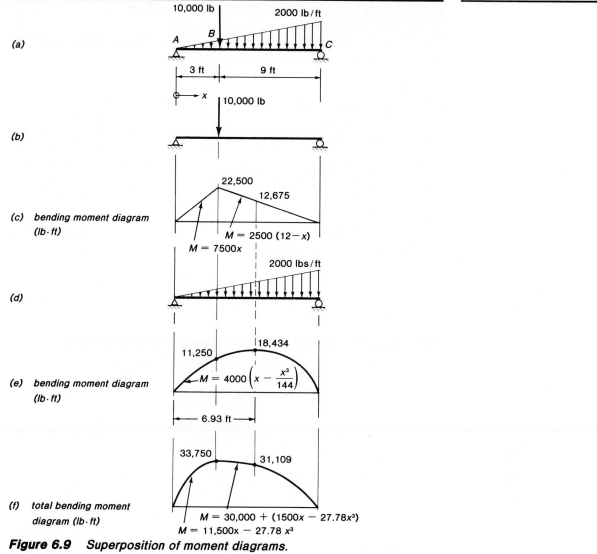

Figure 6.9 *Superposition of moment diagrams.*

6.5 Rules for Construction of Bending Moment Diagrams

Examples 6.5 and 6.6 allow us to draw important conclusions concerning the form of the bending moment diagram.

1. Between concentrated loads the bending moment diagram changes in a linear manner.
2. Under a uniformly distributed load, the moment diagram is a parabolic curve.
3. Under a linearly changing spread load, the moment diagram is a cubic function.
4. The bending moment is zero at pinned supports and roller supports of simple beams, as well as at the free end of cantilever beams, unless a concentrated moment is applied at this location.
5. When the loading includes an externally applied moment, the internal bending moment diagram will exhibit a sudden jump. The magnitude of this sudden change will equal the magnitude of the moment applied.

Figure 6.10 illustrates these concepts and is a useful diagram to refer to in determining the shape of bending moment diagrams. Example 6.7 illustrates the use of these rules.

6.6 Equations for Bending Moment Diagrams

So far, in constructing bending moment diagrams, the equations used have been taken from Appendix B. It is a valid question to ask: Where did these equations come from? In practice, these relationships may be derived by making cuts along the beam at a distance x from some predetermined origin and deriving the moment equations in terms of the distance x. A different cut will need to be made in every region where the loading condition changes. Thus a separate cut will be needed between every point-applied load (moment or force) and wherever the spread load condition changes. Example 6.8 will illustrate a simple case of this procedure where only one free body is required.

equation for loading	loading diagram / moment diagram	equation for bending moment
concentrated forces no other loading (weight of beam neglected)	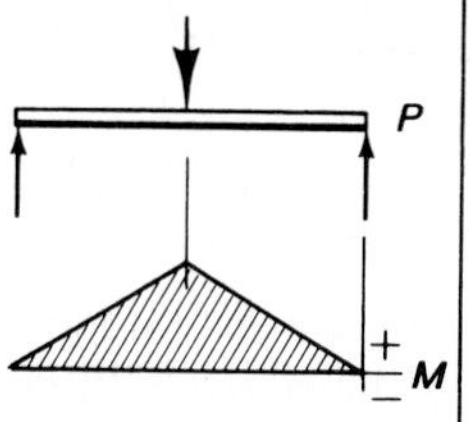	Bending moment varies linearly $M = K \cdot x + C$ slope depends on forces to the left
Uniformly distributed loads w = constant	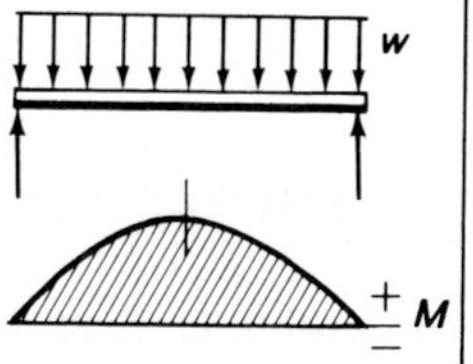	Parabolic moment distribution $M = K \cdot x^2 + C$ slope given by forces to the left
Linearly varying spread load $w = C \cdot x$	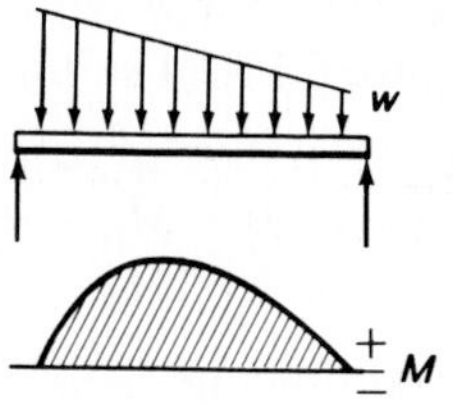	Moment described by cubic function $M = K \cdot x^3 + C$
Concentrated moment acting at point x	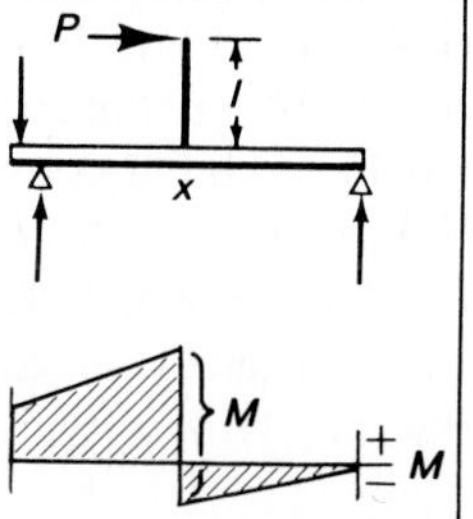	Moment distribution graph will show sudden jump at x. The change will equal the moment applied. $M = P \cdot l$

Figure 6.10 *Relationships between the shape of the loading diagram and the resulting moment diagram.*

Example 6.7

A beam is loaded as shown in Figure 6.11a. The external forces are in equilibrium. Determine the bending moment diagram using the principles discussed.

Solution:

Step 1.   Determination of Reactions. Since the reaction forces were computed already and the beam is in complete equilibrium, the construction of the moment diagram may proceed directly. (If not given, the first step would be the finding of the unknown reactions.)

Step 2.   End Moments. The moment at both cantilevered ends of the beam is zero as long as there are no concentrated moments applied. Accordingly, the moment diagram is known at these points.

Step 3.   Moment Variation Under Unloaded Portions. In the region AB of the beam there is no applied load, and thus the moment diagram will be linear. If a free body is cut at B, the left-hand side of the cut is shown in Figure 6.11c. Taking moments about B gives

$$\Sigma M_B = 0 \; \circlearrowleft +$$
$$- 100 \, (1.2) - M_B = 0$$
$$M_B = -120 \text{ kN} \cdot \text{m}$$

Thus the region AB of the moment diagram shown in Figure 6.11c may be drawn.

Step 4.   Moment Variation Under a Uniform Load. As shown in Figure 6.4, the moment diagram under a uniform load is parabolic and therefore has the approximate shape sketched in the region BC of Figure 6.11b. The diagram may not be completely drawn in this region until the moment at C is known. To find this moment a cut is made in the beam at C. It is easier to consider a free body diagram to the right of this cut (Figure 6.11c), on which the resultant of the linearly varying load is

$$R = \frac{1}{2}(50) \cdot 1.1 = 27.5 \text{ kN}$$

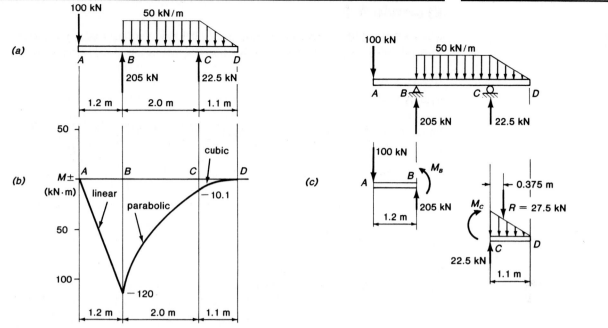

Figure 6.11 *Using rules for construction of bending moment diagrams.*

and it acts at a point $1/3\,(1.1) = 0.367$ meter from C. Formulating the moment equilibrium equation on this free body gives

$$\Sigma\,M_C = 0 \,\circlearrowleft +$$
$$M_C + 0.375 \cdot 27.5 = 0$$
$$M_C = -10.1 \text{ kN} \cdot \text{m}$$

Step 5. Moment Under a Linearly Varying Load. Reference to Figure 6.10 shows that the moment equation is of cubic order under a uniformly varying applied load such as occurs between C and D. Since the end moments (at C and D) of this region are known, the moment diagram may be completed as shown in Figure 6.11b.

Example 6.8

The cantilever beam shown in Figure 6.12a is loaded with a linearly varying spread load. Develop the equation for the internal bending moment.

Solution:

The first step is to determine the equation for the spread load. At the wall its value is w_0. At the free end the loading is zero. Between these two ends it varies linearly. By considering similar triangles we may state that $w_0/L = w_x/(L - x)$. From this relationship the loading at any distance x from the wall is

$$w_x = w_0 \left(1 - \frac{x}{L} \right)$$

There are two ways to proceed with the solution. One is first to determine the reactions at the wall. By writing equilibrium formulas one may find that at the wall

$$\Sigma\, F_V = 0 + \uparrow$$

$$R_A - \frac{w_0 L}{2} = 0$$

$$R_A = \frac{1}{2} w_0 L$$

$$\Sigma\, M_A = 0 \;\curvearrowright +$$

$$M_A + \left(w_0 \frac{L}{2} \right) \cdot \frac{L}{3} = 0$$

$$M_A = -\, w_0 L^2/6$$

The equivalent load representing this spread load and its point of application are given in Chapter 3.

Now consider a free body equilibrium by cutting the beam at location x and examining the portion to the left of this cut. As shown on the drawing, this part of the body carries a greater variety of loading than the other part, to the right side of the cut. On the right side there are no reactions, hence the only thing to contend with is the distributed load. From Figure 6.1 and the corresponding calculations leading to Equations 6.1 and 6.2, we know that the internal bending moment will be the same in magnitude regardless which side of the cut beam is considered as a free body. Therefore, for the sake of simplicity, consider the free body on the right side of the cut.

Equilibrium of the free body to the right of an arbitrarily selected cut at x shows that the internal shear force is

$$\Sigma\, F_V = 0 \uparrow +$$

$$V_x - w_x \left(\frac{L - x}{2} \right) = 0$$

$$V_x = w_x \left(\frac{L - x}{2} \right)$$

$$V_x = \frac{w_0}{2L} (L - x)^2$$

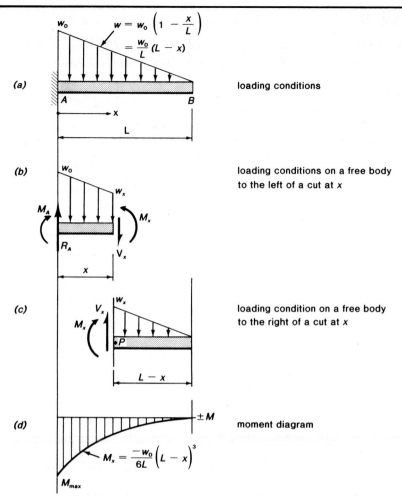

(a) loading conditions

(b) loading conditions on a free body to the left of a cut at x

(c) loading condition on a free body to the right of a cut at x

(d) moment diagram

$$M_x = \frac{-w_0}{6L}(L - x)^3$$

Figure 6.12 Development of the internal bending moment diagram for linearly distributed spread load.

The internal bending moment is given by

$$\Sigma M_p = 0 \; \circlearrowleft +$$

$$M_x + \frac{w_x \cdot (L - x)}{2} \cdot \frac{(L - x)}{3} = 0$$

$$M_x = -\frac{w_x}{6}(L - x)^2$$

Substituting for w_x from the formula developed earlier, we obtain

$$M_x = -\frac{1}{6}\left(\frac{w_0}{L}\right)(L - x)^3$$

This is a similar expression to that given in Appendix B, Case 5, but with a different origin for x.

shear conditions	shear and moment diagram	moment conditions
constant shear force	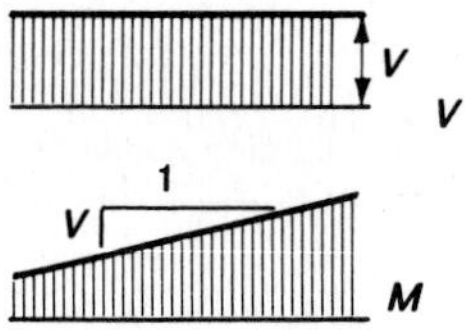	linearly changing moment slope of moment diagram equals shear force
linearly increasing shear	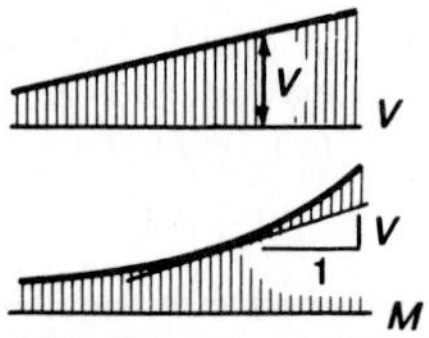	parabolically increasing moment slope of moment diagram at any location equals the shear force at the same location
linearly decreasing shear	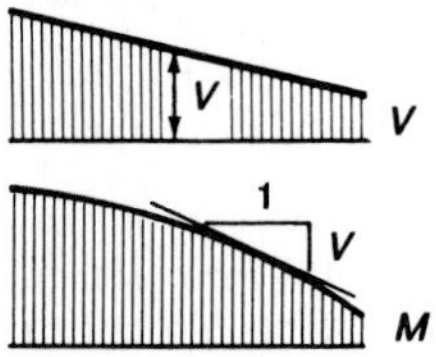	parabolically decreasing moment diagram slope of parabola at any point is given by shear force at that point
zero shear force If shear changes sign, the slope of moment diagram has a peak. If shear goes from + to −, moment has positive maximum value.	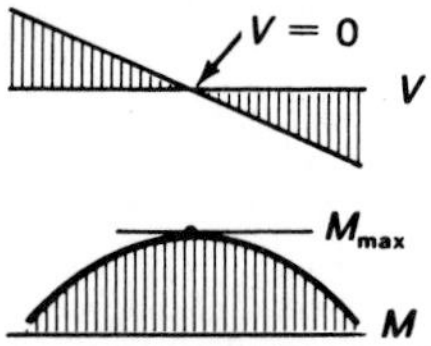	extreme value of moment diagram slope of moment diagram is zero at zero shear value
parabolically increasing shear	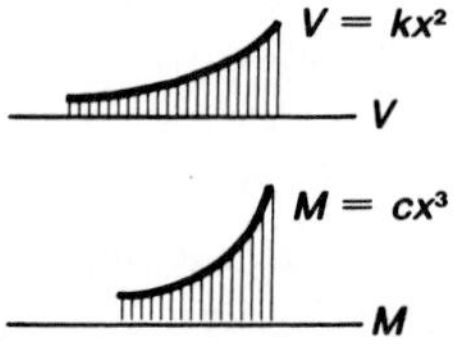	moment diagram increases according to cubic function
negative shear positive shear	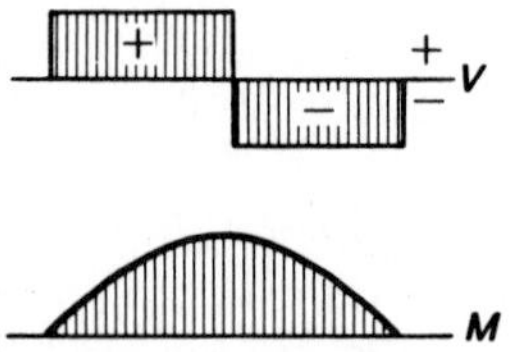	declining moment diagram increasing moment diagram

Figure 6.13 *Functional relationships between shear force diagram and bending moment diagram.*

6.7 Bending Moment Diagrams from Shear Force Diagrams

The easiest and quickest way to construct a bending moment diagram is by using the shear diagram. Figure 5.9 shows the functional relationships between the loading conditions and the corresponding shear diagrams. Figure 6.10 shows similar relationships between loading and moment diagrams. Obviously the two are closely related. A number of the typical relationships between shear force and internal bending moment are shown in Figure 6.13.

It can be shown that *the change in bending moment between any two points on a beam is equal to the area under the shear force diagram between those two points.* Figure 6.14 explains this relationship. Here there are three individual concentrated forces acting on a beam. The shear diagram, constructed individually for each force, is shown in the middle. The shear forces result in three rectangular areas. These areas, with respect to the cut at point A, define the individual moments M_1, M_2, and M_3 of the three forces. With careful consideration of their signs, their sum gives M. This concept is valid for any possible shear diagram.

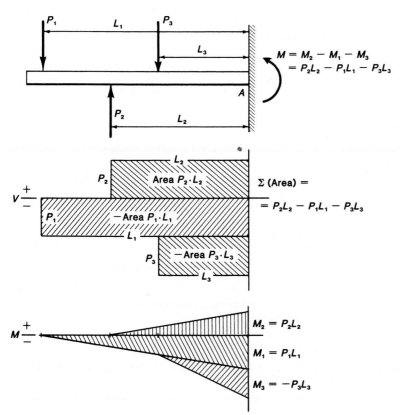

Figure 6.14　*Illustration that the area under the shear force diagram gives the bending moment.*

In the previous chapter we have seen that the same relationship is valid between the shear and the loading diagram, namely, that the change of the internal shear force between two points along the beam equals the area under the loading diagram between the two points considered.

In mathematics, if the value of one variable quantity, say y, depends upon the value of another, say x, then a functional relationship exists between the two variables: y is said to be a function of x, and any change Δx in the value of x

Example 6.9

Construct the shear diagram for the loading shown in Figure 6.15a and develop the bending moment diagram from the shear. (This example repeats Example 6.1.)

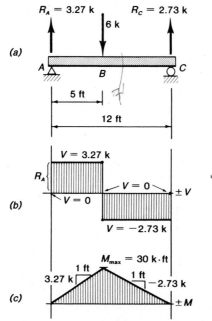

Figure 6.15 *Construction of the bending moment diagram from the shear diagram for Example 6.1.*

will be accompanied by a change Δy in the value of y. The limit of the ratio of Δy to Δx as Δx approaches zero is called the slope of the function (see Appendix D). From this it follows that *the slope of the moment diagram at a point is given by the magnitude of the internal shear force at that point.* Similarly, as we have seen previously, the slope of the shear force diagram at a point equals the negative of the magnitude of the loading at that point. These concepts may be used to prove the validity of the rules given in Section 6.5 for the construction of moment diagrams.

Solution:

Following the principles introduced in the previous chapter, the shear diagram is constructed in Figure 6.15*b*. As shown, the shear between points A and B, is a constant, equal to the value of the reaction at the left support, 3.27 kips. At point B the shear changes abruptly from +3.27 kips to −2.73 kips. The difference of these values is equal to the concentrated force acting at point B. At the ends of the beam beyond the supports, the shear force in the beam is zero.

The moments at both simple supports are zero, as these points are allowed to rotate and hence they do not resist a moment. From the support at point A the moment increases linearly, as proved by the repeated computations made in Example 6.1. The peak moment, at point B, equals the area under the shear force diagram between A and B. Thus

$$M_B = 3.27 \ (\text{kips}) \times 5 \ (\text{ft}) = 16.35 \ \text{kip} \cdot \text{ft}$$

The slope of the line representing the moment between points A and B is equal to the constant value of shear between these points.

The moment diagram between points B and C is, again, represented by a constant slope. The slope of the line equals the shear in this region, −2.73 kips. Since this is a negative value, the moment diagram declines from its peak value at B to zero at the right support.

At the location of the maximum moment, B, the shear force is zero. This proves that the moment has an extreme value at B. Immediately to the left of point B the positive shear corresponds to the positive slope of the moment diagram. To the right of B the negative shear force corresponds to the negative slope of the moment diagram. The discontinuity of the shear diagram at B indicates that the slope of the moment diagram at B changes from a positive value (3.27 k) to a negative value (−2.73 k). Other types of loading distributions produce maxima and minima in the moment diagram (points at which the slope of the diagram is zero) when the shear diagram does not exhibit a sudden jump through zero.

Example 6.10

Construct the moment diagram for the airplane wing shown in Figure 5.10.

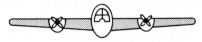

(a)   for loading see figure 5.6

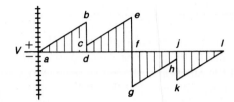

(b)

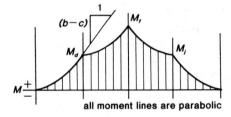

all moment lines are parabolic

M_d = area V (abd)
$M_t = M_d$ + area V (cefd)
(c) $M_j = -$area V (jkl)

Figure 6.16   *Construction of the bending moment diagram for Example 5.4.*

Solution:

First, we produce the shear diagram as shown in Figures 5.10*b* and 6.16*b*. The resulting moment diagram will have three characteristic locations, at points *d*, *f*, and *j*.

M_d equals the area on the shear force diagram enclosed by points *a*, *b*, and *d*, since the moment at the free end of the wing is zero.

M_f, the maximum moment at the base of the wing, equals the area of the shear force diagram enclosed by points *a*, *b*, *c*, *e*, *f*, *d*, and *a*.

For the right wing of the plane, the moment diagram is the mirror image of the left side. The fact that in this portion the areas under the shear force diagram are negative accounts for the negative slope of the moment toward the right wingtip.

Since the shear force diagram is composed of portions that all have the same positive slope (induced by the uniform uplift force throughout the wings), all segments of the moment diagram will be parabolic. Figure 6.14, showing the functional relationships between shear force and moment distribution, confirms this. The only difference between these parabolic segments is in their initial tangents. These are given by the magnitude of the shear force at the initial points. For instance, the shear values at the wing tips are zero, indicating that the parabola representing the moment distribution starts with horizontal tangent. At the left engine, marked by point *d*, the tangent of the parabola from the left side equals the shear described by the distance *bd*, and on the right side, by the distance *cd*. At the fuselage the slope to the left is defined by *ef*, and to the right, by *fg*. The latter is a negative value, showing that to the right of point *f*, the slope of the moment diagram is negative and the magnitude of the moment is decreasing. The values of shear described by the distances *cd* and *jh* are the tangents to the moment diagram on either side.

Example 6.11

Determine the maximum moment and the shape of the moment diagram for the problem solved in Example 6.8 by the use of the shear force diagram.

Solution:

First the reaction at point A is determined. R_A equals the area under the loading distribution,

$$R_A = w_0 L/2$$

The shear force diagram, according to Figure 5.9, is parabolic. The tangent of the parabola at point A equals the negative of w_0; at point B the slope is zero. The value of the shear force at point A equals R_A.

From the geometry of a parabola it is known that the area under a parabolic shear force diagram equals one-third of the area of the rectangle enclosing it; that is,

$$\text{Area of } V = -\frac{1}{3}\left(\frac{w_0 L}{2}\right) L = -\frac{w_0 L^2}{6}$$

Equilibrium conditions show that

$$M_A = -\frac{w_0 L^2}{6}$$

and since the moment at the free end B must be zero.

According to Figure 6.14 the parabolic shear force distribution brings about a cubic change in moment. The slope of this curve at point A equals the magnitude of the shear at the point $w_0 L/2$. At point B the slope of the cubic curve is zero, since the value of the shear force at B is zero. The curve is sketched in Figure 6.17.

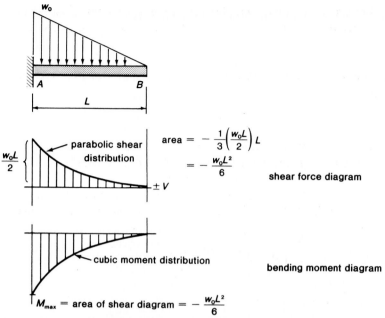

$$\text{area} = -\frac{1}{3}\left(\frac{w_0 L}{2}\right) L$$

$$= -\frac{w_0 L^2}{6}$$

shear force diagram

bending moment diagram

M_{max} = area of shear diagram $= -\dfrac{w_0 L^2}{6}$

Figure 6.17 *Development of the bending moment diagram for Example 6.3 using the shear force diagram.*

Example 6.12

Determine the location and magnitude of the maximum moment on the airframe discussed in Example 5.9.

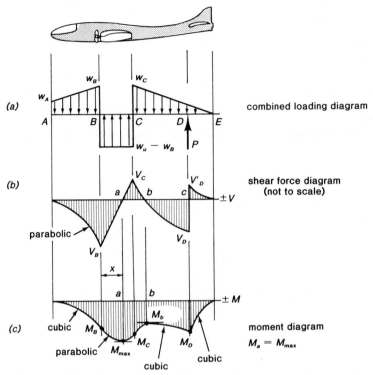

Figure 6.18 *Construction of the bending moment diagram for Example 5.9 using the shear force diagram from Figure 5.12c.*

Solution:

In Figures 6.18*a* and *b* the combined loading and the shear force diagrams are reproduced from Figure 5.12. The shear diagram shows three points where the moment has maxima or minima. These are marked as points *a*, *b*, and *c*. Since the magnitude of the moments may be obtained by the corresponding areas of the shear diagram to either side of a point considered, one may arrive at the following conclusions:

   a. The moment at point *a* equals the area of the shear diagram from the left end to a distance *x* from point *B* since end *A* is a free end. This shear force diagram area is all negative. It consists of a parabolic section from *A* and *B* and a triangular section from *B* to *a*.

   b. The moment at point *b* *must* be less than the moment at point *a* because the area of the shear force diagram between *a* and *b* is positive. Therefore, in summing the shear areas to the left, this area must be subtracted from the negative area, to obtain the moment at b.

c. The moment at point c may be taken as the negative value of the shear area between point c and E. It is a small parabolic section that is rather insignificant when compared to the areas contributing to the moment at point a.

On this basis we may now conclude that the maximum bending moment acting on the fuselage of the airplane will be found at point a under the base of the wings.

To determine the moment at point a we must first locate this point. In order to find the distance x we turn our attention to the loading diagram. At a the shear force is zero, therefore

$$\Sigma\, F_V = 0^+$$

$$\left(\frac{w_A + w_B}{2}\right) L_1 - (w_u - w_B)\, x = 0$$

From this we find that

$$x = \frac{(w_A + w_B)\, L_1}{2\,(w_u - w_B)}$$

Now we are ready to determine the area of the shear force diagram between points A and a. This is equal to

$$\frac{1}{3} L_1 V_B + \frac{1}{2} V_B x = \left(\frac{L_1}{3} + \frac{x}{2}\right) V_B$$

The first term is the area of the parabolic section between A and B; the second is that of the triangle between points B and a. Substituting V_B from Example 5.9, and substituting the value computed for x, one obtains

$$M_{\text{max}} = (w_A + w_B)\, L_1^2 \left[\frac{1}{6} + \frac{1}{8}\left(\frac{w_A + w_B}{w_u - w_B}\right)\right]$$

Without computing the values of the other parts of the moment diagram, we may sketch its shape into the figure according to the principles outlined in this chapter. One may observe that the moment function is cubic between points A and B. Then through a point of inflection at M_B it turns into a parabolic function between points B and C. From C to E the curve consists of two cubic segments. At point b the moment has a localized minimum value. The slope of the line is zero, then it turns negative again. At M_C the transition is smooth, but the concentrated force P at the tail causes a kink to form at M_D.

6.1 Rework Problem 5.1 for finding bending moment to the left and right of the 2,000-pound load.

6.2 Determine the bending moment equation for the beam shown in Figure P 5.2.

6.3–
6.22 Find the bending moment equation and draw a bending moment diagram for each beam shown in Figures P 5.3 through P 5.22.

6.23–
6.27 Find the forces (moments) acting on the beam from the moment diagrams shown in Figures P 6.23 through P 6.27.

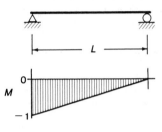

Figure P 6.23

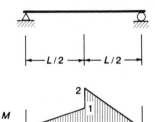

Figure P 6.24

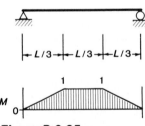

Figure P 6.25

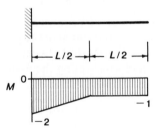

Figure P 6.26

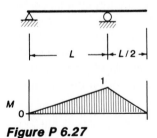

Figure P 6.27

Properties of Cross Sections

7

1. To understand how to find the centroid of any complex shape.

Objectives

2. To be able to find the moments of inertia of any complex shape.

3. To be able to find the partial moment of its area for points on cross section.

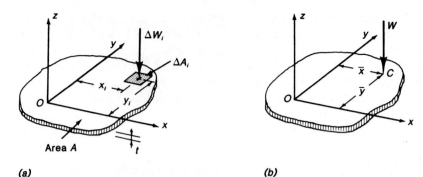

Figure 7.1   Force ΔW_i acting perpendicularly to a small element ΔA_i of a plate of area A.

7.1 Introduction

Why do engineers use I-shaped steel members in constructing buildings? Why do producers of the pieces of machines and buildings make pieces with such a wide variety of cross-sectional shapes—L-shapes, I-shapes, box-shapes, etc.? Why are the shafts in machines of circular cross section? The answers to these questions and other similar questions may be found in the geometric properties of the cross-sectional shapes themselves. For this reason it is very important that the relevant properties of a cross section are understood clearly, and that they may be readily found.

It is difficult at this stage to recognize which properties of a cross section are important. In subsequent chapters of this book, however, various properties of a cross section are used extensively. In order to avoid discussion of the derivation of these properties as they are required, this chapter gathers together all of the required properties for consideration at one time.

7.2 Center of Mass and Centroid

Consider the flat plate of Figure 7.1a, which is in the xy-plane. The plate has a constant thickness t and is of one material with a density ρ. The very small area ΔA_i thus has a weight ΔW_i, which is given by the expression

(7.1) $$\Delta W_i = \rho t \Delta A_i$$

The weight of this small area acts in the negative z-direction. The total weight W of the plate must therefore also act in the negative z-direction, and the value of W must be equal to the sum of the weights of all the small areas like the area ΔA_i, which when taken together would spread over the entire area A of the plate, i.e.,

(7.2) $$W = \Sigma \, \Delta W_i$$

or

(7.3) $$W = \Sigma_i \rho t \Delta A_i$$

The weight W is thus the resultant of the small weight forces ΔW_i acting on all the small areas ΔA_i. It is now necessary to know where this resultant acts on the plate. Figure 7.1b shows the resultant weight force W acting at the point C, with coordinates $(\bar{x}, \bar{y})$. The point C is known as the *center of mass* of the plate. The center of mass of any body is defined as that point on the body through which the resultant weight force acts.

If the plate is of constant thickness and of uniform density throughout, the point C is also the *centroid* of the area A. A centroid is a geometric property of a cross section, bearing no relation to the thickness of the plate or the density of the material.

The finding of the centroid of a section is a relatively simple matter. Suppose that the weight of the plate is W, and it is to be placed in such a position on the plate that the external effects of the weight force are identical to the effects of the small weight forces ΔW_i acting on all the small areas ΔA_i of which the plate is composed. One of the evidences that the force W is indeed equivalent to the sum of the effects of the small weight forces ΔW_i is that the moment of W about any axis must be the same as the sum of the moments of the small forces ΔW_i about the same axis. Thus, taking moments about the y-axis

$$\bar{x} \cdot W = \Sigma x_i \Delta W_i$$

or

(7.4)
$$\bar{x} = \frac{\Sigma x_i \Delta W_i}{W} = \frac{\Sigma x_i \Delta W_i}{\Sigma \Delta W}$$

Since the plate thickness is constant and the density uniform,

$$\Sigma \Delta W_i = \Sigma \rho t \Delta A_i = \rho t \, \Sigma \Delta A_i$$

and

$$\Sigma x_i \Delta W_i = \Sigma x_i \rho t \Delta A_i = \rho t \, \Sigma x_i \Delta A_i$$

Thus

$$\bar{x} = \frac{\rho t \, \Sigma x_i \Delta A_i}{\rho t \, \Sigma A_i}$$

or

(7.5)
$$\bar{x} = \frac{\Sigma x_i \Delta A_i}{\Sigma \Delta A_i}$$

In a similar manner it may be shown that

(7.6)
$$\bar{y} = \frac{\Sigma y_i \Delta A_i}{\Sigma \Delta A_i}$$

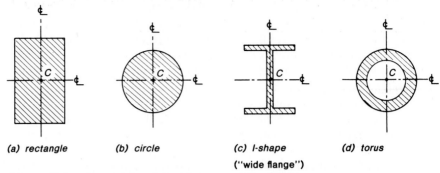

(a) rectangle   (b) circle   (c) I-shape   (d) torus

("wide flange")

Figure 7.2   *Centroids of doubly symmetric sections.*

The location of the centroid C $(\bar{x}, \bar{y})$ is thus determined. It is noted that if the thickness of the plate varies, or the density is nonuniform, the point C will still be the centroid of the plate area A, although it will no longer be the center of mass of the plate.

In locating the centroids of cross sections it may be shown that the *centroid* of a shape *is always on any axis of symmetry* that the shape possesses. In certain circumstances this fact alone is enough to locate the centroid. The rectangular, circular, I-shaped, and toroidal areas of Figure 7.2 all have two axes of symmetry, and the centroid of the area must therefore be at the point where the axes of symmetry meet. Such sections are called doubly symmetric sections (although the circle and the torus have an infinite number of axes of symmetry).

The location of the centroid of some common simple sections is shown in Figure 7.3. The use of these simple shapes for which the areas are known, and for each of which the location of the centroid is known, permits the finding of the area and location of the centroid of a more complex section. This process of determining the properties of a complex section by considering it to be made up of a number of simple sections is referred to as *analyzing a section "by parts."*

It is appropriate to note that the process of replacing a distributed force by a resultant force acting at the centroid of the area has already been mentioned in this book. In Table 3.4 the distributed loads representing spread loadings on a beam were replaced by a resultant force R, which acted at a position on the beam such that the force R had the same moment about one end of the beam as did the distributed load. The resultant force R must be acting at the centroid of the loading.

The centroid of sections with holes in them may be easily found using the same procedure. The area of the hole is taken to be a negative area for the purposes of the calculations. If the entire area of the section is within the first quadrant, the moment of the negative area about either axis will thus be negative.

The summation $\Sigma\, y_i A_i$ is known as the *first moment of area* about the x-axis, and the summation $\Sigma\, x_i A_i$ is thus the first moment of area about the y-axis. This term will be used in later sections of this chapter.

Figure 7.3 Common areas and centroids.

shape	area	$\bar{x}$	$\bar{y}$	
(a) rectangle	bh	$\dfrac{b}{2}$	$\dfrac{h}{2}$	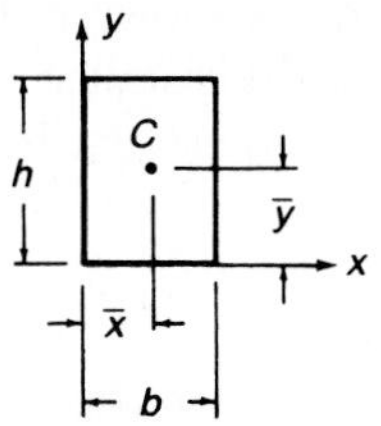
(b) triangle	$\dfrac{bh}{2}$	$\dfrac{b}{3}$	$\dfrac{h}{3}$	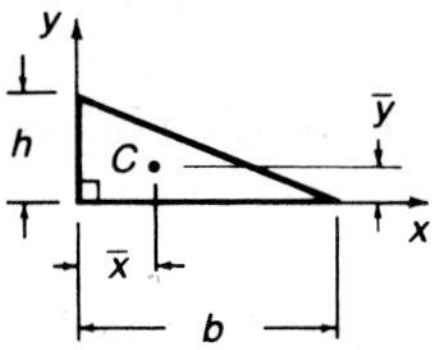
(c) circle	πr^2	r	r	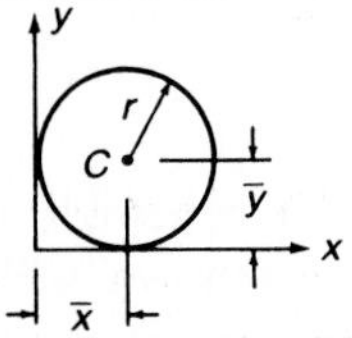
(d) semicircle	$\dfrac{\pi r^2}{2}$	r	$\dfrac{4r}{3\pi}$	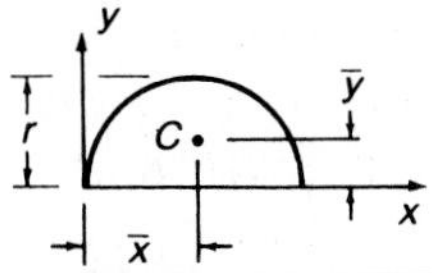
(e) quadrant	$\dfrac{\pi r^2}{4}$	$\dfrac{4r}{3\pi}$	$\dfrac{4r}{3\pi}$	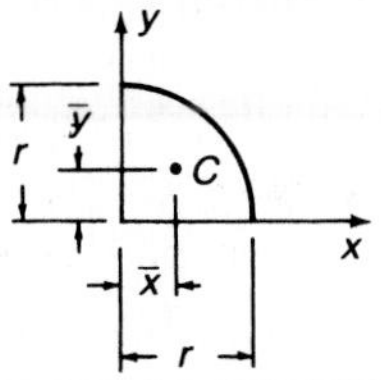
(f) circular sector (α in radians)	αr^2	$\dfrac{2 r \sin \alpha}{3\alpha}$	0	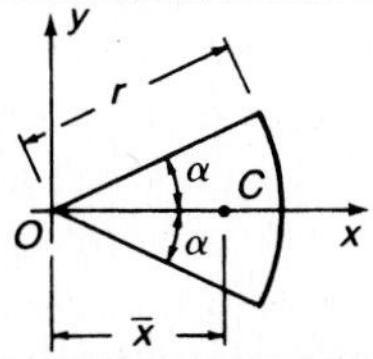
(g) parabola (i) above	$\dfrac{2ab}{3}$	$\dfrac{3a}{8}$	$\dfrac{3b}{5}$	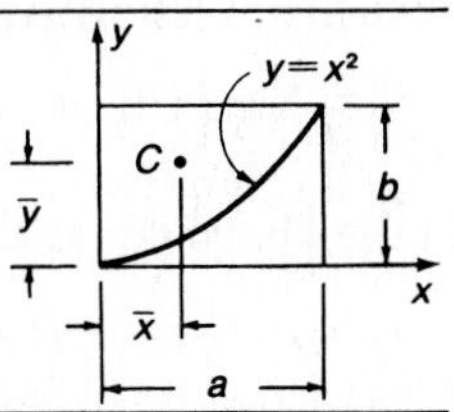
(ii) below	$\dfrac{ab}{3}$	$\dfrac{3a}{4}$	$\dfrac{3b}{10}$	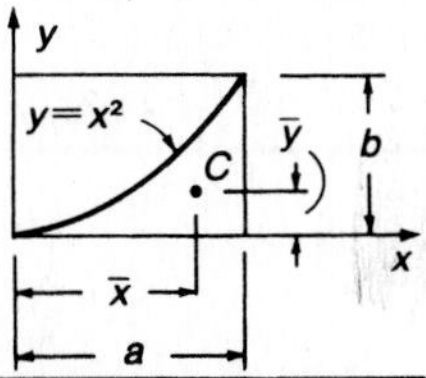

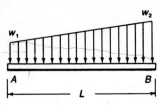

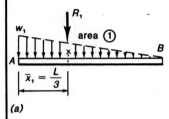

$$\bar{x}_1 = \frac{L}{3}$$

(a)

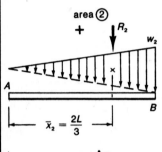

$$\bar{x}_2 = \frac{2L}{3}$$

(b)

Figure 7.4

Example 7.1

A beam is loaded as shown in Figure 7.4a. Find the resultant force R acting on the beam and the distance $\bar{x}$ from the left-hand end A to the point where the resultant force must act. If $w_1 = 3$ kips/ft, $w_2 = 5$ kips/ft, and $L = 12$ ft, find R and $\bar{x}$.

Solution:

As shown in Figure 7.4b the distributed force given may be replaced by two triangular segments. The area and location of a triangular segment is given in Figure 7.3b. The values of the resultant forces R_1 and R_2, from the areas A_1 and A_2, respectively, are given by

$$R_1 = \frac{w_1 L}{2}$$

$$R_2 = \frac{w_2 L}{2}$$

The locations of the centroids of these triangles, $\bar{x}_1$ and $\bar{x}_2$ for areas 1 and 2, respectively, are also shown in Figure 7.4b. The resultant force R on the beam is thus given by

$$R = R_1 + R_2 = \frac{L}{2}(w_1 + w_2)$$

The distance $\bar{x}$ is given by Equation 7.5 as

$$\bar{x} = \frac{\Sigma x_i A_i}{\Sigma A_i} = \frac{\bar{x}_1 R_1 + \bar{x}_2 R_2}{R_1 + R_2}$$

$$= \frac{\frac{L}{3} \cdot \left(\frac{w_1 L}{2}\right) + \frac{2L}{3}\left(\frac{w_2 L}{2}\right)}{\frac{L}{2}(w_1 + w_2)}$$

$$= \frac{L}{3}\left(\frac{w_1 + 2w_2}{w_1 + w_2}\right)$$

This is exactly the expression given for this loading in Table 3.4c.

If $w_1 = 3$ kips/ft, $w_2 = 5$ kips/ft, and $L = 12$ ft, then

substitution in the above equations yields

$$R = 48 \text{ kips}$$
$$\bar{x} = 6.5 \text{ ft}$$

Example 7.2

Find the centroid of the shape shown in Figure 7.5a.

Solution:
It is first noted that the shape has an axis of symmetry along the line $x = 25$ mm. Since the centroid of any shape must be somewhere on its line of symmetry (if it has one), then the centroid of this shape must be somewhere on the line $x = 25$ mm, i.e., $\bar{x} = 25$ mm.

To find the value of $\bar{y}$, the distance from the x-axis to the centroid, the shape is divided into two simpler areas as shown in Figure 7.5b. For each of these two areas, a triangle and a rectangle, the area and the location of the centroid are given in Figure 7.3. The necessary properties of each of these simple areas are now found:

For area 1:

$$A_1 = bh = 50 \cdot 20 = 1,000 \text{ mm}^2$$

$$\bar{y}_1 = \frac{h}{2} = \frac{20}{2} = 10 \text{ mm}$$

For area 2:

$$A_2 = \frac{bh}{2} = \frac{50 \cdot 30}{2} = 750 \text{ mm}^2$$

$$\bar{y}_2 = 20 + \frac{h}{3} = 20 + \frac{30}{3} = 30 \text{ mm}$$

To find the required products for calculating the location of the centroid it is often useful to organize the results as shown in Table 7.1.

Table 7.1

Part (i)	A_i (mm²)	$\bar{y}_i$ (mm)	$A_i \cdot \bar{y}_i$ (mm³)
1	1,000	10	10,000
2	750	30	22,500
Total	1,750		32,500

Thus,

$$\bar{y} = \frac{\Sigma A_i \cdot \bar{y}_i}{\Sigma A_i} = \frac{32,500}{1,750} = 18.6 \text{ mm}$$

The location of the centroid $C(\bar{x}, \bar{y})$ is then given by

$$\bar{x} = 25 \text{ mm}$$
$$\bar{y} = 18.6 \text{ mm}$$

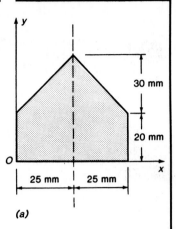

(a)

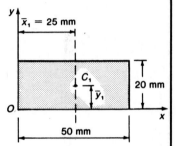

area ①

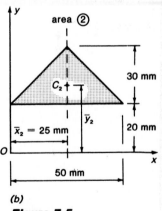

(b)
Figure 7.5

Example 7.3

Find the centroid and area of the shape shown in Figure 7.6a.

Solution:

As shown in Figure 7.6b, the area given is broken into four simple shapes, for each of which the area and centroid are easily found from Figure 7.3. For each of these shapes, the area and the coordinates of its centroid with respect to the x- and y-axes are now computed.

For area 1:

$$A_1 = 150 \cdot 20 = 3{,}000 \text{ mm}^2$$

$$\bar{x}_1 = \frac{150}{2} = 75 \text{ mm}$$

$$\bar{y}_1 = \frac{20}{2} = 10 \text{ mm}$$

For area 2:

$$A_2 = 180 \cdot 30 = 5{,}400 \text{ mm}^2$$

$$\bar{x}_2 = 80 + \frac{30}{2} = 95 \text{ mm}$$

$$\bar{y}_2 = 20 + \frac{180}{2} = 110 \text{ mm}$$

For area 3:

$$A_3 = \frac{45 \cdot 30}{2} = 675 \text{ mm}^2$$

$$\bar{x}_3 = 110 + \frac{30}{3} = 120 \text{ mm}$$

$$\bar{y}_3 = 115 + \frac{2}{3}(45) = 145 \text{ mm}$$

For area 4:

$$A_4 = \frac{\pi \cdot 30^2}{2} = 1{,}414$$

$$\bar{x}_4 = 110 \text{ mm}$$

$$\bar{y}_4 = 200 + \frac{4 \cdot 30}{3\,\pi} = 212.7 \text{ mm}$$

To find the products necessary to find the area and centroid of the complex shape, Table 7.2 is formed.

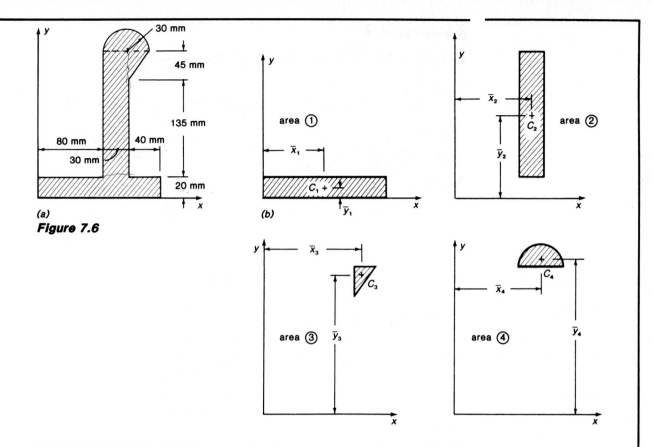

Figure 7.6

Table 7.2

Part (i)	A_i (mm²)	$\bar{x}_i$ (mm)	$\bar{y}_i$ (mm)	$A_i \cdot \bar{x}_i$ (mm³)	$A_i \cdot \bar{y}_i$ (mm³)
1	3,000	75	10	225,000	30,000
2	5,400	95	110	513,000	594,000
3	675	120	145	81,000	97,000
4	1,414	110	212.7	155,500	300,800
Total	10,489			974,500	1,021,800

Thus,

$$A = \Sigma A_i = 10{,}489 \text{ mm}^2$$

$$\bar{x} = \frac{\Sigma x_i A_i}{\Sigma A_i} = \frac{974{,}500}{10{,}489} = 92.9 \text{ mm}$$

$$\bar{y} = \frac{\Sigma y_i A_i}{\Sigma A_i} = \frac{1{,}021{,}800}{10{,}489} = 97.5 \text{ mm}$$

Example 7.4

Find the centroid of the area of the shape shown in Figure 7.7a.

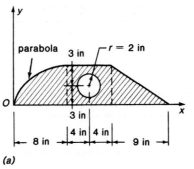

(a)

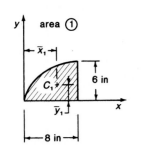

area ①

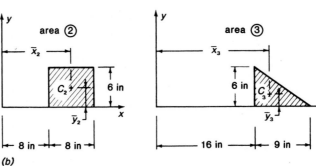

area ②

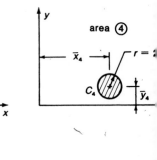

area ③

area ④

(b)

Figure 7.7

Solution:

The complex area of Figure 7.7a is divided into four simpler areas as shown in Figure 7.7b. The area and centroid of each of these simpler areas may be readily found from Figure 7.3, although some care will need to be taken to ensure that the correct shape is being used, and that the centroid is being measured from the correct place on the section. It is also noted that the area of the hole is to be taken as a negative area since it is an area to be subtracted from a solid shape. The coordinates of the centroid of each simpler area are computed with respect to the x- and y-axes.

For area 1:

$$A_1 = \frac{2ab}{3} = \frac{2 \cdot 8 \cdot 6}{3} = 32 \text{ in}^2$$

$$\bar{x}_1 = a - \frac{3a}{8} = \frac{5a}{8} = \frac{5 \cdot 8}{8} = 5.0 \text{ in}$$

$$\bar{y}_1 = b - \frac{3b}{5} = \frac{2b}{5} = \frac{2 \cdot 6}{5} = 2.4 \text{ in}$$

Table 7.3

Part (*i*)	A_i (in²)	$\bar{x}_i$ (in)	$\bar{y}_i$ (in)	$A_i \cdot \bar{x}_i$ (in³)	$A_i \cdot \bar{y}_i$ (in³)
1	32	5	2.4	160	76.8
2	48	12	3.0	576	144
3	27	19	2.0	513	54
4	−12.57	12	3.0	−150.8	−37.7
Total	94.43			1,098.2	237.1

For area 2:

$$A_2 = 8 \cdot 6 = 48 \text{ in}^2$$

$$\bar{x}_2 = 8 + \frac{8}{2} = 12.0 \text{ in}$$

$$\bar{y}_2 = \frac{6}{2} = 3.0 \text{ in}$$

For area 3:

$$A_3 = \frac{6 \cdot 9}{2} = 27.0 \text{ in}^2$$

$$\bar{x}_3 = 16 + \frac{9}{3} = 19.0 \text{ in}$$

$$\bar{y}_3 = \frac{6}{3} = 2.0 \text{ in}$$

For area 4:

$$A_4 = -\pi \cdot r^2 = -\pi \cdot 2^2 = -12.57 \text{ in}$$

$$\bar{x}_4 = 12.0 \text{ in}$$

$$\bar{y}_4 = 3.0 \text{ in}$$

To find the location of the centroid of the complex shape, the necessary products are assembled in Table 7.3.
Thus,

$$\bar{x} = \frac{\Sigma A_i x_i}{\Sigma A_i} = \frac{1{,}098.2}{94.43} = 11.53 \text{ in}$$

$$\bar{y} = \frac{\Sigma A_i y_i}{\Sigma A_i} = \frac{237.1}{94.43} = 2.51 \text{ in}$$

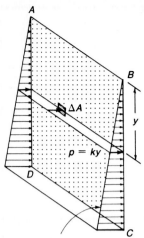

pressure of water on plate

Figure 7.8
Hydrostatic pressure variation on a submerged plate.

7.3 Moment of Inertia

In Chapter 10 the strength of beams in bending is considered. At that point it will be necessary to use the *moment of inertia* of a section, which is now derived. The moment of inertia is a property of a cross section, and yet it cannot be measured by any measuring device. It is purely a numerical concept that helps make many calculations easier.

To help in understanding how the moment of inertia of a section is derived, consider the plate *ABCD* shown in Figure 7.8, which is suspended vertically in water with the top edge *AB* just at the water surface. Those who are divers will know that the pressure of water on any object gets greater as the object gets further beneath the water surface. In fact, the pressure is proportional to the depth, or $p = k \cdot y$, where k is a constant. The horizontal force ΔF acting on a small area ΔA of the plate at a depth y below the water surface is thus given by

(7.7)
$$\Delta F = p \cdot \Delta A$$
$$= (ky) \cdot \Delta A$$

The moment ΔM of this small force about the horizontal axis *AB* at the water surface is thus given by

(7.8)
$$\Delta M = \Delta F \cdot y$$
$$= (k \cdot y \cdot \Delta A) \cdot y$$
$$= k \cdot y^2 \cdot \Delta A$$

Figure 7.9 *Properties of cross sections.*

shape of area	$\bar{I}_x$	Z_x	k_x
(a) rectangle 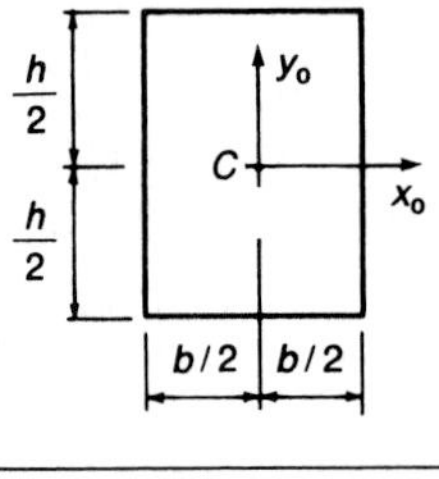	$\dfrac{bh^3}{12}$	$\dfrac{bh^2}{6}$	$0.289h$
(b) triangle 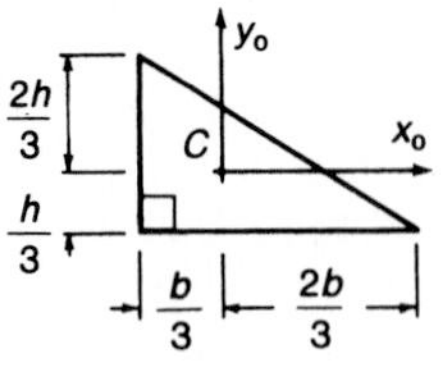	$\dfrac{bh^3}{36}$	$\dfrac{bh^2}{24}$	$0.236h$

The sum of all the little moments ΔM acting on all the little areas ΔA, of which the plate $ABCD$ (with total area A) is composed, is thus

(7.9)
$$\begin{aligned} M &= \Sigma\,\Delta M \\ &= \Sigma\,k \cdot y^2 \cdot \Delta A \\ &= k \cdot \Sigma\,y^2 \cdot \Delta A \end{aligned}$$

The summation $\Sigma\,y^2 \cdot \Delta A$ is defined as the moment of inertia of the area A about the axis AB. The moment of inertia is usually given the symbol I and is written

(7.10)
$$I = \Sigma\,y^2 \cdot \Delta A$$

It is a matter of great importance to know the axis about which the moment of inertia is given. In order to standardize this the moments of inertia of common sections are usually given about centroidal axes. The moments of inertia about centroidal axes (i.e., about axes through the centroid) for simple sections is given in Figure 7.9. The moment of inertia about the x_0-axis through the centroid is given the symbol $\bar{I}_x$, while the moment of inertia about the y_0-axis through the centroid is given the symbol $\bar{I}_y$.

$\bar{I}_y$	Z_y	k_y	J
$\dfrac{hb^3}{12}$	$\dfrac{hb^2}{6}$	$0.289b$	$\dfrac{bh(b^2 + h^2)}{12}$
$\dfrac{hb^3}{36}$	$\dfrac{hb^2}{24}$	$0.236h$	

shape of area	$\bar{I}_x$	Z_x	k_x
(c) circle 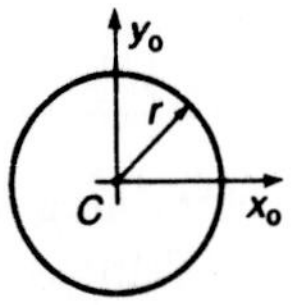	$\dfrac{\pi r^4}{4}$	$\dfrac{\pi r^3}{4}$	$\dfrac{r}{2}$
(d) semicircle 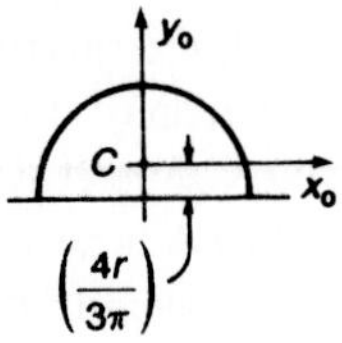	$r^4\left(\dfrac{\pi}{8} - \dfrac{8}{9\pi}\right)$ $= 0.110r^4$	$0.191r_3$	$0.264r$
(e) ring 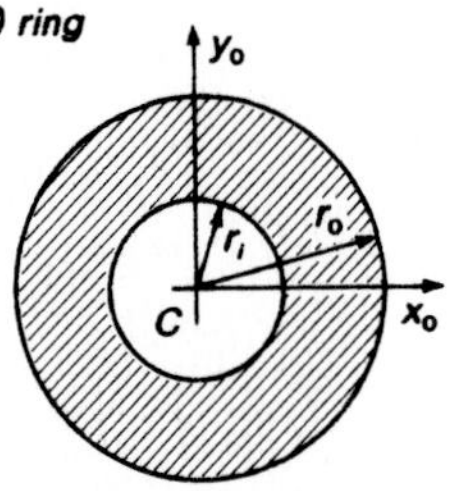	$\dfrac{\pi}{4}\left(r_0^4 - r_i^4\right)$	$\dfrac{\pi(r_0^4 - r_i^4)}{4r_0}$	$\dfrac{\sqrt{r_0^2 + r_i^2}}{2}$
(f) hollow rectangle 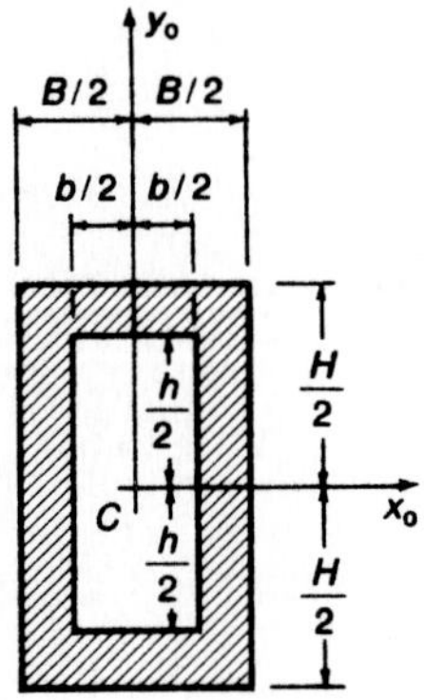	$\dfrac{1}{12}\left(BH^3 - bh^3\right)$	$\dfrac{1}{6H}\left(BH^3 - bh^3\right)$	$\sqrt{\dfrac{BH^3 - bh^3}{12\,(BH - bh)}}$

Figure 7.9 (Continued)

$\bar{I}_y$	Z_y	k_y	J
$\dfrac{\pi r^4}{4}$	$\dfrac{\pi r^3}{4}$	$\dfrac{r}{2}$	$\dfrac{\pi r^4}{2}$
$\dfrac{\pi r^4}{8}$ $= 0.393 r^4$	$\dfrac{\pi r^3}{8}$ $= 0.393 r^3$	$\dfrac{r}{2}$	
$\dfrac{\pi}{4}\,(r_o{}^4 - r_i{}^4)$	$\dfrac{\pi(r_o{}^4 - r_i{}^4)}{4 r_o}$	$\sqrt{\dfrac{r_o{}^2 + r_i{}^2}{2}}$	$\dfrac{\pi}{2}\,(r_o{}^4 - r_i{}^4)$
$\dfrac{1}{12}\,(HB^3 - hb^3)$	$\dfrac{1}{6B}\,(HB^3 - bh^3)$	$\sqrt{\dfrac{HB^3 - hb^3}{12\,(BH - bh)}}$	

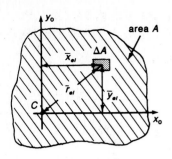

Figure 7.10
An elementary area
on a plane.

In Figure 7.10 a general section of area A is shown. The x_0- and y_0-axes are drawn through the centroid C of the section. A small area ΔA of the section is at a distance $\bar{y}_{el}$ from x_0-axis and at a distance $\bar{x}_{el}$ from the y_0-axis. The moment of inertia about the x_0-axis is the summation over the entire area A of

(7.11)
$$\bar{I}_x = \Sigma \, \bar{y}_{el}^2 \cdot \Delta A$$

In a similar manner the moment of inertia about the y_0-axis may be shown to be

(7.12)
$$\bar{I}_y = \Sigma \, \bar{x}_{el}^2 \, \Delta A$$

The moment of inertia is sometimes called the *second moment of area* of a section.

Three other properties of a section that may readily be derived from a moment of inertia of a section are also given in Figure 7.9. The first is the *section modulus, Z*. The section modulus is defined as the moment of inertia of a section about a given axis, divided by the maximum distance that a point on a section may have from that axis. Thus for the rectangular section of Figure 7.9a the section modulus about the x_0-axis, given the symbol Z_x, is given by

(7.13)
$$Z_x = \frac{\bar{I}_x}{\left(\dfrac{h}{2}\right)} = \frac{\left(\dfrac{bh^3}{12}\right)}{\left(\dfrac{h}{2}\right)} = \frac{bh^2}{6}$$

This value of Z_x applies only to rectangular sections when the moment of inertia is derived about the centroidal x_0-axis. Fortunately, this is the axis about which the section modulus is most commonly required.

Another common property of a cross section that may be found from the moment of inertia is the *radius of gyration, r*. The radius of gyration is defined by the relationship

(7.14)
$$r = \sqrt{\frac{I}{A}}$$

where A is the area of the cross section. The radius of gyration is also dependent on the axis about which the moment of inertia is taken. The radius of gyration r_x associated with the moment of inertia $\bar{I}_x$ for the rectangular section in Figure 7.9a, for example, is found to be

(7.15)
$$r_x = \sqrt{\frac{\bar{I}_x}{A}} = \sqrt{\frac{\left(\dfrac{bh^3}{12}\right)}{bh}} = 0.289 \, h$$

For some sections another property—the *polar moment of inertia, J*—is also given. It refers to the moment of a section about its centroid rather than one of its axes. It is important in calculation of the twisting strength of sections (see Chapter 9). It is defined in terms of a summation over the area of a section as

(7.16)
$$J = \Sigma \, \bar{r}_{el^2} \cdot \varDelta A$$

where $\bar{r}_{el}$ and $\varDelta A$ are defined in Figure 7.10. Since this property is not used commonly for a large number of sections, its value is not tabulated for all sections. However, if the value of J is required for more complex sections, it may be shown that the value of J for any section may be found by adding the moments of inertia about the x_0- and y_0-axes, i.e.,

(7.17)
$$J = I_x + I_y$$

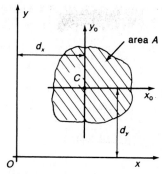

Figure 7.11
An arbitrary area in the xy-plane.

These values, together with the dimensions of the sections, are tabulated for the cross sections of some common engineering shapes in Appendix A. The sections listed in Appendix A are not at all exhaustive but they are a good sample, showing the large range of properties which may be expected in their values.

7.4 Transfer of Axes

It has previously been mentioned that it is very important to know the axes about which the moments of intertia are being given. It is standard practice to tabulate the moments of inertia about axes that pass through the centroid of the shape. However, it often happens that the moment of inertia of a section is required about an axis that does not pass through the centroid. In order to "transfer" the given moments of inertia from the centroidal axes to different axes, the parallel axis theorem is used.

Consider that the moment of inertia $\bar{I}_x$ about the x_0-axis through the centroid C for the area A shown in Figure 7.11 is known. It is desired to find the moment of inertia about the x-axis, which is parallel to the x_0-axis but does not go through the centroid C. Instead, the x-axis is at a distance d_y from the x_0-axis. It may be shown that the moment of inertia of the area A about the x-axis is given by

(7.18)
$$I_x = \bar{I}_x + A \cdot d_y^2$$

where I_x is the moment of inertia about the x-axis, and $\bar{I}_x$ is the moment of inertia about the x_0-axis. In a similar way the moment of inertia about the y-axis, I_y, may be found in terms of the moment of inertia about the y_0-axis ($\bar{I}_y$) as

(7.19)
$$I_y = \bar{I}_y + A \cdot d_x^2$$

In using Equations 7.18 and 7.19, care must be taken to make sure that the transfer of axes is being made between parallel axes (e.g., between the x- and x_0-axes, or between the y- and y_0-axes). Notice also that the moment of inertia about a centroidal axis is always included in Equations 7.18 and 7.19. These equations *cannot* be used to transfer the moment of inertia between two axes when neither axis goes through the centroid.

Properties of Cross Sections   195

The main use of the parallel axis theorem is to permit the finding of the moments of inertia of complex sections. Recall that a complex section was divided into a series of simpler sections in order to find the centroid of the complex section. In the same way the moments of inertia of a complex shape may be found from the moments of inertia of simpler sections. In general, however, the moments of inertia of a complex shape are required about axes that pass through the centroid of the complex shape, and thus the moments of inertia of the simpler shapes need to be transferred to coincide with these axes.

Example 7.5

Find the moments of inertia about axes through the centroid of the T-section given in Figure 7.12a parallel to the x- and y-axes.

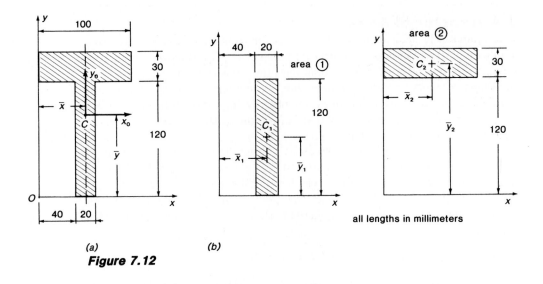

(a)   (b)

Figure 7.12

Solution:

The first step in finding the moment of inertia is to locate the centroid. As the T-section has a vertical axis of symmetry, the centroid must lie on this axis. Thus $\bar{x} = 5.0$ centimeters. To find the $\bar{y}$ value locating the centroid, the T-section is divided into two rectangular sections as shown in Figure 7.12b. For each of these sections the area, centroid location, and moments of inertia about their own centroid, $\bar{I}_{xi}$ and $\bar{I}_{yi}$, may all be found using Figures 7.3 and 7.9. The calculations to locate the centroid are done in the same manner as previously demonstrated.

When summed over a number of simple sections, Equations 7.18 and 7.19 become

(7.20)
$$I_x = \Sigma \, (\bar{I}_{xi} + A_i d_y{}^2)$$

(7.21)
$$I_y = \Sigma \, (\bar{I}_{yi} + A_i \cdot d_y{}^2)$$

where $\bar{I}_{xi}$ and $\bar{I}_{yi}$ are the moments of inertia for a simple section i about axes parallel to the x- and y-axes through its own centroid.

In the same way as for centroids, holes in sections are treated as negative areas. They also have negative moments of inertia about their own centroids.

Table 7.4

Part (i)	A_i (mm²)	$\bar{y}_i$ (mm)	$A_i \cdot \bar{y}_i$ (mm³)
1	2,400	60	144,000
2	3,000	135	405,000
Total	5,400		549,000

For area 1:

$$A_1 = 120 \cdot 20 = 2,400 \text{ mm}^2$$
$$\bar{x}_1 = 50 \text{ mm}$$
$$\bar{y}_1 = \frac{h}{2} = \frac{120}{2} = 60 \text{ mm}$$

For area 2:

$$A_2 = b \cdot h = 100 \cdot 30 = 3,000 \text{ mm}^2$$
$$\bar{x}_2 = 50 \text{ mm}$$
$$\bar{y}_2 = 120 + \frac{h}{2} = 120 + \frac{30}{2} = 135 \text{ mm}$$

The figures for calculating the centroid coordinate are collected in Table 7.4. Thus,

$$\bar{y} = \frac{\Sigma \, A_i \cdot \bar{y}_i}{\Sigma \, A_i} = \frac{549,000}{5,400} = 101.7 \text{ mm}$$

With the location of the centroid of the complex section known the finding of the moments of inertia may now proceed. The moments of inertia of each individual section about its own centroid are first found and then these values are transferred to the centroid of the complex section, using Equations 7.20 and 7.21.

Example 7.5 (Continued)

Table 7.5

Part (i)	$\bar{I}_{xi}$ (mm⁴)	A_i (mm²)	$d_y = \bar{y} - \bar{y}_i$ (mm)	$A_i \cdot d_y^2$ (mm⁴)	$\bar{I}_{xi} + A_i \cdot d_y^2$ (mm⁴)
1	2.88×10^6	2,400	$101.7 - 60$ $= 41.7$	4.17×10^6	7.05×10^6
2	22.5×10^4	3,000	$101.7 - 135$ $= -33.3$	3.33×10^6	3.35×10^6
Total					1.040×10^7

For area 1:

$$\bar{I}_{x1} = \frac{bh^3}{12} = \frac{20 \cdot 120^3}{12} = 2.88 \cdot 10^6 \text{ mm}^4$$

$$\bar{I}_{y1} = \frac{hb^3}{12} = \frac{120 \cdot 20^3}{12} = 8.0 \cdot 10^4 \text{ mm}^4$$

For area 2:

$$\bar{I}_{x2} = \frac{bh^3}{12} = \frac{100 \cdot 30^3}{12} = 22.5 \cdot 10^4 \text{ mm}^4$$

$$\bar{I}_{y2} = \frac{hb^3}{12} = \frac{30 \times 100^3}{12} = 2.50 \cdot 10^6 \text{ mm}^4$$

The calculation of the centroidal moment of inertia of the complex section about the x_0-axis is facilitated by using a table such as Table 7.5. Care must be taken to ensure that the correct distance—given the symbol d_y in Equation 7.20—is being used to shift the x-axes.

Thus,

$$\bar{I}_x = \Sigma (\bar{I}_{xi} + A_i \cdot d_y^2) = 1.04 \times 10^7 \text{ mm}^4$$

Table 7.6

Part (i)	$\bar{I}_{yi}$ (mm⁴)	A_i (mm²)	$d_x = \bar{x} - \bar{x}_i$ (mm)	$A_i \cdot d_x^2$ (mm⁴)	$\bar{I}_{yi} + A_i \cdot d_x^2$ (mm⁴)
1	8.0×10^4	240	0	0	8.0×10^4
2	2.50×10^6	300	0	0	2.50×10^6
Total					2.58×10^6

The moment about the y_0-axis is found in an exactly similar way. This time, however, both the complex section and the simple sections have their centroids on the line $x = 5$ cm. The distance d_x, which represents the shift in the y-axis from the centroids of the simple sections to that of the complex, is therefore zero. Again, Table 7.6 shows this best.

Thus,

$$\bar{I}_y = \Sigma\,(\bar{I}_{yi} + A_i \cdot d_x^2) = 2.58 \times 10^6 \text{ mm}^4$$

The polar moment of inertia, $\bar{J}$, about the point C is found using Equation 7.17, i.e.,

$$\bar{J} = \bar{I}_x + \bar{I}_y$$
$$= 1.04 \times 10^7 + 2.58 \times 10^6$$
$$= 1.30 \times 10^7 \text{ mm}^4$$

Example 7.6

From the tables of Appendix A select the wide-flange shape W 16×100. List the values of moment of inertia about the x_0- and y_0-axes through the centroid of this shape. Then, approximating the wide-flange shape by rectangular sections, calculate the same values for the section. Compare your answers with these shown in the tables of Appendix A.

Solution:

From the tables, the values of the moments of inertia for the W 16×100 section are

$$I_x = 1,490 \text{ in}^4$$

and

$$I_y = 186 \text{ in}^4$$

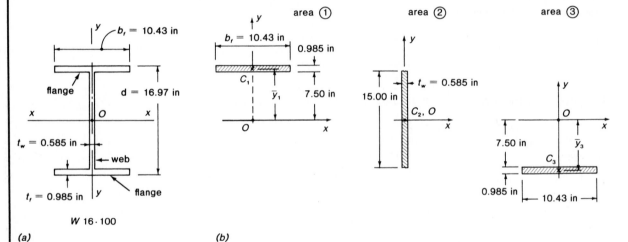

(a)

(b)

Figure 7.13

Since the average thicknesses of the web and flanges are also given in the tables of Appendix A, the section may be approximated by the straight-sided shape of Figure 7.13a. This shape is doubly symmetric and there is therefore no need to find the centroid of the section. The centroid must be at the intersection of the two axes of symmetry, i.e., at the point O.

The area is now divided into three areas as shown in Figure 7.13b. For these areas the following properties are relevant.

For area 1:

$$A_1 = b \cdot h = 10.43 \cdot 0.985 = 10.27 \text{ in}^2$$

$$\bar{y}_1 = 7.50 + \frac{d}{2} = 7.50 + \frac{0.985}{2} = 7.50 + 0.49 = 7.99 \text{ in}$$

$$\bar{I}_{x1} = \frac{bh^3}{12} = \frac{10.43(0.985)^3}{12} = 0.831 \text{ in}^4$$

$$\bar{I}_{y1} = \frac{hb^3}{12} = \frac{0.985(10.43)^3}{12} = 93.13 \text{ in}^4$$

$$d_y = \bar{y} - \bar{y}_1 = 0.0 - 7.99 = -7.99 \text{ in}$$

For area 2:

$$A_2 = b \cdot h = 0.585 \cdot 15.00 = 8.775 \text{ in}$$

$$\bar{y}_2 = 0$$

$$\bar{I}_{x2} = \frac{bh^3}{12} = \frac{0.585(15)^3}{12} = 164.5 \text{ in}^4$$

$$\bar{I}_{y2} = \frac{hb^3}{12} = \frac{15(0.585)^3}{12} = 0.25 \text{ in}^4$$

$$d_y = \bar{y} - \bar{y}_2 = 0.0$$

For area 3:

$$A_3 = b \cdot h = 10.43 \cdot 0.985 = 10.27 \text{ in}^2$$

$$\bar{y}_3 = 7.50 - \frac{h}{2} = -7.50 - \frac{0.985}{2} = -7.99 \text{ in}$$

$$\bar{I}_{x3} = \frac{bh^3}{12} = \frac{10.43(0.985)^3}{12} = 0.831 \text{ in}^4$$

$$\bar{I}_{y3} = \frac{hb^3}{12} = \frac{0.985(10.43)^3}{12} = 93.13 \text{ in}^4$$

$$d_y = \bar{y} - \bar{y}_3 = 0.0 - (-7.99) = 7.99 \text{ in}$$

The calculations for $\bar{I}_x$ are collected in Table 7.7.

Example 7.6 *(Continued)*

Table 7.7

Part (*i*)	$\bar{I}_{xi}$ (in⁴)	A_i (in²)	d_y (in)	$A_i \cdot d_y{}^2$ (in⁴)	$\bar{I}_{xi} + A_i \cdot d_y{}^2$ (in⁴)
1	0.831	10.27	−7.99	656.3	657.1
2	164.5	8.775	0	0	164.5
3	0.831	10.27	7.99	656.3	657.1
Total					1,478.7

Thus,

$$\bar{I}_x = \Sigma\,(\bar{I}_{xi} + A_i \cdot d_y{}^2) = 1,478.7 \text{ in}^4$$

For $\bar{I}_y$, the calculations appear in Table 7.8.

Example 7.7

Find the moments of inertia of the shape shown in Figure 7.14*a* about axes through the centroid of the area.

Solution:

The shape has been redrawn in Figure 7.14*b*, in which the reference axes x and y have been shown going through the origin O. Also shown are the x_0- and y_0-axes going through the centroid C of the area. Before the moments of inertia may be found, the location of the centroid must be found. This is specified as being at distances $\bar{y}$ and $\bar{x}$ from the x- and y-axes, respectively. The area is shown divided into four simple areas in Figure 7.14*c*. The area 4 represents the hole and will be taken as negative. The following properties of each simple area are now found:

For area 1:

$$A_1 = \frac{b \cdot h}{2} = \frac{30 \cdot 60}{2} = 900 \text{ mm}^2$$

$$\bar{x}_1 = \frac{2b}{3} = \frac{2}{2} \cdot 30 = 20 \text{ mm}$$

$$\bar{y}_1 = \frac{h}{3} = \frac{60}{3} = 20 \text{ mm}$$

Table 7.8

Part (i)	$\bar{I}_{yi}$ (in⁴)	A_i (in²)	d_x (in)	$A_i \cdot d_x^2$ (in⁴)	$\bar{I}_{yi} + A_i \cdot d_x^2$ (in⁴)
1	93.13	10.27	0	0	93.13
2	0.25	8.775	0	0	0.25
3	93.13	10.27	0	0	93.13
Total					186.51

Thus,

$$\bar{I}_y = \Sigma\,(\bar{I}_{yi} + A_i \cdot d_x^2) = 186.5 \text{ in}^4$$

When these values are compared to the values taken from the tables of Appendix A, it is noted that the agreement is good but not exact. This is because the flanges of a wide-flange member are typically not parallel, as was assumed in the calculations. Furthermore the corners are always rounded slightly. The values tabulated have been corrected for these facts, and are therefore more accurate than the values calculated.

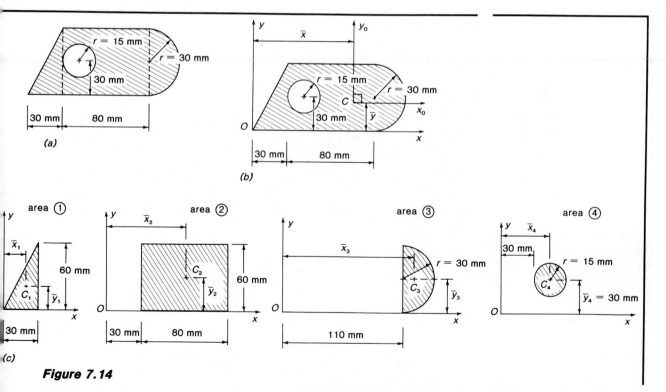

Figure 7.14

Example 7.7 *(Continued)*

Table 7.9

Part (*i*)	A_i (mm²)	$\bar{x}_i$ (mm)	$\bar{y}_i$ (mm)	$A_i \cdot \bar{x}_i$ (mm³)	$A_i \cdot \bar{y}_i$ (mm³)
1	900	20	20	18,000	18,000
2	4,800	70	30	336,000	144,000
3	1,413.7	122.7	30	173,461	42,411
4	−706.9	45	30	−31,811	−21,207
Total	6,406.8			495,650	183,204

For area 2:

$$A_2 = b \cdot h = 80 \cdot 60 = 4,800 \text{ mm}^2$$

$$\bar{x}_2 = 30 + \frac{b}{2} = 30 + \frac{20}{2} = 70 \text{ mm}$$

$$\bar{y}_2 = \frac{h}{2} = \frac{60}{2} = 30 \text{ mm}$$

For area 3:

$$A_3 = \frac{1}{2}\pi r^2 = \frac{1}{2}\pi \cdot 30^2 = 1,413.7 \text{ mm}^2$$

$$\bar{x}_3 = 110 + \frac{4r}{3\pi} = 110 + \frac{4 \cdot 30}{3 \cdot \pi} = 122.7 \text{ mm}$$

$$\bar{y}_3 = r = 30 \text{ mm}$$

For area 4:

$$A_4 = -\pi r^2 = -\pi \cdot 15^2 = -706.9 \text{ mm}^2$$

$$\bar{x}_4 = 30 + r = 30 + 15 = 45 \text{ mm}$$

$$\bar{y}_4 = 30 \text{ mm}$$

The usual method—of Table 7.9—is now followed to find the centroid coordinates.

Thus,

$$\bar{x} = \frac{\Sigma A_i \cdot \bar{x}_i}{\Sigma A_i} = \frac{495,650}{6,406.8} = 77.36 \text{ mm}$$

and

$$\bar{y} = \frac{\Sigma A_i \cdot \bar{y}_i}{\Sigma A_i} = \frac{183,204}{6,406.8} = 28.59 \text{ mm}$$

Extra properties are now required to find the moments of inertia of the complex section about its own centroid. These include the moments of inertia of each simple section about its own centroid, and the distance the centroid of each simple section must be moved (in order for it to coincide with the centroid of the complex section).

For area 1:

$$I_{x1} = \frac{bh^3}{36} = \frac{30(60)^3}{36} = 180,000 \text{ mm}^4$$

$$I_{y1} = \frac{hb^3}{36} = \frac{60(30)^3}{36} = 45,000 \text{ mm}^4$$

$$d_{y1} = \bar{y} - \bar{y}_1 = 28.59 - 20 = 8.59 \text{ mm}$$

$$d_{x1} = \bar{x} - \bar{x}_1 = 77.36 - 20 = 57.36 \text{ mm}$$

For area 2:

$$I_{x2} = \frac{1}{12}bh^3 = \frac{1}{12} \cdot 80(60)^3 = 1,440,000 \text{ mm}^4$$

$$I_{y2} = \frac{1}{12}hb^3 = \frac{1}{12} \cdot 60(80)^3 = 2,560,000 \text{ mm}^4$$

$$d_{y2} = \bar{y} - \bar{y}_2 = 28.59 - 30 = 1.41 \text{ mm}$$

$$d_{x2} = \bar{x} - \bar{x}_2 = 77.36 - 70 = 7.36 \text{ mm}$$

For area 3:

$$\bar{I}_{x3} = 0.393 \cdot r^4 = 0.393(30)^4 = 318,330 \text{ mm}^4$$

$$\bar{I}_{y3} = 0.110 \cdot r^4 = 0.110(30)^4 = 89,100 \text{ mm}^4$$

$$d_{y3} = \bar{y} - \bar{y}_3 = 28.59 - 30 = -1.41 \text{ mm}$$

$$d_{x3} = \bar{x} - \bar{x}_3 = 77.36 - 122.67 = -45.31 \text{ mm}$$

For area 4:

$$\bar{I}_{x4} = \bar{I}_{x4} = -\frac{\pi r^4}{4} = -\frac{\pi(15)^4}{4} = -39,761 \text{ mm}^4$$

$$d_{y4} = \bar{y} - \bar{y}_4 = 28.59 - 30 = -1.41 \text{ mm}$$

$$d_{x4} = \bar{x} - \bar{x}_4 = 77.36 - 45 = 32.36 \text{ mm}$$

The calculations for the moment of inertia about the x_0-axis appear in Table 7.10.

Example 7.7 (Continued)

Table 7.10

Part (i)	$\bar{I}_{xi}$(mm⁴)	A_i(mm²)	d_y(mm)	$A_i \cdot d_y^2$ (mm⁴)	$\bar{I}_{xi} + A_i \cdot d_y^2$ (mm⁴)
1	180,000	900	8.59	66,409	246,409
2	1,440,000	4,800	−1.41	9,543	1,539,543
3	318,330	1,413.7	−1.41	2,811	321,141
4	−39,761	−706.9	−1.41	−1,405	−41,166
Total					2,065,927

Thus,

$$\bar{I}_x = 2.065 \times 10^6 \text{ mm}^4$$
$$= 2.065 \times 10^{-6} \text{ m}^4$$

A similar table (Table 7.11) may be derived for the moment of inertia about the y_0-axis.

7.5 Partial Moments of Area

In the calculation of certain sorts of shear stresses (see Chapter 11) another property is required; this is called the *partial moment of area* and given the symbol Q. To be strictly accurate, the partial moment of area is not a property of a cross section but a property of a portion of or a *point on a cross section*. In order to find Q, both the shape of the section and the point on the section at which the value of Q is required will need to be known. The partial moment of area is always derived with respect to the centroidal axes of the section.

Consider the cross section shown in Figure 7.15. The location of the centroid is assumed to be known so the x_0- and y_0-axes are shown. It is required to find the partial moment of area with respect to the x_0-axis of the point A on the cross section where the area is cut. The partial moment of area is defined as the product of the area farther away from the x_0-axis than the point A (i.e., area A_p) and the distance from the centroid of this area to the centroid of the whole section (i.e., $\bar{y}_A$).
This may be written

(7.22)
$$Q_A = A_p \cdot \bar{y}_A$$

In a similar calculation, the partial moment of area with respect to the x_0-axis of the point B may be written

(7.23)
$$Q_B = A_p' \cdot \bar{y}_B$$

Table 7.11

Part (i)	$\bar{I}_{yi}$ (mm⁴)	A_i (mm²)	d_x (mm)	$A_i \cdot d_x^2$ (mm⁴)	$\bar{I}_{yi} + A_i \cdot d_x^2$ (mm⁴)
1	45,000	900	57.36	2,961,153	3,006,153
2	2,560,000	4,800	7.36	260,014	2,820,014
3	89,100	1,413.7	−45.31	2,902,321	2,991,421
4	−39,761	−706.9	32.36	−740,244	−780,005
Total					8,037,583

Thus,

$$\bar{I}_y = 8.038 \times 10^6 \text{ mm}^4$$
$$= 8.04 \times 10^{-6} \text{ m}^4$$

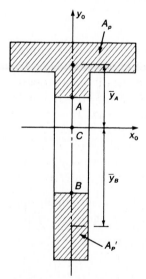

Figure 7.15 *Partial areas of a cross section.*

Note that in this case the area farther away from the x_0-axis than the point B is the area A_p', which is below the point B.

Example 7.8

The cross section of a W 18 × 50 wide flange beam has been approximated by the shape shown in Figure 7.16a. Find the partial moments of inertia with respect to the x_0-axis of the points A, B, and C on the cross section.

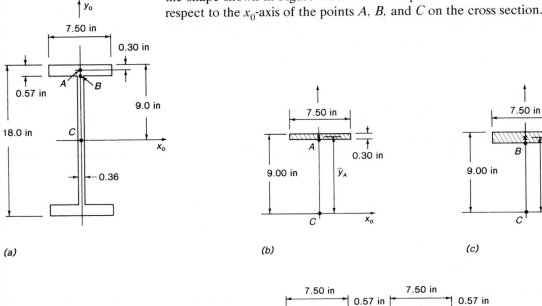

(a)

(b)

(c)

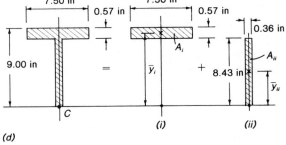

(d)

Figure 7.16

Solution:

Since the shape is doubly symmetric, the centroid may be located without any calculations. The centroid is at the point C, which is shown as the origin of the x_0- and y_0-axes.

The partial area associated with the point A is shown in Figure 7.16b, for which the following calculations are valid

$$A_p = 7.50 \cdot 0.30 = 2.25 \text{ in}^2$$

$$\bar{y}_A = 9.00 - \frac{0.30}{2} = 8.85 \text{ in}$$

and therefore

$$Q_A = 2.25 \cdot 8.85 = 19.91 \text{ in}^3$$

For the point B the calculations are based on the area of Figure 7.16c.

$$A_p = 7.50 \cdot 0.57 = 4.275 \text{ in}^2$$

$$\bar{y}_B = 9.00 - \frac{0.57}{2} = 8.715 \text{ in}$$

and therefore

$$Q_B = 4.275 \cdot 8.715 = 37.26 \text{ in}^3$$

For point C the area shown in Figure 7.16d must be divided into two areas (i) and (ii) as shown. It is not necessary to find the centroid of the T-shaped section, for the partial moment of this area about the x_0-axis is equal to the sum of the moments of the areas A_i and A_{ii} about the same axis. The following quantities are then calculated.

$$A_i = 7.50 \cdot 0.57 = 4.275 \text{ in}^2$$

$$\bar{y}_i = 9.00 - \frac{0.57}{2} = 8.715 \text{ in}$$

$$A_{ii} = 8.43 \cdot 0.36 = 3.035 \text{ in}^2$$

$$\bar{y}_{ii} = \frac{8.43}{2} = 4.215 \text{ in}$$

Therefore,

$$Q_C = A_i \cdot \bar{y}_i + A_{ii} \cdot \bar{y}_{ii} = 4.275 \cdot 8.715 + 3.035 \cdot 4.215$$
$$= 50.05 \text{ in}^3$$

Problems

7.1–
7.9   Find the centroids of the given built-up sections or composite shapes shown in Figures P 7.1 to P 7.9.

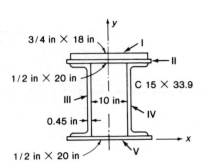

Figure P 7.1

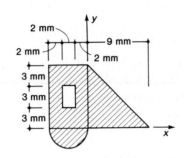

Figure P 7.2

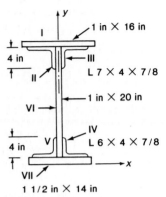

Figure P 7.3

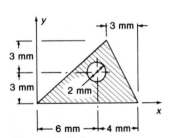

Figure P 7.4

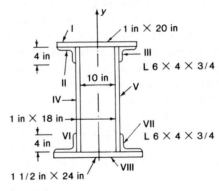

Figure P 7.5

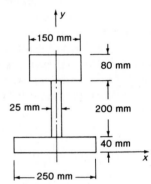

Figure P 7.6

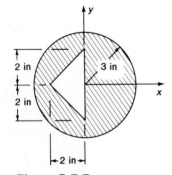

Figure P 7.7

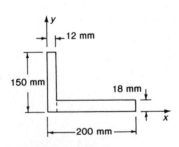

Figure P 7.8

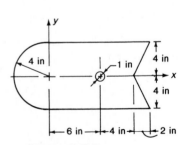

Figure P 7.9

210   Chapter Seven

7.10–
7.18 Find I_{xx}, I_{yy} $\bar{I}_{xx}$, $\bar{I}_{yy}$ for the given built-up sections or composite shapes shown in Figure P 7.2, P 7.1, P 7.4, P 7.3, P 7.6, P 7.5, P 7.8, P 7.7, and P 7.10, respectively.

7.19–
7.22 For these problems, take $L^* = (SS + 50)$ in and $P^* = (SS + 50)^2$ lb, where SS is either the last two digits of your Social Security number, or two digits assigned by the instructor. Find $\bar{I}_{xx}$, $\bar{I}_{yy}$ for the sections shown in Figures P 7.19 to P 7.22.

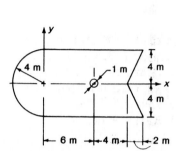

Figure P 7.10

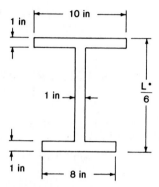

Figure P 7.19

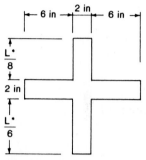

Figure P 7.20

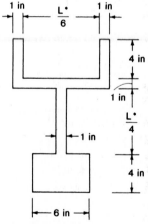

Figure P 7.21

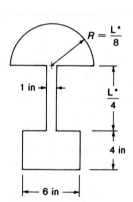

Figure P 7.22

Axial Stress and Strain 8

1. *To understand how stresses and strains occur within a body, how they are calculated, and what shape a stress-strain diagram is likely to have for a given material.*

2. *To comprehend certain technical terms important in characterizing the behavior of material, such as modulus of elasticity, Poisson's ratio, and ductility.*

3. *To calculate stresses and deflections in an axially loaded member.*

4. *To understand the need for factors of safety and allowable stresses, and to be able to derive the required cross-sectional areas of axially loaded members so that allowable stress criteria are not violated.*

5. *To be able to analyze the behavior of bodies subjected to uniform stresses in two perpendicular directions.*

6. *To be able to predict the effects of changes in temperature or of geometrical discontinuities in an axially loaded member.*

Objectives

8.1 Definition of Stress and Strain

In the previous chapters of this book discussion has been centered on the forces that may be applied to the parts of a structure or a machine. Structures such as trusses and beams were analyzed to find the internal forces within such systems.

However, it has not yet been determined whether the truss or beam is able to support those forces without collapsing. Clearly some materials are stronger than others, and a truss or beam made of stronger materials can support larger forces than can one made of other materials. To be able to design a structure or machine component, it is therefore very important that the behavior of the material under stress is understood.

There are many ways that the behavior of a material may be characterized, but before the behavior of two materials may be compared, the behavior of each material must somehow be standardized. In order to help with this, two important properties of a loaded member are defined. The first of these is *stress,* which is defined as the force per unit area on the part, i.e.,

$$\text{(8.1)} \qquad \text{stress } (\sigma) \triangleq \frac{\text{force } (F)}{\text{area } (A)}$$

(The notation $\triangleq$ means "equal by definition.")

It is important to note the units of stress. American Customary Units of stress are pounds per square inch and kips per square inch; metric units are pascals and megapascals. (Remember that 1 pascal (Pa) is 1 newton per square meter.) Stress therefore has the same units as pressure. The stress in a structural element, then, represents the force per unit area to which the part is being subjected.

The next standard measure of performance for an engineering material involves the tendency of a material to stretch or compress when subjected to a force. It is easy to see that a piece of steel stretches very little when pulled with the hands. On the other hand, a piece of rubber of the same size may stretch a great deal when pulled in the same manner. The relative deformation of a material is the *strain,* which is defined as the deformation per unit length of the member, i.e.,

$$\text{(8.2)} \qquad \text{strain } (\epsilon) \triangleq \frac{\text{deformation } (\Delta)}{\text{original length } (l)}$$

Once again it is important to note the units of strain. Since both the deformation and the original length will have units of length (inches or feet or meters), strain will have no unit. In other words, strain is dimensionless and does not depend on which system is used in the calculations. (It is, however, often written as in/in or m/m.)

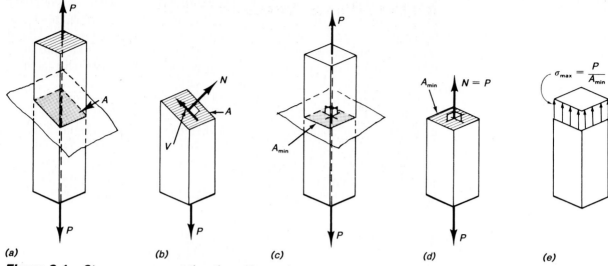

Figure 8.1 *Stresses on a section through a rod.*

8.2 Calculation of Axial Stress

Consider now one of the members of a truss. When the truss is loaded, the principles of Chapter 4 may be used to analyze the truss and find the axial force in each member. As noted previously, the forces are axial forces: the force is directed along the longitudinal axis of the member. One such member from a truss is shown in Figure 8.1*a*. Assume now that the member is cut as shown in Figure 8.1*b*. The cut is at an angle other than 90° to the longitudinal axis of the member. A free body diagram of the lower portion of the member is shown. It may be readily shown that for equilibrium there must be a force on the area A with components both parallel to the area A (i.e., the force V) and normal to the area (i.e., the force N). The force N acts normal to the area and is said to cause a *normal* or *axial* stress. On the other hand, the force V is a shear force on the area A, since it acts in the plane of the area A. This shear force V is said to cause a *shear* stress on the area A. Using Equation 8.1, the forces N and V may be converted to stresses. Thus

(8.3)
$$\text{normal stress } (\sigma) = \frac{N}{A}$$

and

(8.4)
$$\text{shear stress } (\tau) = \frac{V}{A}$$

It is usual to differentiate between the normal stress and the shear stress by giving the axial stress the symbol sigma (σ) and the shear stress the symbol tau (τ). It is easy to see that the axial stress σ is inversely proportional to the area. Thus if the cross-sectional area of the member is reduced, the value of the axial stress is increased. In order to find the maximum axial stress that exists in the member subjected to axial force, it is therefore necessary to find the area of the minimum cross section which is present in the member. The minimum cross-sectional area A_{min} is clearly shown in Figure 8.1c, where the member is cut by a plane perpendicular to the longitudinal axis of the member. As shown in Figure 8.1d, the forces on the cut with area A_{min} have reduced to an axial force N only. There is now no shear force on the cross section, and by equilibrium it may be shown that the value of the axial force N at the section is equal to the axial force P applied to the member. If Equation 8.1 is now applied to this situation, the value of the maximum normal stress σ_{max} is

$$(8.5) \qquad\qquad \sigma_{max} = \frac{N}{A_{min}} = \frac{P}{A}$$

As indicated in Figure 8.1e, this may be visualized as a stress of uniform intensity spread over the area A_{min}. Since there is no shear stress on the area A_{min}, stress σ_{max} is known as a *principal* stress. Principal stresses will be further dealt with in later chapters.

Care must be taken to ensure that the member is indeed subject only to axial force before Equation 8.5 is applied to find the value of axial stress on a cross section. Take for example the open link of Figure 8.2a. At first glance it would appear that the link is under axial forces only, since the applied forces P

Example 8.1

The circular rod of Figure 8.3a is rigidly attached to a vertical wall at A. The cross-sectional area of the rod between A and C is 4 square inches, and between C and E it is 6 square inches. If the rod is acted upon by the axial forces shown at B, D, and E, find the maximum axial stress in the rod.

Solution:

It is first necessary to find the reaction force at the support A. If the reaction force is R, then Figure 8.3a may be converted to a free body diagram by inclusion of this force as shown. Application of the equilibrium equation to forces acting in the x-direction yields

$$\Sigma F_x = 0 \overset{+}{\rightarrow}$$
$$- R - 20 + 60 + 36 = 0$$
$$R = 76 \text{ kips}$$

The value of axial stress must be calculated wherever the value of axial force in the member changes or where the cross-sectional area of the member changes. Thus, four cuts must be made in the rod, and the situation at each cut examined separately. These cuts are labelled as cuts 1, 2, 3, and 4 in Figure 8.3a. The precise locations of each of these cuts is not specified, although

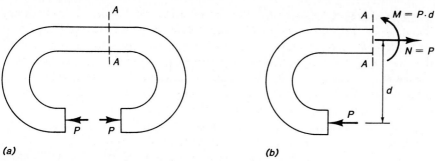

Figure 8.2 Free-body diagram for a cut through A-A.

have the same line of action and are applied in opposite directions. However, if a cut is taken at section $A-A$ and a free body is drawn of part of the link, as shown in Figure 8.2b, then a different situation emerges. At section $A-A$ the link is subject to an axial force N acting at the centroid of the cut part, as well as a bending moment M. Bending moments cause different types of stress in a cross section (see Chapter 10), and so the stresses on the cross section may not be completely determined by using Equation 8.5.

In order to find the maximum normal stress existing in an axially loaded member, it is essential, as suggested above, to consider the stress on an area perpendicular to the longitudinal axis of the member. In addition it is necessary to recalculate the value of the stress wherever the value of the applied axial load changes or where the value of the cross-sectional area changes.

it is known that cut 1 is somewhere between points A and B, cut 2 between B and C, cut 3 between C and D, and cut 4 is between D and E. Free body diagrams of the portions of the rod to the left of each cut have been drawn in Figures 8.3b, c, d, and e.

At cut 1: (Figure 8.3b)

$$\Sigma F_x = 0 \overset{+}{\rightarrow}$$

$$N_1 - R = 0$$

$$N_1 = R = 76 \text{ kips}$$

Thus the axial stress is given by

$$\sigma_1 = \frac{N_1}{A_1} = \frac{76}{4} = 19 \text{ ksi}$$

At cut 2: (Figure 8.3c)

$$\Sigma F_x = 0 \overset{+}{\rightarrow}$$

$$N_2 - 20 - R = 0$$

$$N_2 = 96 \text{ kips}$$

Example 8.1 *(Continued)*

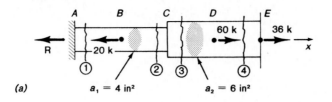

(a)

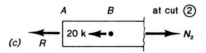

at cut ①

(b)

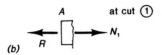

at cut ②

(c)

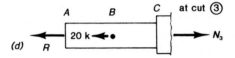

at cut ③

(d)

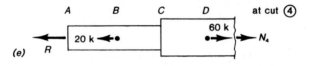

at cut ④

(e)

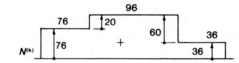

(f) axial force diagram

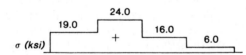

(g) axial stress diagram

Figure 8.3

The axial stress is then given by

$$\sigma_2 = \frac{N_2}{A_1} = \frac{96}{4} = 24 \text{ ksi}$$

At cut 3: (Figure 8.3d)

$$\Sigma F_x = 0 \overset{+}{\rightarrow}$$

$$N_3 - 20 - R = 0$$

$$N_3 = N_2 = 96 \text{ kips}$$

The axial stress at cut 3 is

$$\sigma_3 = \frac{N_3}{A_2} = \frac{96}{6} = 16 \text{ ksi}$$

Note that although the axial force at cut 3 is the same value as the axial force at cut 2, the value of the axial stress is different at the two cuts. This is because the cross-sectional area is different at the two cuts.

At cut 4: (Figure 8.3e)

$$\Sigma F_x = 0 \overset{+}{\rightarrow}$$

$$N_4 + 60 - 20 - R = 0$$

$$N_4 = 36 \text{ kips}$$

The axial stress at cut 4 is

$$\sigma_4 = \frac{N_4}{A_2} = \frac{36}{6} = 6 \text{ ksi}$$

The same values of axial stress would have been derived at each cut if free bodies had been considered to the right of each cut. Note that the positive value of each stress indicates that it is a tensile stress, in the same way that a positive value of axial force indicates that the force is a tensile force.

Based on the analysis of each cut, it is possible to construct an axial force diagram as shown in Figure 8.3f. This diagram shows the variation of axial force along a rod in the same way that a shear force diagram shows the variation of shear force along a beam. As indicated, the axial force diagram "jumps" at each location where an axial load is applied. In the same manner it is also possible to construct an axial stress diagram, showing the variation of axial stress along the rod. Such a diagram is shown in Figure 8.3g. Examination of the axial stress diagram shows that the maximum axial stress is given by

$$\sigma_{max} = 24 \text{ ksi}$$

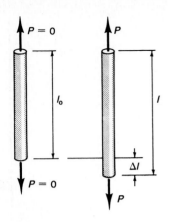

$P = 0$ P

l_0 l

$P = 0$ Δl

P

(a)   (b)

Figure 8.4
Axial stretching of a rod under axial stress.

8.3 Stress-Strain Diagram

Thus far the calculation of axial stress has been discussed. If now the definition of axial strain is recalled from Equation 8.2, the measurement of axial strain may be examined. If the rod of Figure 8.4a has a length of l_0 before any axial load is applied to it, and then it stretches to a length l when the axial force P is applied (see Figure 8.4b) then the strain of the rod is given as

(8.6)
$$\text{axial strain } (\epsilon) = \frac{\text{change in length}}{\text{original length}} = \frac{\Delta l}{l_0} = \frac{l - l_0}{l_0}$$

With this definition it is now possible to derive a standard test that characterizes the load-carrying behavior of many materials. Such tests have been specified by the American Society for Testing and Materials (ASTM) as well as other groups. In each test a specimen of the material to be tested must be made in some specified fashion. Examples of such specimens, known as *tension test coupons,* are shown in Figure 8.5. Whether round or flat, the specimens have common features. At each end of each specimen is a thickened portion, which enables the specimen to be held in the testing machine. Near the center of each specimen is a smaller machined section over which the cross-sectional area is essentially constant. It is the behavior of the central section of the specimen that is to be examined. It is pulled in the testing machine; the extremities of this machined portion of the test specimen are marked and the distance between the marks is called the gage length of the specimen. A typical gage length for a metal specimen is two inches.

The gage length is carefully measured before any axial force is applied to the specimen and then the axial force P is increased very slowly. At specific values of axial force P, the new length of the gage length of the specimen is measured, and thus the strain in the material corresponding to that load may be computed using Equation 8.6. Consider first a mild steel, a very common material. The diagrams of Figure 8.6 illustrate what happens as the applied load P is increased on the tensile test coupon. Before any force is applied to the coupon, the initial gage length, l_0 and the initial cross-sectional area of the reduced portion of the specimen, A_0, are carefully measured. In a properly machined specimen, the area A_0 should be constant over the entire gage length of the specimen. The unloaded specimen is shown in Figure 8.6a.

As the axial force on the specimen is increased, the specimen gets longer as the material stretches, or "strains." At an axial force of P_1 the new length of the reduced portion of the specimen is l_1 and the cross-sectional area of this portion has been reduced to A_1. The strain in the specimen is then computed from Equation 8.6. The stress at this load may also be computed, but care must be taken with this calculation. The *true stress* at this load is given by

(8.7)
$$\sigma_{\text{true}} = \frac{P_1}{A_1}$$

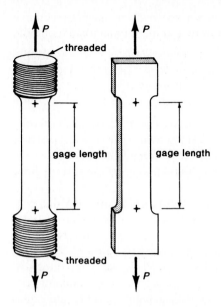

(a) round specimen   (b) flat specimen

Figure 8.5   *Tension test coupons.*

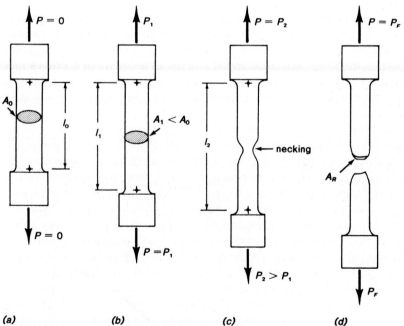

(a)   *(b)*   *(c)*   *(d)*

Figure 8.6   *Axial stress on a tensile coupon.*

However, only rarely is the testing stopped in order to measure the cross-sectional area A_1 accurately. For convenience, therefore, it is more common to compute the *engineering stress,* which is given by

(8.8)
$$\sigma_{\text{eng}} = \frac{P_1}{A_0}$$

This calculation is much easier, since the cross-sectional area was carefully measured before the experiment was started. At low axial loads, the value of true stress, calculated using Equation 8.7, is very close to the value of engineering stress calculated using Equation 8.8, but at high loadings these results differ considerably, as will be demonstrated.

For a large portion of its load range, a steel tension coupon looks very much as shown in Figure 8.6b. It gets longer as more force is applied to it, but it does not change shape noticeably. Eventually, however, the axial force P gets close to the force that will break the specimen. As this load is approached, several interesting events occur. At an applied axial force of P_2 (Figure 8.6c), the cross-sectional area of the reduced portion of the specimen no longer remains constant: it starts to decrease markedly within a localized region, as illustrated. This process is known as necking and is a strong indication that the specimen is about to fracture. As indicated in Figure 8.6d, the fracture of the specimen finally occurs within the necked area at a fracture load P_F. The reduced area of the fracture surface at failure of the specimen, A_R, is easily measured after the test is completed.

If the values of force P and gage length l are measured at frequent intervals as the force P is increased, then the values of engineering stress and strain may be readily computed using Equations 8.8 and 8.6 respectively. If these results are then plotted, the result is a *stress-strain diagram.* The shape of a stress-strain diagram for a typical mild steel is shown in Figure 8.7. It is clear from this diagram that the material behaves differently at different values of applied axial force P.

The stress-strain diagram starts with a straight portion in which the engineering stress is proportional to the strain. The slope of this portion of the curve is called the *modulus of elasticity* of the material and is given the symbol E. However, as the force on the specimen is increased, there comes a point on the curve—designated as the point 1 in Figure 8.7—where the curve starts to bend away from the straight line that has characterized its behavior up to this stage. This point is known as the *proportional limit* of the material. For stresses above the proportional limit, the stress-strain diagram is no longer a straight line.

As the applied stress is further increased, the diagram soon reaches the point 2, which is called the *elastic limit.* If the applied stress is less than or equal to the elastic limit, then the specimen may be unloaded and the specimen will return to its original length. If, however, the specimen has been loaded beyond its elastic limit, there will always be some permanent deformation of the specimen; i.e., if unloaded, the specimen will remain longer than the length l_0 measured before the test was commenced. Close to the point 2 on the stress-strain diagram are the points 3 and 4, usually very close together. At loadings

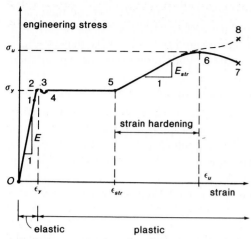

Figure 8.7 *Typical stress-strain diagram for a mild steel.*

between these points the material is said to have reached its *yield stress.* At the yield stress (designated σ_y) extensive permanent straining of the specimen starts to occur with very little further increase in the axial force P on the specimen. The yield strain associated with the yield stress is designated ϵ_y. In many cases the points 1, 2, 3, and 4 occur so close together on the stress-strain diagram for a mild steel that it is difficult to separate them.

In the region between points 4 and 5 on the stress-strain diagram the curve is almost horizontal, indicating that a great amount of strain (and therefore a great amount of extension) is taking place with very little increase in the force applied to the specimen. For obvious reasons, this portion of the curve is sometimes called the *yield plateau.* Once the curve has passed the point 5, the specimen appears to get somewhat stronger again, indicated by the fact that the curve once again has a positive slope. In this region the material is said to undergo *strain hardening.* The slope of this portion of the curve, between points 5 and 6, is almost constant; this slope is called the modulus of strain hardening and given the symbol E_{str}. At point 6 the material has reached the greatest stress it can possibly sustain. The stress at this point is called the *ultimate stress, σ_u,* and the corresponding strain is the *ultimate strain, ϵ_u.* At point 6, necking of the specimen usually starts (as illustrated in Figure 8.6d), and the specimen usually traces the curve from point 6 to point 7 extremely rapidly. Point 7 represents the point at which the specimen breaks, and the stress at this point is known as the *fracture stress.*

It is in the region between points 6 and 7 that it makes a great deal of difference whether the engineering stress or the true stress is being plotted. The dashed line between points 6 and 8 represents the behavior of the specimen if true stress is being plotted instead of engineering stress. It is higher than the engineering stress curve because the cross-sectional area of the specimen has been substantially reduced by necking.

When the stress is below the yield stress, it is said that the material is behaving elastically, or that the material is in the *elastic range* of material behavior. Once the maximum axial stress has reached the yield stress, the material is within the *plastic* range of material behavior and there will be some permanent deformation.

For a typical mild steel, some of the variables mentioned above have values of the order listed below:

$$E \ = 30 \times 10^6 \text{ psi} = 30{,}000 \text{ ksi} \ (207 \text{ GPa})$$
$$\sigma_y \ = 36 \text{ to } 50 \text{ ksi} \ (250 \text{ to } 350 \text{ MPa})$$
$$\epsilon_y \ = 0.0012 \text{ to } 0.0018$$
$$\epsilon_{str} \ = \text{about } 0.02$$
$$\sigma_u \ = 60 \text{ to } 100 \text{ ksi} \ (400 \text{ to } 700 \text{ MPa})$$
$$\epsilon_u \ = \text{about } 0.20$$

These values are not exact because the properties of steels vary greatly, but they do indicate the expected order of the results. One particularly noteworthy fact about a mild steel is the large strain ϵ_u that it is able to withstand before it fractures. It is noted that a strain of 0.2 is the same as an increase of 20 percent in the length of a member. Thus a steel truss member that was initially 10 feet in length would be about 12 feet long before it fractured. This fact indicates that steel is a very *ductile* material, a term that will be better defined in the next section.

Before leaving stress-strain diagrams, it must be recognized that Figure 8.7 is the form of the stress-strain diagram for a mild steel and that other materials may not have stress-strain diagrams of this form. Some materials do not exhibit a clearly defined yield stress, where the material behavior changes from elastic to plastic behavior. Copper and aluminum fall into this category, and the stress-strain diagrams for these materials have a shape like that shown in Figure 8.8. It can be seen from this curve that there is no readily defined yield stress for the material. Because engineers have found the yield stress to be a convenient number on which to base a large number of calculations, an artificial yield stress is often defined for such materials. This yield stress is sometimes called the *0.2 percent offset stress,* because of the way it is derived. As indicated, a strain of 0.002 is arbitrarily chosen as the starting point of a line that travels up to the curve at a slope equal to the original slope of the curve. The point at which this straight line crosses the stress-strain diagram for the material is defined as the yield stress of the material. This definition is made for convenience only, and it should be remembered that materials with a stress-strain diagram of this shape do not have a true yield stress. Typical properties of some metals, ceramics, and plastics are tabulated in Appendix C.

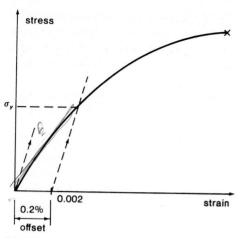

Figure 8.8 *Stress-strain diagram for copper or aluminum.*

8.4 Definitions

Some of the useful properties of a material may be found directly from the stress-strain diagram for the material. Other properties must be found by different tests. There are a number of standard properties of materials that typically interest engineers. Some of the more useful properties of a material are defined now.

Modulus of Elasticity, or Young's Modulus, E. As defined in Section 8.3, the modulus of elasticity of a material is the slope of the initial, elastic portion of the stress-strain diagram for the material. Since the stress is proportional to the strain in this portion of the diagram, the very important equation relating stress and strain may be written.

(8.9)
$$\sigma = E \cdot \epsilon$$

The units of the modulus of elasticity need to be carefully determined. Since the strain is a dimensionless quantity, the modulus of elasticity must have the same units as the stress (e.g., psi or MPa). The modulus of elasticity for some typical engineering materials is given in Table 8.1 and also in Appendix C.

Ductility. A material is said to be *ductile* if it will undergo a large tensile strain before it fractures. On the other hand, a material is said to be *brittle* if it fractures at very small strains. Steel is generally considered to be a ductile material, whereas both glass and concrete are usually considered to be brittle materials. The ductility is usually defined as the maximum elongation (expressed as a percentage of the original length) obtained in a tensile test on a specimen, prior to breaking the specimen.

Toughness. For some applications, such as those in which large forces will be applied suddenly (such as in blasting or in impact situations), it is important that a material exhibit a high degree of *toughness*. The *modulus of toughness* of a material is defined as the area under the stress-strain curve of the material.

Malleability. A material is *malleable* if it will deform appreciably before it ruptures. This property is important in such operations as rolling, hammering, or forging.

Hardness. The resistance of a material to wear or surface penetration is measured by its *hardness*. Usually a stronger material will have greater hardness. There are several standard tests for hardness, such as the Rockwell hardness test, in which small balls or cones are pushed into the surface of the material with a given force and their penetration is measured.

8.5 Calculation of Axial Deflections

For materials that remain within the elastic range of material behavior it is possible to readily calculate the deflection of an element subjected to axial forces. The relationship between axial force and axial extension or contraction may be derived from the equations already presented.

First, recall the definition of strain (Equation 8.6):

$$\epsilon = \frac{\Delta L}{L}$$

from which the change in length of an axially loaded element (Figure 8.4) is found to be

(8.10) $$\Delta L = \epsilon \cdot L$$

Example 8.2

The rod *AB* shown in Figure 8.9*a* is rigidly fixed to a vertical wall at *A* and acted on by the 12 kilonewton force at the free end *B*. If the cross-sectional area of the rod is constant, and equal to 1,000 square millimeters, and the rod is made of aluminum with a modulus of elasticity of 70 gigapascals, find the horizontal movement *Δ* of the point *B*. The rod is 0.3 meter long.

Solution:

As in past examples, the first step is to find the reactions of the support using the equations of equilibrium. All the possible elements of reaction at a fixed support have been shown in the free body diagram of Figure 8.9*b*. However, application of the equations of equilibrium yields:

$$H_A = 12 \text{ kN}$$
$$V_A = 0$$
$$M_A = 0$$

Table 8.1 Approximate moduli of elasticity

Material	Modulus of elasticity, E		
	psi		**GPa**
Steel	30×10^6		207
Aluminum	10×10^6		70
Copper	17×10^6		120
Brick	2×10^6	(compression)	14
Wood	$\approx 1 \times 10^6$	(varies with species)	≈ 12.4
Cast iron	11×10^6	(varies)	77
Glass	9×10^6		63
Brass	15×10^6		105

Now, substituting Equation 8.9:

(8.11)
$$\Delta L = \left(\frac{\sigma}{E} \right) \cdot L$$

The stress, σ, within an axially loaded element is given by Equation 8.5. Placing this in Equation 8.11 gives

(8.12)
$$\Delta L = \frac{PL}{AE}$$

Thus, the extension of an axially loaded element is related to the applied force on the element (P), the length of the element (L), the cross-sectional area of the element (A), and the modulus of elasticity of the material (E). Note that the axial force, area, and modulus of elasticity must all be constant over the length of L.

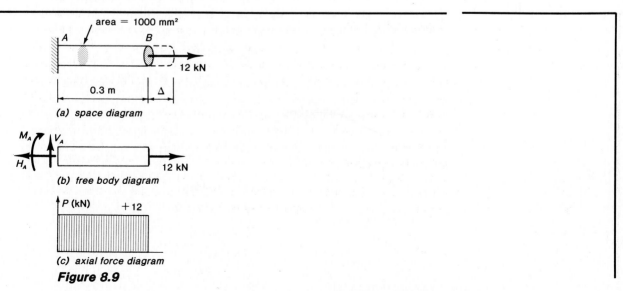

(a) space diagram

(b) free body diagram

(c) axial force diagram

Figure 8.9

Example 8.2 (Continued)

It is then possible to draw an axial force diagram, showing the variation of axial force throughout the member. As the diagram of Figure 8.9c shows, the axial force is constant throughout the length of the member AB in this example. Since the area, the axial force, and the modulus of elasticity are all constant between A and B, Equation 8.12 may be applied to the whole distance AB. Clearly the movement Δ of the point B is due totally to the extension of the rod AB between A and B.

Thus,

$$\Delta = \Delta L_{AB}$$

$$= \frac{P_{AB} \cdot L_{AB}}{A_{AB} \cdot E_{AB}}$$

$$A_{AB} = 1{,}000 \text{ mm}^2 = 1{,}000 \times 10^{-6} \text{ m}^2$$

$$= 10^{-3} \text{ m}^2$$

$$E = 70 \times 10^3 \text{ MPa} = 70 \times 10^6 \text{ kPa}$$

Example 8.3

The rod $ABCD$ is acted upon by the axial forces shown in Figure 8.10a. The rod is of variable section, as shown, and is manufactured from a steel with Young's modulus of 207 GPa. Find the movement of the point D and draw an axial deflection diagram showing the horizontal movement of any point on the rod.

Solution:

A free body diagram of the entire rod is first drawn to find the support reaction H_A. As in the previous example, the force H_A is the only nonzero element of reaction at A. Using equilibrium equations,

$$H_A = 100 \text{ kN}$$

With the support forces known, it is now possible to find the internal forces within the rod. Cuts are made at locations 1, 2, and 3 as shown in Figure 8.10b. The free bodies formed by the portion of the rod to the left of cuts 1, 2, and 3 are shown in Figures 8.10b, c, and d, respectively. Applying the horizontal force equilibrium equations to each of these free bodies gives the three unknown internal forces as

$$N_1 = 100 \text{ kN}$$

$$N_2 = -150 \text{ kN}$$

$$N_3 = 50 \text{ kN}$$

Substituting in Equation 8.12,

$$\varDelta = \frac{12 \times 0.3}{10^{-3} \times 70 \times 10^6}$$

$$= 5.14 \times 10^{-5} \text{ m}$$

$$= 0.0514 \text{ mm}$$

The positive sign indicates that rod AB extends; i.e., the point B moves to the right.

The negative sign of N_2 indicates that the region BC of the rod is in compression. This region of the rod will therefore get shorter as it is being compressed. The resulting axial force diagram is shown in Figure 8.10f. The deflection of the point D is equal to the sum of the extensions of the various parts of the rod; i.e.,

$$\varDelta_D = \varDelta_{AB} + \varDelta_{BC} + \varDelta_{CD}$$

$$= \left(\frac{PL}{AE} \right)_{AB} + \left(\frac{PL}{AE} \right)_{BC} + \left(\frac{PL}{AE} \right)_{CD}$$

In substitution, units of meters and newtons are used. Thus

$$\varDelta_D = \frac{(100 \times 10^3) \times 2}{0.001 \times (207 \times 10^9)} + \frac{(-150 \times 10^3 \times 1)}{0.002 \times (207 \times 10^9)}$$

$$+ \frac{(50 \times 10^3) \times 1.5}{0.001 \times (207 \times 10^9)}$$

$$\varDelta_D = 9.66 \times 10^{-4} - 3.62 \times 10^{-4} + 3.62 \times 10^{-4} \text{ m}$$

$$= 9.66 \times 10^{-4} \text{ m}$$

$$= 0.966 \text{ mm}$$

Note that in these calculations, the extension or contraction of the three parts of the rod has been found to be

$$\varDelta_{AB} = +0.966 \text{ mm}$$

$$\varDelta_{BC} = -0.362 \text{ mm}$$

$$\varDelta_{CD} = +0.362 \text{ mm}$$

Example 8.3 (Continued)

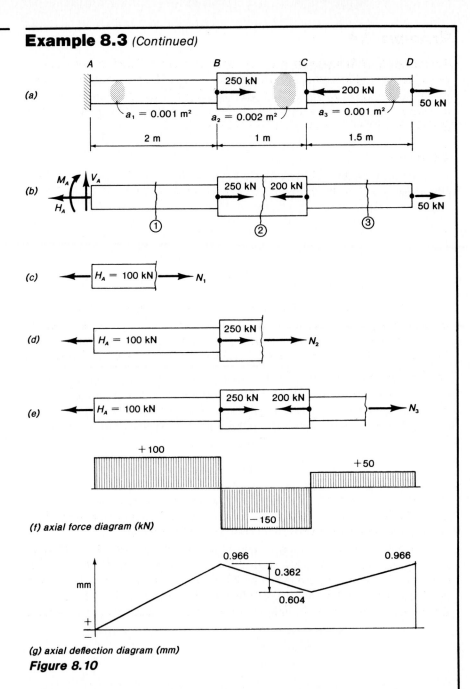

(a)

(b)

(c)

(d)

(e)

(f) axial force diagram (kN)

(g) axial deflection diagram (mm)

Figure 8.10

The negative sign of Δ_{BC} indicates that the part BC has become shorter, as had been anticipated by the fact that this region of the rod is in compression. Using the deflections of each part given above, the axial deflection diagram of Figure 8.10g may be derived. This diagram gives the axial deflection of any point of the rod. The diagram is composed of straight lines because the deflection of each part is proportional to the length of that part, as shown in Equation 8.12.

Example 8.4

The rod *ABCDE* is rigidly fixed to the vertical wall at *A*. The rod is of variable cross section and loaded as shown in Figure 8.11*a*. The rod is made of a steel with a modulus of elasticity of 30×10^6 pounds per square inch. Draw the axial deflection diagram for the rod and use the diagram to find the horizontal movement of a point midway between *D* and *E*.

Solution:

A free body diagram of the whole rod is drawn to find the unknown support reactions. Applying the equilibrium equations to this free body diagram, shown in Figure 8.11*b*, gives:

$$H_A = -6 \text{ kips}$$
$$V_A = 0$$
$$M_a = 0$$

Free bodies are now formed by cuts at four locations, as shown in Figures 8.11*c*, *d*, *e*, and *f*. It is necessary to form a separate free body diagram wherever the loading condition on the rod changes. Applying the horizontal force equilibrium condition to each free body gives the following values for the unknown internal forces:

$$N_1 = -6 \text{ kips}$$
$$N_2 = 8 \text{ kips}$$
$$N_3 = 10 \text{ kips}$$
$$N_4 = 12 \text{ kips}$$

These values of internal force are used to draw the axial force diagram of Figure 8.11*g*. Note that the negative sign of N_1 indicates that the region *AB* of the rod is in compression and is therefore expected to shorten.

In many instances it is a good idea to plot a cross-sectional area diagram, showing the variation of area along the rod. Such a diagram, given in Figure 8.11*h*, is compiled from the given information. The value of this diagram lies in the fact that it readily shows where the area changes. It is recalled that Equation 8.12 may be applied only over regions where the modulus of elasticity,

Example 8.4 *(Continued)*

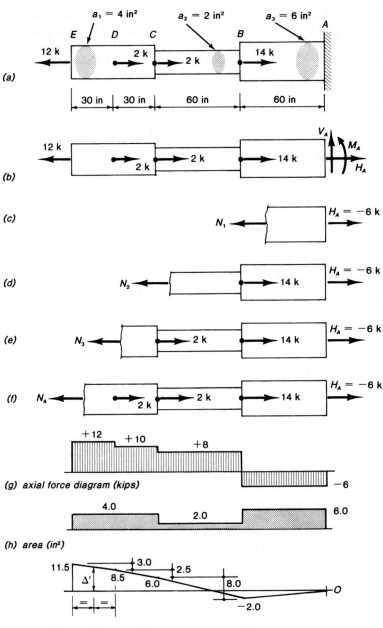

(a)

(b)

(c)

(d)

(e)

(f)

(g) axial force diagram (kips)

(h) area (in²)

(i) axial deflection (in × 10⁻³)

Figure 8.11

the axial force, and the area are all constant. The horizontal deflection of the point E is now the sum of the extensions or contractions of each of the parts of the rod. Thus

$$E = 30 \times 10^6 \text{ psi} = 30 \times 10^3 \text{ ksi}$$

$$\Delta_E = \Delta_{AB} + \Delta_{BC} + \Delta_{CD} + \Delta_{DE}$$

$$= \left(\frac{PL}{AE}\right)_{AB} + \left(\frac{PL}{AE}\right)_{BC} + \left(\frac{PL}{AE}\right)_{CD} + \left(\frac{PL}{AE}\right)_{DE}$$

$$= \frac{(-6) \times 60}{6 \times (30 \times 10^3)} + \frac{8 \times 60}{2 \times (30 \times 10^3)} + \frac{10 \times 30}{4 \times (30 \times 10^3)}$$

$$+ \frac{12 \times 30}{4 \times (30 \times 10^3)}$$

$$= -2.0 \times 10^{-3} + 8.00 \times 10^{-3} + 2.5 \times 10^{-3} + 3.0 \times 10^{-3}$$

$$= 11.5 \times 10^{-3} \text{ in}$$

Using these calculations, the axial deflection diagram of Figure 8.11i may be drawn. The deflection of a point midway between the points D and E may be found from the axial deflection diagram and is labelled Δ' in Figure 8.11i, whence

$$\Delta' = 1.5 \times 10^{-3} \text{ in}$$
$$= 1.5 \times 10^{-3} \text{ in}$$

8.6 Allowable Stresses and Factors of Safety

In the discussion thus far, stresses have been calculated using, for example, Equation 8.3 or 8.4. In order to find the stress it was necessary to be given both the force P and the area A. However, in engineering design it is usual that the area A is not given. Instead, the force P is usually found by some analysis and then the designer must find the necessary area A such that the part is strong enough to withstand the stress. If Equation 8.3 is rearranged in the form

(8.13)
$$A = \frac{N}{\sigma}$$

to reflect the fact that the area A is unknown, then it is clear that some value of stress must be assumed in order to find the area A. At first, one might consider using the ultimate strength of the material, σ_u (see Figure 8.7), as the stress σ in Equation 8.13 since this would give the minimum area A and the part would be just on the point of breaking when the load P was applied. However, this condition would be unsafe for the following reasons:

a. If the applied force was increased by a small fraction—say 1 percent—the part would break. In practice, engineers rarely know the true force that will be placed on a part to within 10 percent, and thus it is quite likely that the part will have a force larger than P exerted on it at some time during its useful life.

b. If the material from which the part was made was for some reason slightly weaker than the expected strength σ_u, then the part would break. There are usually small variations in the strength of materials supplied by manufacturers.

c. If the part was exposed to a corrosive environment and it rusted a little, then the area A of the part would be slightly reduced, and the part would break.

d. Maybe the part could indeed support the force P but the strain would be so large that the part would stretch too much to be of any use. A good example of this might be a truss member that is nominally 10 feet long stretching to a length of 12 feet when the design force P was acting in the member. Clearly this is unacceptable, because the truss, while still complete, would be so distorted as to be of no use.

For all of these reasons, it is necessary to introduce some *factors of safety* into the calculations. This is done by defining an *allowable stress* as

(8.14*a*) $$\text{allowable stress } (\sigma_{all}) = \frac{\text{ultimate strength } (\sigma_u)}{\text{factor of safety (F.S.)}}$$

The factor of safety may also be applied to the material yield stress or to the shear strength of the material, depending on which is more appropriate. (The factor of safety will have a different value depending on which stress value it is being applied to.)

It may therefore be appropriate to define an allowable stress as

(8.14*b*) $$\text{allowable stress } (\sigma_{all}) = \frac{\text{yield strength } (\sigma_y)}{\text{factor of safety (F.S.)}}$$

or

(8.14*c*) $$\text{allowable shear stress } (\tau_{all}) = \frac{\text{yield strength in shear } (\tau_y)}{\text{factor of safety (F.S.)}}$$

In general, design codes give the engineer the appropriate allowable stress that may be used in a given situation. In most cases the writers of the design code have attempted to introduce a factor of safety with a value somewhere between 2 and 5. It is, however, very difficult to be sure about the size of the factor of safety actually achieved, and the factor of safety might more properly be referred to as a factor of ignorance, since it reflects the uncertainty about many aspects of engineering design.

If the allowable stress is used in Equation 8.13, then the required area of the part, A_{req} is given by

(8.15) $$A_{req} = \frac{P}{\sigma_{all}}$$

Example 8.5

The two rods shown in Figure 8.12a support a weight W. The rod AB is made of aluminum with an ultimate strength of 400 megapascals and having a cross-sectional area of 50 square millimeters. The rod BC is made of steel having an ultimate strength of 600 megapascals and a cross-sectional area of 30 square millimeters. If it is desired to have a factor of safety against rupture of 4 for the aluminum rod and a factor of safety of 3 for the steel rod, find the maximum weight W that can be supported by the rods and the stresses in the rods when this maximum weight is hanging from the point B.

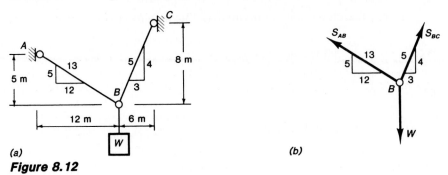

(a)
(b)

Figure 8.12

Solution:

It is first necessary to find the forces in the rods in terms of the unknown force W hanging from the point B. To do this, a free body diagram of the point B is drawn in the same manner as was done for the analysis of trusses by the method of joints. This diagram is shown in Figure 8.12b. The horizontal and vertical force equilibrium equations are now applied to this free body:

$$\Sigma F_H = 0 \overset{+}{\rightarrow}$$

$$S_{BC} \cdot \left(\frac{3}{5}\right) - S_{AB} \cdot \left(\frac{12}{13}\right) = 0$$

$$\Sigma F_V = 0 \uparrow +$$

$$S_{AB} \cdot \left(\frac{5}{13}\right) + S_{BC} \cdot \left(\frac{4}{5}\right) - W = 0$$

Solving these gives:

$$S_{BC} = \frac{20}{21} \cdot W$$

$$S_{AB} = \frac{13}{21} \cdot W$$

Consider the rod AB. Using Equation 8.14 the allowable stress, σ_{all}, may be found:

$$\sigma_{all} = \frac{\sigma_u}{F.S.} = \frac{400}{4} = 100 \text{ MPa}$$

Rearranging Equation 8.15 gives the allowable axial force in the member, P_{all} (all units are in newtons and meters):

$$P_{all} = \sigma_{all} \cdot A = (100 \times 10^6)(50 \times 10^{-6}) = 5,000 \text{ N}$$
$$= 5.0 \text{ kN} = S_{AB_{all}}$$

Now

$$S_{AB} = \left(\frac{13}{21}\right) W$$

$$W_{max} = \left(\frac{21}{13}\right) \cdot S_{AB_{all}} = \left(\frac{21}{13}\right) \cdot 5.0 = 8.07 \text{ kN}$$

Thus if the rod AB has its maximum allowable stress in it, a force W of 8.07 kilonewtons will be supported at point B. However, at present it is not clear whether this force is permissible without exceeding the allowable stress in rod BC.

Consider the rod BC:

$$\sigma_{all} = \frac{\sigma_u}{\text{F.S.}} = \frac{600}{3} = 200 \text{ MPa}$$

Thus,

$$P_{all} = \sigma_{all} \cdot A = (200 \times 10^6) \times (30 \times 10^{-6}) = 6,000 \text{ N}$$
$$= 6.0 \text{ kN} = S_{BC_{all}}$$

Now

$$S_{BC} = \left(\frac{20}{21}\right) W$$

and

$$W_{max} = \left(\frac{21}{20}\right) \cdot S_{BC_{all}} = \left(\frac{21}{20}\right) \cdot 6.0 = 6.3 \text{ kN}$$

Thus the maximum force W that may be supported by the rods is 6.3 kN (if a force of 8.07 kN was suported the stress in rod BC would be greater than the allowable stress). When W is 6.3 kilonewtons,

$$\sigma_{BC} = 200 \text{ MPa} = \sigma_{BC_{all}}$$

The force in rod AB is given by

$$S_{AB} = \frac{13}{21} W = \frac{13}{21} \cdot 6.3 = 3.9 \text{ kN}$$

therefore

$$\sigma_{AB} = \frac{S_{AB}}{A_{AB}} = \frac{3.9 \times 10^3}{50 \times 10^{-6}} \text{ Pa} = 78 \times 10^6 \text{ Pa}$$

$$= 78 \text{ MPa} < \sigma_{all} = 100 \text{ MPa}$$

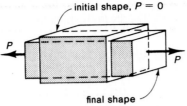

initial shape, $P = 0$

final shape

Figure 8.13 *Change in shape of a block under axial stress.*

8.7 Poisson's Ratio

The equations developed in this chapter have primarily involved the relationship between axial load and axial deflection, or between axial stress and axial strain. However, when a member is subject to an axial force it not only strains in a direction parallel to the applied force, but it also strains in directions perpendicular to the line of action of the applied force. This is illustrated by the block shown in Figure 8.13. When the force P is applied to the block, it lengthens in a direction parallel to the force P, but it shortens in the perpendicular directions. The relationship between the axial strain and the perpendicular strain is defined as Poisson's ratio, v, which is given by

$$(8.16) \qquad \text{Poisson's ratio}\,(v) \triangleq \left| \frac{\text{lateral strain}}{\text{axial strain}} \right|$$

The modulus signs (vertical bars) on this expression indicate that Poisson's ratio is always a positive number, even though the lateral strain is usually of the opposite sign to the axial strain. (This is because the member must shrink in directions perpendicular to the force if it extends parallel to the force.) For most engineering materials, Poisson's ratio is in the range of 0.2 to 0.35. It may, however, extend outside this range, as the values of Table 8.2 show.

8.8 Hooke's Law in Two Dimensions

The relationship expressed by Equation 8.9 linking stress and strain is often called Hooke's law, after an English scientist who discovered a similar equation while experimenting with springs in the late seventeenth century. However, Equation 8.9 shows how axial strain will occur when a member is subjected to one axial stress, but does not show how an element will behave if it is subject to stresses in two perpendicular directions. Consider the small plate shown in Figure 8.14 subject to stresses σ_x and σ_y and in the x- and y-directions, respectively. It may be shown that the resulting strains ϵ_x and ϵ_y in the x- and y-directions are given by:

$$(8.17) \qquad \epsilon_x = \frac{\sigma_x}{E} - v\frac{\sigma_y}{E}$$

$$(8.18) \qquad \epsilon_y = \frac{\sigma_y}{E} - v\frac{\sigma_x}{E}$$

Table 8.2   Poisson's Ratio

Material	Poisson's Ratio (υ)
Steel	0.29
Stainless steel	0.31
Iron	0.27
Aluminum	0.33
Brass	0.34
Copper	0.35
Concrete	0.1-0.2
Titanium	0.34
Rubber	0.50

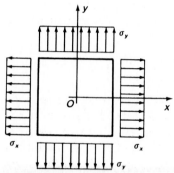

Figure 8.14   *Stresses in two directions.*

It is noted that Poisson's ratio enters both of these equations, as this ratio indicates how a stress in one direction influences the strain in a perpendicular direction. In these expressions a stress is considered to be positive if it is a tensile stress, and positive strain is indicative of a stretching of the material.

Example 8.6

A steel plate 10 feet by 12 feet by 3/8 inch thick is subjected to uniformly distributed edge loads, p_x and p_y, as shown in Figure 8.15. Under these loads the plate stretches 0.096 inch in the x-direction and 0.0864 inch in the y-direction. If the modulus of elasticity of the steel is 30×10^6 pounds per square inch and Poisson's ratio is 0.25, find the values of the edge loads p_x and p_y in kips per foot.

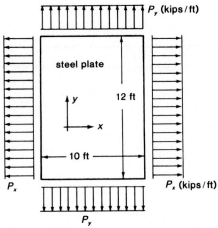

Figure 8.15

Solution:

In order to use Equations 8.17 and 8.18, the stresses and strains in perpendicular directions must be found; if the edge loads are p_x and p_y in kips per foot then the forces acting in the x- and y-directions—P_x and P_y, respectively—are given by

$$P_x = 12\,p_x \text{ kips}$$

and

$$P_y = 10\,p_y \text{ kips}$$

The stresses on these faces of the plate are then found by using Equation 8.3. Care must be taken that the right cross-sectional area is being used in calculation of the stresses. If A_x and A_y are the areas of the plate normal to the x- and y-axes, respectively, then

$$A_x = (12 \cdot 12) \cdot 0.375 = 54 \text{ in}^2$$

and

$$A_y = (10 \cdot 12) \cdot 0.375 = 45 \text{ in}^2$$

The expressions for the stresses are then

$$\sigma_x = \frac{P_x}{A_x} = \frac{12p_x}{54} = 0.222\, p_x \text{ ksi}$$

and

$$\sigma_y = \frac{P_y}{A_y} = \frac{10p_y}{45} = 0.222\, p_y \text{ ksi}$$

The strains in the two perpendicular directions may also be computed by using Equation 8.6. Let ϵ_x and ϵ_y be the strains in the x- and y-directions, respectively;

$$\epsilon_x = \frac{\Delta_x}{L_x} = \frac{0.096}{10 \times 12} = 0.0008$$

and

$$\epsilon_y = \frac{\Delta_y}{L_y} = \frac{0.0864}{12 \times 12} = 0.0006$$

Thus, Equations 8.17 and 8.18 become

$$0.0008 = \frac{0.222\, p_x}{30 \times 10^3} - \frac{(0.25) \cdot 0.222\, p_y}{30 \times 10^3}$$

and

$$0.0006 = \frac{0.222\, p_y}{30 \times 10^3} - \frac{(0.25) \cdot 0.222\, p_x}{30 \times 10^3}$$

Recall that $E = 30 \times 10^6$ psi $= 30 \times 10^3$ ksi. Solving these equations gives

$$p_x = 136.8 \text{ kips/ft}$$
$$p_y = 81.0 \text{ kips/ft}$$

8.9 Thermal Strains

In the previous sections of this chapter we have considered only strains that are caused by the application of forces that cause stresses in the material. However, in addition to strains resulting from forces, strains are also produced when a body is subjected to a change in temperature. For example, when the temperature of a body rises, the body swells in all directions. In other words, the rise in temperature produces a positive strain in all directions within the body. The magnitude of this strain is proportional to the change in temperature and is given by

(8.19)
$$\epsilon_x = \epsilon_y = \epsilon_z = \alpha \cdot \Delta T$$

where ϵ_x, ϵ_y, and ϵ_z, are the strains in the x-, y-, and z-directions respectively (i.e., the strain is the same in all directions); ΔT is the change in temperature (in °F or °C); and α is the coefficient of linear thermal expansion (1/°F or 1/°C).

The coefficient of linear thermal expansion is determined experimentally and is different for different materials. It has been found that this coefficient is approximately a constant for many materials over the range of temperatures for which an engineer usually wants to use them. A selection of coefficients of linear thermal expansion is given in Table 8.3.

It is important to note the units of the coefficient of linear thermal expansion. Since strain is always dimensionless, the units must be (1/°F) or (1/°C). If the body has an initial length of L_0 in one direction before the temperature changes, then after a temperature change of ΔT the change in length ΔL will be

(8.20)
$$\Delta L = \alpha \cdot \Delta T \cdot L_0$$

Example 8.7

A brass rod 12 inches long and having a cross-sectional area of 2 square inches is rigidly held between two surfaces as shown in Figure 8.16a. After the rod has been fixed in this position, it experiences a temperature rise of 40° F. If the modulus of elasticity of the rod is 16×10^6 pounds per square inch and the coefficient of thermal expansion is 1.1×10^{-5} (1/°F), find the resulting axial stress in the rod.

Solution:

Note that the rod is effectively prevented from expanding in an axial direction. There will thus be no final axial strain due to the temperature change. There will, however, be axial stresses introduced because the axial strain has been prevented.

Table 8.3   Coefficients of Linear Thermal Expansion

Material	Coefficient of Linear Thermal Expansion	
	1/°F	**1/°C**
Steel	6.5×10^{-6}	1.17×10^{-5}
Cast iron	6.1×10^{-6}	1.10×10^{-5}
Wood	3.0×10^{-6}	5.4×10^{-6}
Brick	5.0×10^{-6}	9.0×10^{-6}
Aluminum	1.3×10^{-5}	2.34×10^{-5}
Copper	9.3×10^{-6}	1.67×10^{-5}
Concrete	6.0×10^{-6}	1.08×10^{-5}
Bronze	1.0×10^{-5}	1.8×10^{-5}
Brass	1.1×10^{-5}	1.8×10^{-5}

Thus the new length L will be

(8.21)
$$L = L_0 + \Delta L = L_0 (1 + \alpha \, \Delta T)$$

A somewhat unexpected feature of thermal strains is that the strain is not necessarily accompanied by any stresses. In fact, if a body is completely unrestrained then a change in temperature will produce thermal strains but no stresses. On the other hand, if a body is completely restrained (i.e., is unable to expand or contract as the temperature changes), then a change in temperature produces stresses inside the body but no strains. Thus, *thermal stresses result only when thermal strains are prevented.*

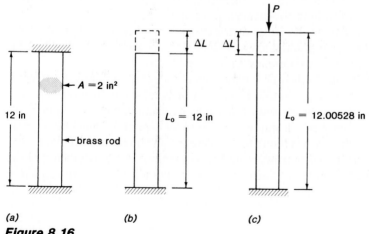

(a)   (b)   (c)

Figure 8.16

Example 8.7 *(Continued)*

The first step in the solution is to assume that the rod is completely free to expand. The rod is free to expand if the upper boundary is removed as shown in Figure 8.16b. If the rod extends a distance ΔL as shown, then this extension is given by Equation 8.20:

$$\Delta L = a \cdot \Delta T \cdot L_0$$
$$= (1.1 \times 10^{-5}) \times 40 \times 12 = 5.28 \times 10^{-3} \text{ inch}$$

Thus if there was no restraint at the end of the rod, it would lengthen by 0.00528 inch because of the temperature rise, and the new length of the rod would be 12.00528 inches.

To account for the fact that the rod cannot expand axially, force P is applied to the end of the rod as shown in Figure 8.16c. If this force compresses the rod so that its length is once again 12 inches, then it has the same effect as if it had been present all during the heating of the rod. The strain required in the rod is now compressive and is given by

8.10 Stress Concentration

In Equation 8.3 the expression for axial stress in an axially loaded member indicated that the axial stress is constant over the whole cross section of the member (see Figure 8.1e). This is indeed the case if the cross-sectional area of the member is constant. A problem arises, however, if the area of the member changes. Two common cases where this situation occurs are shown in Figures 8.17a and b. In Figure 8.17a, a hole has been drilled in the member, maybe as a bolt hole. In Figure 8.17b two plates of the same thickness, but of different widths, have been joined somewhere along the axially loaded member. Figures 8.17c and d show the stress distributions across critical sections of the members at the discontinuities shown in Figures 8.17a and b, respectively. (This problem was ignored in Examples 8.1, 8.3, and 8.4.)

It is clear from Figure 8.17 that for both types of discontinuity considered, the maximum stress, σ_{max}, at the critical section is substantially larger than the average stress, σ_{av}, at the location. Since it is the maximum stress that controls the design of a member, it is important to find the size of the maximum stress. In practice it is often very difficult to calculate the maximum stress theoretically, and so use of stress concentration graphs is common. If it is assumed that the maximum stress is related to the average, or "nominal," stress by the equation

(8.22)
$$\sigma_{max} = K \cdot \sigma_{av}$$

$$\epsilon = -\frac{\Delta L}{L_0} = -\frac{0.00528}{12.00528} \approx -\frac{0.00528}{12.00}$$

$$= -4.40 \times 10^{-4}$$

The stress needed to produce this strain is

$$\sigma = E\epsilon = (16 \times 10^6) \times (-4.40 \times 10^{-4})$$
$$= -7,040 \text{ psi} = -7.04 \text{ ksi}$$

The negative sign indicates compressive stress. Note that it was not necessary to know the cross-sectional area to find the axial stress. However, if the axial force had been requested it would have been given by

$$P = \sigma \cdot A = -7.04 \times 2 = -14.08 \text{ kips}$$

When the bar has been returned to its original length, there is no axial thermal strain, but there is a significant thermal stress instead.

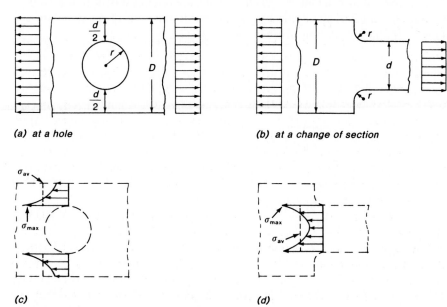

(a) at a hole

(b) at a change of section

(c)

(d)

Figure 8.17 Stress concentrations in a plate under tension (cross section is rectangular).

then the maximum stress can easily be calculated provided the constant K is available. The basis of the nominal stress is usually given with the location where the average stress is to be determined. It may be shown that the value of K is determined by the "sharpness" of the discontinuity and the graph of Figure 8.18 may be used to find its size. To use the graph of Figure 8.18 the fraction r/d (as defined in Figure 8.17) must be calculated from the known geometry of the part before the graph is entered along the horizontal axis. From the correct point on the horizontal axis, a straight line is plotted vertically upward until the correct curve is reached. At the point where the vertical line reaches the curve, the value of K may be read from the vertical axis. In derivation of the curves shown in Figure 8.18 it has been assumed that the thickness of the plate has remained constant, and that the discontinuity has been caused solely by the hole or the change in width of the member. (A different set of curves would be needed to include the effects of changing member thickness.)

Example 8.8

The flat plate shown in Figure 8.19a has a constant thickness of 30 millimeters. If an axial load P of 20 kilonewtons is applied to the ends of the member as shown, find the maximum stress in the member.

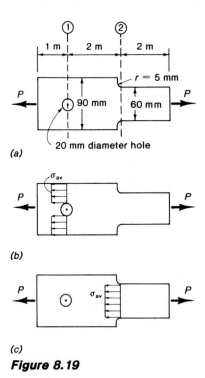

(a)

(b)

(c)

Figure 8.19

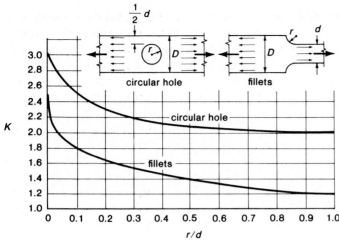

Figure 8.18 *Stress concentration graph.*

Source: Focht, M. M. "Factors of Stress Concentration Photoelastically Determined."
Transactions of ASME 57 (1936): A-67.

Solution:

There are two possible critical locations within the member at which the maximum stress may occur—at the hole and at the change of width—and the stress condition must be investigated at both of these locations. These locations are indicated as locations 1 and 2 on Figure 8.19a.

At location 1. In order to use the graph of Figure 8.18, the value of r/d must first be found:

$$\frac{r}{d} = \frac{10}{70} = 0.143$$

Note that in this case r is the radius of the hole and d is the width of plate left after the hole has been taken out of the plate. From the graph of Figure 8.18 the value of K is found to be 2.35.

The average stress at location 1 must now be calculated. Figure 8.19b clearly shows the area over which the average stress is being determined. Once again it is the reduced area of the cross section after the hole has been made in the plate. Thus

$$\sigma_{av} = \frac{P}{A} = \frac{(20 \times 10^3)}{(30 \times 10^{-3})(70 \times 10^{-3})}$$

$$= 9.52 \times 10^6 \text{ Pa}$$

In this calculation all forces are expressed in newtons and all lengths, in meters.

Now, from Equation 8.22 the maximum stress at location 1 may be found:

$$\sigma_{max_1} = K \cdot \sigma_{av} = 2.35 \times 9.52 \times 10^6$$
$$= 2.24 \times 10^7 \text{ Pa}$$
$$= 22.4 \text{ MPa}$$

Example 8.8 (Continued)

At location 2. Once again the value of r/d must be found:

$$\frac{r}{d} = \frac{5}{60} = 0.0833$$

In this case the value of r is the radius of the fillet joining the two pieces and d is the width of the narrower of the two parts being joined. From the graph of Figure 8.18, the value of K is found to be 1.8.

The calculation of the average stress—distributed as shown in Figure 8.19c—is comparatively easy for this case. Once again units of newtons and meters are used in the calculations.

$$\sigma_{av} = \frac{P}{A} = \frac{20 \times 10^3}{(30 \times 10^{-3})(60 \times 10^{-3})}$$

$$= 11.1 \times 10^6 \text{ Pa}$$

$$= 11.1 \text{ MPa}$$

From Equation 8.22, the maximum stress at location 2 is given by

$$\sigma_{\text{max}_2} = K \cdot \sigma_{\text{av}} = 1.8 \times 11.1 \times 10^6$$
$$= 2.0 \times 10^7 \text{ Pa}$$
$$= 20.0 \text{ MPa}$$

Thus the maximum stress in the part is 22.4 MPa, and it occurs at location 1. Figure 8.17c indicates that the maximum stress occurs at the edges of the hole.

Problems

8.1 For the stress-strain curve given in Figure P 8.1, determine graphically σ_y, ϵ_y, E, and ϵ_{st}.

8.2 Rework Problem 8.1 for the stress-strain curve given in Figure P 8.2.

8.3 A steel rod with a constant cross section of 2 inches is subjected to the axial forces shown in Figure P 8.3. Find the axial deflection at point A and the axial stress at point B. ($E = 30 \times 10^6$ psi.)

8.4 For the rod in Figure P 8.4, with constant cross-sectional area of 30 square millimeters, determine the axial deflection at point A and the axial stress at point B. ($E = 100$ GPa.)

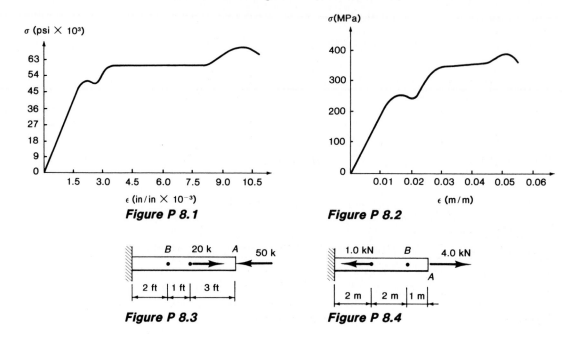

Figure P 8.1

Figure P 8.2

Figure P 8.3

Figure P 8.4

8.5   The brass rod shown in Figure P 8.5 undergoes a temperature change from 45° F to 90° F. Find the change in length caused by this temperature change. ($a = 11 \times 10^{-6}$ in/in/° F.)

8.6   The copper rod shown in Figure P 8.6 is subjected to a temperature increase of 50° C. Determine the stress developed in the rod. ($a = 20 \times 10^{-6}$ m/m/° C; $E = 120$ GPa.)

8.7   For the rod shown in Figure P 8.7 determine the axial deflection at point A and axial stress at point B. ($E = 30 \times 10^{6}$ psi; ignore stress concentrations.)

8.8   Rework Problem 8.7 for the rod shown in Figure P 8.8. ($E = 100$ GPa; ignore stress concentrations.)

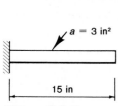

Figure P 8.5

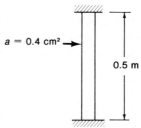

Figure P 8.6

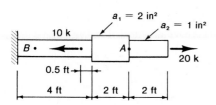

Figure P 8.7

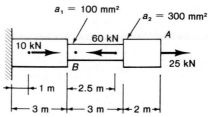

Figure P 8.8

8.9 Rework Problem 8.7 for the rod shown in Figure P 8.9. ($E = 20 \times 10^6$ psi; ignore stress concentrations.)

8.10 Rework Problem 8.7 for the rod shown in Figure P 8.10. ($E = 120$ GPa; ignore stress concentrations.)

8.11 If the ultimate stress for the rod shown in Figure P 8.11 is 200 kips per square inch, determine the factor of safety when the rod is acted upon by the force shown. (Ignore stress concentrations.)

8.12 If the ultimate stress for the rod shown in Figure P 8.12 is 500 megapascals, determine the factor of safety when the rod is acted upon by the force shown. (Ignore stress concentrations.)

8.13 Rework Problem 8.11 for the rod shown in Figure P 8.13 if the ultimate stress is 190 kips per square inch. (Ignore stress concentrations.)

8.14 For a factor of safety of 3, what is the minimum ultimate stress required of the material from which the rod shown in Figure P 8.14 is made? (Ignore stress concentrations.)

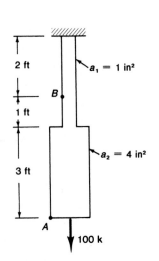

Figure P 8.9

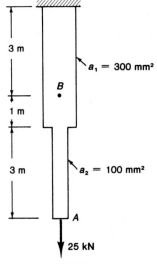

Figure P 8.10

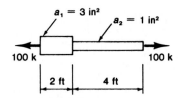

Figure P 8.11

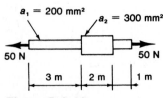

Figure P 8.12

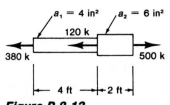

Figure P 8.13

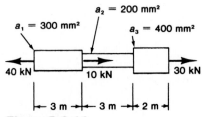

Figure P 8.14

8.15 If the ultimate stress for the rod shown in Figure P 8.15 is 60 kips per square inch and a factor of safety equal to 3 is required, determine the maximum force, P, which may be applied to the rod.

8.16 For the rod shown in Figure P 8.16 the allowable stress is 50 megapascals. Determine the maximum force P.

8.17 If the allowable stress for the rod shown in Figure P 8.17 is 50 kips per square inch, determine the maximum force P. (Ignore stress concentrations.)

8.18 Determine the maximum force, P, for the rod shown in Figure P 8.18 if the ultimate stress is 750 megapascals and a factor of safety of 3 is assumed. (Ignore stress concentrations.)

8.19 Determine the force P that, when applied to the rod shown in Figure P 8.19, produces an elongation of 0.25 inch. ($E = 30 \times 10^6$ psi; ignore stress concentrations.)

8.20 Determine the force P required to compress the rod shown in Figure P 8.20 by 15 millimeters. ($E = 50$ GPa; ignore stress concentrations.)

8.21 The rod in Figure P 8.21 is elongated 0.5 inch at point X as a result of the given forces. Determine the forces P and $2P$ that produce this elongation. ($E = 20 \times 10^6$ psi; ignore stress concentrations.)

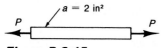

Figure P 8.15

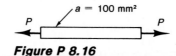

Figure P 8.16

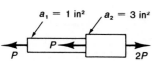

Figure P 8.17

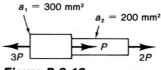

Figure P 8.18

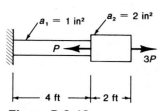

Figure P 8.19

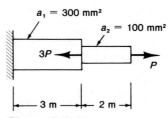

Figure P 8.20

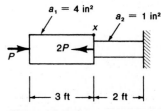

Figure P 8.21

8.22 The rod in Figure P 8.22 is elongated 5 millimeters at point X because of the given forces. Determine the force P that produces this elongation. ($E = 50$ GPa; ignore stress concentrations.)

8.23 The 1-inch-thick steel bar in Figure P 8.23 has a 30-kip force applied to it. Determine the maximum stress in the bar.

8.24 The 5-millimeter-thick steel bar shown in Figure P 8.24 is subjected to a force, P. The allowable stress is 100 megapascals. Determine the maximum force, P.

8.25 The 2-inch-thick bar shown in Figure P 8.25 has an allowable stress of 30,000 pounds per square inch. Determine the maximum force, P.

8.26 The 0.1-meter-thick steel plate shown in Figure P 8.26 is subjected to a 25-kilonewton force. Determine the maximum stress.

8.27 For the 2-inch-thick steel bar shown in Figure P 8.27, which has allowable stress of 30,000 pounds per square inch, determine the maximum force, P.

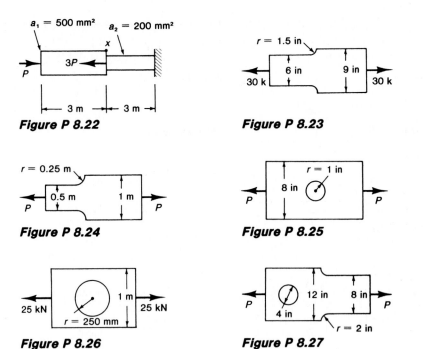

$a_1 = 500$ mm² $a_2 = 200$ mm²

Figure P 8.22

Figure P 8.23

Figure P 8.24

Figure P 8.25

Figure P 8.26

Figure P 8.27

8.28   Determine the maximum force *P* that may be applied to the
        3-centimeter-thick steel plate shown in Figure P 8.28. The ultimate
        stress of the plate material is 400 megapascals, and a safety factor
        of 4 is required.

8.29   Determine the maximum force *P* that may be applied to the 1-inch-
        thick steel bar shown in Figure P 8.29 if the allowable stress is
        20,000 pounds per square inch.

8.30   Determine the maximum force *P* that may be applied to the
        2-centimeter-thick bar shown in Figure P 8.30 if the allowable stress
        is 300 megapascals.

8.31   Take *L** as defined in Problem 7.19. If the rod shown has a
        thickness of 2 inches, and is made of steel with a Young's modulus
        of 30×10^6 pounds per square inch, determine the change in length
        of the rod. (Neglect stress concentrations.)

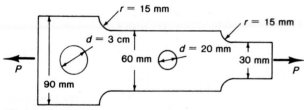

Figure P 8.28

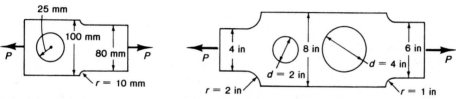

Figure P 8.29

Figure P 8.30

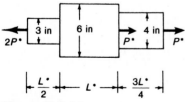

Figure P 8.31

Shear, Torsion, and Twist

9

Objectives

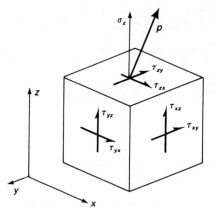

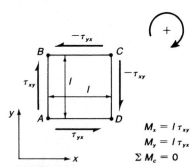

Figure 9.1
Notation for shear stress
components.

Figure 9.2
The equivalence of shear
stresses acting in perpendicular
directions.

9.1 Shear Stress and Shear Strain

Stresses that are parallel with the surface on which they act are called *shear stresses*. While a normal stress may be defined by the direction in which it acts, a shear stress must be defined by two terms: the plane on which it acts and its direction. In a rectangular coordinate system a plane may be defined in two ways. Either the two coordinates in which the plane lies are given, or the coordinate that is perpendicular to the plane is specified. We will use the latter method. In this manner the direction and plane of application of a shear stress may be defined by using two subscripts. The first will specify the plane, the second will specify the direction of the shear stress. Figure 9.1 shows examples for such notations.

When a stress p acts in an arbitrary direction, as shown in Figure 9.1, it may be separated into three components. These consist of the component normal to the plane, σ_z, and the two perpendicular shear stress components, τ_{zx} and τ_{zy}.

Considering the static equilibrium of a very small rectangle subjected to shear on all sides as shown in Figure 9.2, we may draw a few important conclusions. Shear stresses come in pairs: their magnitudes on adjacent parallel planes are equal and their signs are opposite. Given the sides of the small rectangle as l, as the two parallel shear forces give rise to two moments around point C that are

(9.1)
$$M_x = l \cdot \tau_{xy}$$

and

$$M_y = -l \cdot \tau_{yx}$$

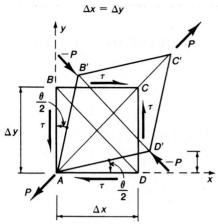

Figure 9.3 *Deformation of an elementary small square caused by shear.*

By static equilibrium,

$$\Sigma M = M_x + M_y = 0$$

which can be true only if the magnitudes of the shear stresses are equal; that is,

(9.2)
$$|\tau_{xy}| = |\tau_{yx}|$$

In general terms, this conclusion states that *at any and all points of a structure, when shear exists on a certain plane, the same shear exists in a plane perpendicular to the first one.*

Another conclusion to be drawn from the foregoing is that the two pairs of shear stresses acting on a small rectangular element are such that they balance each other's rotation. At all corner points of the rectangle, the two shear stresses acting are both pointing toward the corner or away from it.

Let us now investigate the *deformation* of an elementary small square under the influence of shear stresses acting on its four sides. If the original positions of the corner points are *A, B, C,* and *D,* in Figure 9.3, the shear stresses will deform the square into a figure with new corner points *A, B', C',* and *D'.* Looking at the figure carefully, one will notice that the two opposite corners, *A* and *C,* toward which the shear stresses point, will move diagonally farther from each other. The other two corners, *B* and *D,* will move diagonally closer to each other. These are the points where the shear stresses point away from the corner. The diagonals of the square are important orientations. If there are no normal stresses acting on the small square element, this is called *pure shear.*

The deformation under shear, shown in Figure 9.3, may be produced by another means. Rather than applying shear, the same deformation may be attained by applying compressive stresses, $-p$, between points *B* and *D* and tensile stresses, p, between points *A* and *C.* The idea may be better visualized

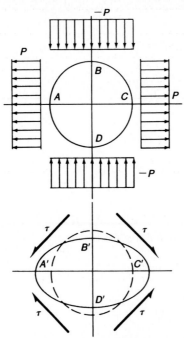

Figure 9.4   *The equivalence of deformation of an elementary circle into an ellipse under perpendicular compressive and tensile stresses and under pure shear.*

by considering the deformation of the elementary circle shown in Figure 9.4 under tension in the horizontal direction and compression in the vertical direction. The resulting deformation will cause the circle to become an ellipse, as shown in the lower drawing. There the identical deformation is shown to be caused by two pairs of shear stresses diagonally oriented, under which the element is subjected to pure shear. The understanding of this concept will be useful when combined stresses are considered in Chapter 12.

In Figure 9.3, the angular deformation of the element is denoted by $\theta/2$ on each side. This angular deformation under the influence of shear stresses is comparable to the linear deformation ϵ due to normal stresses. The latter was

Table 9.1 Shear modulus

Material	Shear modulus, G	
	psi	GPa
Steel	12×10^6	83
Magnesium alloy	2.4×10^6	17
Cast iron, gray	6×10^6	41
Cast iron, malleable	12×10^6	83
Aluminum alloy (2024-T4)	4×10^6	27.6
Polyethylene	0.8×10^6	5.5

an important result of the previous chapter. The magnitude of the angular deformation depends on the magnitude of the shear stresses and the elastic resistance of the material. While linear strain causes volumetric changes in the material, shear strain causes changes in shape. For small deformations the shear stress and the angle of deformation are linearly related to each other. This linear relationship,

(9.3)
$$\tau = G\theta$$

is *Hooke's law for shear.* The angle θ is expressed in radians. The factor of proportionality G is called *shear modulus,* or torsional modulus; it is a property of the material. For the most common materials its values are listed in Table 9.1. Since the angle θ is dimensionless, the dimension of G must be the same as that of the shear stress.

For isotropic materials, for which the elastic constants are the same in all orientations, the value of the shear modulus is related to Young's modulus and Poisson's ratio. The relationship is given by the formula

(9.4)
$$G = \frac{E}{2(1 + v)}$$

For most structural materials the magnitude of G is approximately 38 to 40 percent of the Young's modulus E.

Example 9.1

A 2 by 6-inch wooden plank is spliced with a 30-degree glued scarf splice that has a maximum shear strength of 60 pounds per square inch. Determine the maximum axial tensile force the plank may resist.

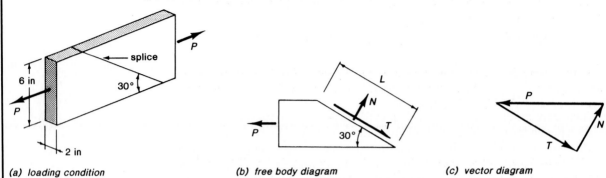

(a) loading condition   (b) free body diagram   (c) vector diagram

Figure 9.5   *Shear on a spliced connection.*

Solution:

Figure 9.5 shows the geometry of the problem. First the area of the glued surface is computed. The length of the splice is

$$L = \frac{6 \text{ in}}{\sin 30°} = 12 \text{ in}$$

Example 9.2

The clevis shown in Figure 9.6 is held with a pin 15 millimeters in diameter. All parts are made of steel that has an allowable tensile stress of 165 megapascals and an allowable shear stress of 100 megapascals. Determine the maximum tensile force that may be supported by the device. Ignore stress concentration.

Solution:

Figures 9.6b, c, and d show the detailed dimensions of the three components. The loading on the pin is shown in Figure 9.6d.

First we review the drawings to locate those sections where the cross-sectional areas resisting the tensile force P are smallest. These are the cross sections where potential failure may occur. We find that the smallest cross section of the bar is that marked as ab and cd across the centerline of the hole. The clevis is composed of two plates, each equal in dimension to the bar; hence the tensile stresses in the clevis present no critical locations. The minimal area in the bar is

$$A_{bar, \, min.} = 2(10 \text{ mm} \times 7.5 \text{ mm}) = 150 \text{ mm}^2 = 1.5 \times 10^{-4} \text{ m}^2$$

The area of the glued splice is then

$$A = 12 \times 2 = 24 \text{ in}^2$$

For a maximum allowable shear stress of 60 pounds per square inch, the shear force T acting in the glued area is

$$T = 60 \text{ psi } (24 \text{ in}^2) = 1,440 \text{ lb}$$

The free body diagram shown indicates that for force T to balance the axial force P, there must be a force N acting normal to the splice. The vector diagram shown in Figure 9.5 indicates that

$$N = T \tan 30° = 1,440 \ (0.577) = 831.4 \text{ lb}$$

As the forces T and N acting on the splice are known, the axial force pulling the plank may be computed by the Pythagorean theorem as

$$P = \sqrt{N^2 + T^2} = \sqrt{831.4^2 + 1,440^2} = 1,663 \text{ lb}$$

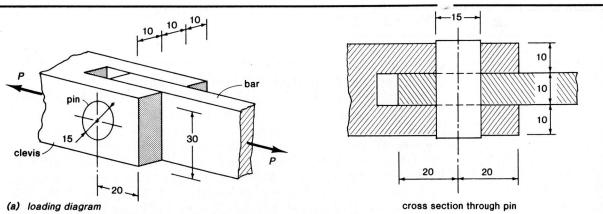

(a) loading diagram

cross section through pin

Figure 9.6 *Double shear in a pin (all dimensions are in mm).*

Example 9.2 (Continued)

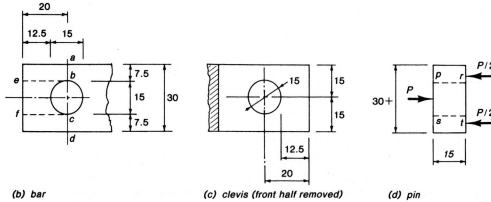

(b) bar　　　　　　　**(c) clevis (front half removed)**　　**(d) pin**

Figure 9.6 (Continued)

In light of the allowable tensile stress of 165 megapascals, the pulling force P may not exceed

$$P = 165 \text{ MPa } (1.5 \times 10^{-4} \text{ m}^2) = 24.75 \text{ kN}$$

The shear in the pin occurs at both sides as it enters the clevis. The cross-sectional area of the pin is

$$A_{pin} = \frac{\pi D^2}{4} = \frac{3.14 \,(15)^2}{4} = 176 \text{ mm}^2 = 1.76 \times 10^{-4} \text{ m}^2$$

As shown in Figure 9.6d, only one half of P acts on this area. (Cases like this are referred to as *double shear.*) Hence, for the allowable shear stress of 100

9.2 Torsional Transmission of Forces

One of the most common ways to transfer forces is by *torsion.* It means the transmission of a moment of the couple in the direction perpendicular to the plane in which the couple acts. Its application ranges from the simple screwdriver or doorknob to the shafts of electric motors and automobiles, pumps and turbines. Let us consider the example shown in Figure 9.7. It shows a simple frame consisting of two parallel *V*-members and a perpendicular connecting bar that is loaded by a couple. The couple results in a torsional moment that twists the structure around the axis connecting the points *G* and *H.* As in any other truss system discussed in Chapter 4 the moment must be transferred to the base of the structure by compressive and tensile stresses in the bars of the *V*-members. Close observation of the structure will enable the reader to visualize that there will be compressive stresses in bars *AB* and *DF.* The other two bars,

megapascals, the shear force in the pin is

$$A_{pin} \tau_{allow} = (1.76 \times 10^{-4}) \, 100 = 17.6 \, \text{kN}$$

$$= \frac{P}{2}$$

Therefore, the maximum tensile force that may be resisted by the pin in shear cannot exceed

$$P = 2 \, (17.6) = 35.2 \, \text{kN}$$

which is more than the bar would be able to support in tension.

Other critical cross sections for shear are the two planes in the bar denoted as eb and fc. The combined shearing surfaces jointly resist the pulling force P. The area of plane fc is

$$A_{fc} = 20 \, \text{mm} \, (10 \, \text{mm}) = 200 \, \text{mm}^2 = 2 \times 10^{-4} \, \text{m}^2$$

At the allowable shear stress of 100 megapascals, the shear in the bar is

$$A_{fc} \tau_{allow} = (2 \times 10^{-4}) \, 100 = 20 \, \text{kN}$$

$$= \frac{P}{2}$$

Therefore the maximum pulling force P may not exceed

$$P = 2 \, (20 \, \text{kN}) = 40 \, \text{kN}$$

Reviewing our results, we conclude that the tensile resistance of the bar will control the design. The pulling force therefore may not exceed 24.75 kilonewtons, and the device is adequate to resist shear.

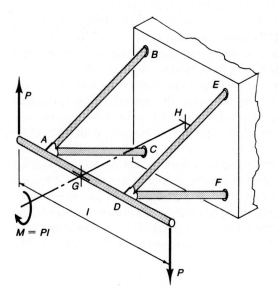

Figure 9.7
Simple frame in pure torsion (moment is transferred by tensile and compressive forces).

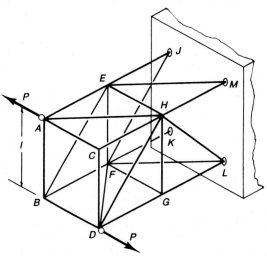

Figure 9.8 *Transfer of twisting moment by compressive and tensile stresses in the components of the structure.*

AC and *DE,* will be under tension. Another example for the transmission of torque is shown in Figure 9.8. It is a boxlike frame loaded by a torsional moment that is the product of *l* and the force *P.* One may visualize the flow of tensile and compressive forces through the members. The bars that allow the structure to carry the torque to the wall are the diagonal ones. The force at point *A* is carried in tension through points *A, H,* and *L.* The one at point *D* creates tension through points *D, F,* and *J.* Other members balance these tensile forces by compressive ones. In general, one may, again, realize that *the transfer of pure torsion involves compressive and tensile stresses.*

If the structure or machine component is composed of a solid piece instead of bars like the two examples discussed above, the same general concept still holds. Torsional moments are transferred by compressive and tensile stresses. Take a thin-walled circular tube, for instance, as shown in Figure 9.9. If the tube is fixed at one end and a twisting moment is applied, the compressive and tensile stresses will flow through the wall of the tube in a spiral manner. The direction of these two stresses are perpendicular to each other at all points of the tube's wall. As tension in one direction is coupled by compression in the other direction, the magnitudes of the two stresses are the same. Because of this fact, required by virtue of static equilibrium, the positioning of these stresses must be symmetrical, hence they are oriented at 45 degrees with respect to the axis of the tube.

Let us consider this compressive and tensile stress distribution at one point of the wall. In Figure 9.9 the stress conditions at point *A* are shown. If a small element of the tube at that point was circular before the deformation of the tube because of the twisting moment, it will be distorted into an ellipse as shown. The major axis of the ellipse is oriented 45 degrees from the axial direction of the tube. This corresponds to the direction of the tensile stress in the tube's wall. The

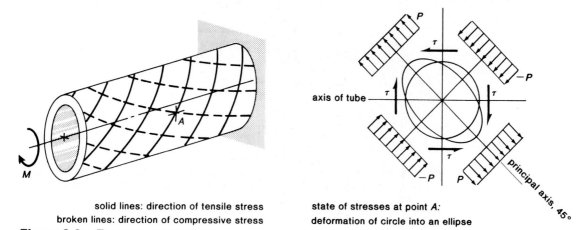

solid lines: direction of tensile stress
broken lines: direction of compressive stress

state of stresses at point A:
deformation of circle into an ellipse

Figure 9.9 *Transfer of twisting moment in a tube by compressive and tensile stresses acting along diagonal, spiral directions in the wall.*

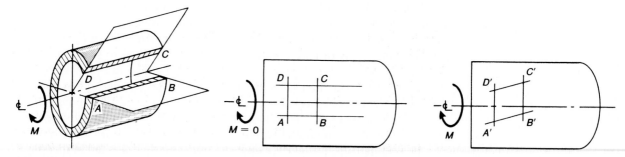

(a) *twisted cylindrical tube with element ABCD removed*

(b) *rectangular grid element ABCD drawn on cylindrical tube before twisting*

(c) *distortion of rectangular grid element after torsion*

Figure 9.10 *Transfer of twisting moment in a cylindrical tube by pure shear (by nature of pure shear, shear stresses act equally on radial and axial planes).*

minor axis is the direction of the compressive stress. This sketch is identical with the one shown in Figure 9.4. From this, one may conclude that the two axes of the ellipse are directions of the principal normal stresses, and that the deformation can be also construed to have been caused by pure shear in the tube. The shear stresses transferring the torsional moment lie in two directions: in a plane bisecting the tube perpendicularly, and in planes defined by the tube's axis and any radial direction. Figure 9.10a shows such an arrangement.

If a rectangle grid is drawn on the surface of a cylindrical tube and one of its elemental rectangles is marked as *ABCD*, as indicated in Figure 9.10b, then, after the torsional moment is applied the element will be deformed into the shape *A'B'C'D'* indicated on Figure 9.10c. One of the fundamental properties of cylindrical bars and tubes under torsion is that *plane sections perpendicular*

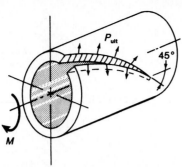

Figure 9.11 *Spiral failure in a cylindrical tube due to tension caused by twisting moment* **M** *(there is* no shear *on the failure surface).*

to the axis remain plane. In other words, the vertical grid lines shown in Figure 9.10*b* will remain vertical in Figure 9.10*c*. In other shapes, particularly in those with corners, like rectangular bars and structural shapes, the vertical cuts will warp when torsion is applied.

Failure of a cylindrical tube under pure torsion will occur when either one of the principal normal stresses or one of the shear stresses exceeds its limiting value. For most common structural materials *three failure modes* may occur:

a. The diagonally oriented principal tensile stress reaches its limiting value. The failure surface appears on the 45-degree spiral line as shown in Figure 9.11. The maximum tensile stress is perpendicular to this line. This type of failure is known to oil drillers when the drilling pipe gets accidentally overtwisted. The same occurs in the leg of a falling skier whose bindings do not release in time. Twisting a piece of chalk results in a similar type of failure. Glass and many plastics also fail in this way.

b. When the tensile strength of a material exceeds its shear strength, failure occurs in a plane perpendicular to the axis. This is a normal shear failure.

c. In some composite materials where the ultimate shear strength in the axial direction is significantly less than in the plane perpendicular to the axis, failure may occur in the axial direction. For example, in a fiber-reinforced plastic material where the fibers are oriented axially, the axial shear may tend to unravel the fibers. The same may happen to reeds and bamboo.

The detailed analysis of combined stresses in Chapter 12 will shed light on an important basic fact that must be mentioned here. This is that *the magnitude of both the principal and the shear stresses is the same throughout a body when it is subject to a pure torsional loading.* That is, at point *A,* for example, shown in Figure 9.9, the absolute values of *p,* −*p,* and τ are identical. As the twisting moment creating the pure shear increases, they all increase uniformly. Failure will occur when any one of the ultimate strength values of the material is reached by these stresses.

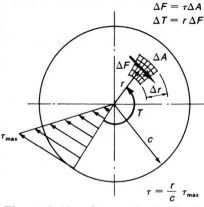

$$\Delta F = \tau \Delta A$$
$$\Delta T = r \, \Delta F$$

$$\tau = \frac{r}{c} \, \tau_{max}$$

Figure 9.12   *Stress distribution in a circular bar subjected to torsion.*

It follows from this same fact that to obtain all of the characteristic stresses shown in Figure 9.4, determining only one is sufficient. In the following section the determination of the shear stress distribution in a cut perpendicular to the axis will be shown.

9.3 The Torsion Formula

One of the fundamental assumptions concerning the torsion of cylindrical bars is that the magnitude of shearing stresses increases linearly with the radius, as shown in Figure 9.12. At the center the radius is zero and the shear stress is also zero; the maximum value of the shear is reached at the circumference of the bar. The shear stress at any point defined by r is

(9.5)
$$\tau = \left(\frac{r}{c} \right) \tau_{max}$$

where c is the outside radius of the bar; $c = r_{max}$. Such linearity in the distribution of shear stresses under torsion has been proven valid for circular bars and tubes as long as the stresses are within the elastic range.

The relationship between the twisting moment T and the shear stresses in a twisted bar may be found by the application of the principle of static equilibrium. The external moment must be balanced by the internal moments. The internal moments are considered at each elementary part of the surface area of the bar that is cut to form a free body. On each small surface element the tangential shear force equals

(9.6)
$$\Delta F = \tau \, \Delta A$$

Substituting Equation 9.5, one obtains

(9.7)
$$\Delta F = \left(\frac{r}{c} \right) \tau_{max} \, \Delta A$$

Example 9.3

A 50-millimeter diameter steel shaft is subjected to a twisting moment of 5,250 newton meters. Determine the maximum shear stress developed at the extreme fibers of the shaft.

Solution:
The radius of the shaft is 0.025 meter. This is c in Equation 9.12. The polar moment of inertia J may be computed from Equation 9.10 with $D = 0.05$ m. Hence

$$J = \frac{\pi (0.05)^4}{32} = 6.14 \times 10^{-7} \, m^4$$

Substituting into Equation 9.12, we have

$$\tau_{max} = \frac{5{,}250 \, (N \cdot m) \times 0.025 \, (m)}{6.14 \times 10^{-7} \, (m^4)}$$

$$= 2.14 \times 10^8 \, N/m^2 = 214 \, MPa$$

The moment in a small surface element is a part of the resisting internal moment and can be expressed by the formula

(9.8)
$$\Delta M = r \cdot \Delta F = \left(\frac{r^2}{c} \right) \tau_{max} \, \Delta A$$

Summing the contributions of all surface elements leads to the total internal resisting moment. This, by Equation 3.3, must be equal to the twisting moment T. In mathematical form, the total twisting moment is equal to

(9.9)
$$T = \Sigma \, \Delta M = \left(\frac{\tau_{max}}{c} \right) \Sigma \, r^2 \, \Delta A$$

In this equation the maximum shear stress and the radius of the extreme fibers are taken outside of the summation sign as they have constant values. One may then recognize that the geometric term in the summation sign is nothing else but the polar moment of inertia J, defined by Equation 7.16.

The polar moment of inertia for circular shapes is given by

(9.10)
$$J = \frac{\pi \, c^4}{2} = \frac{\pi \, D^4}{32}$$

Example 9.4

A hollow shaft with an outside diameter of 4 inches and an inside diameter of 3 inches has an allowable shearing stress of 7,500 pounds per square inch. Determine the maximum torsional moment the shaft may transmit.

Solution:

The polar moment of inertia computed by Equation 9.11 is

$$J = \frac{\pi}{32}(4^4 - 3^4) = 0.098\,(256 - 81) = 17.15\,in^4$$

Equation 9.12 may now be used with $c = 2$ inches. Substituting all known values, one obtains

$$7,500 = \frac{T\,(2)}{17.15}$$

from this the torsional moment will be

$$T = 7,500\,(17.15)/2 = 64,312\,lb \cdot in$$

where D is the diameter of the circle. For tubular shapes the polar moment of inertia equals J for the outside diameter minus J for the inside diameter, or in mathematical form,

(9.11)
$$J = \frac{\pi}{32}(D^4 - d^4)$$

in which d represents the inside diameter of the tube.

By substituting the polar moment of inertia and rearranging to express the maximum shear stress, Equation 9.9 will take the form

(9.12)
$$\tau_{max} = \frac{T\,c}{J}$$

This is the *basic design equation for torsion* of cylindrical members. For non-cylindrical bars under torsion, Equation 9.12 is not applicable.

Example 9.5

A round polyethylene bar is 4 inches in diameter. Polyethylene has an ultimate tensile strength of 1,300 pounds per square inch and an ultimate shear strength of 2,200 pounds per square inch. Determine the maximum torque the bar may transmit.

Solution:

In Section 9.2 three possible failure modes under torsion were explained. These were failure under pure tension and failure under pure shear either in the axial or in the tangential direction. It was also noted that in pure torsion all stresses at one point are numerically equal. Hence the maximum shear stress under torsion in the case of polyethylene should be its ultimate *tensile* strength, 1,300 pounds per square inch. The polar moment of inertia of the 4-inch diameter bar is

$$J = \left(\frac{\pi}{32} \right) 4^4 = 25.13 \text{ in}^4$$

Substituting into the torsional formula, one obtains

$$1,300 = \frac{T(2)}{25.13}$$

and from this

$$T = 16,334 \text{ lb} \cdot \text{in}$$

In machines, power is often transmitted by rotating shafts. These shafts are subjected to torque. In industry, power is most commonly expressed in terms of *horsepower* (hp). In English units the formula of conversion between torque T and horsepower is

(9.13) $$T \text{ (lb} \cdot \text{in)} = 63,000 \text{ hp}/N$$

where N is the speed of rotation in revolutions per minute. The conversion formula to the S.I. system is

(9.14) $$T \text{ (N} \cdot \text{m)} = 119 \text{ hp}/f$$

where f is the frequency of rotation in revolutions per second (hertz). In the S.I. system power is measured in watts and rotational speed in hertz.

Gears or pulleys transmitting torque to a shaft are usually attached by *keys*. To receive the key a keyway is machined into the shaft, as shown in Figure 9.13. The keyway considerably decreases the torsional strength of the shaft, and designers take this into account by reducing the allowable shear stress by about 25 percent.

Example 9.6

Determine the required shaft diameter for a 12-horsepower motor operating at 35 hertz. The allowable shearing stress for the shaft is 50 megapascals.

Solution:

Using Equation 9.14 the torque in question is

$$T = 119\,(12)/35 = 40.8\text{ N}\cdot\text{m}$$

From Equation 9.12,

$$\frac{J}{c} = \frac{T}{\tau_{max}} = \frac{40.8}{50 \times 10^6} = 0.816 \times 10^{-6}\text{ m}^3$$

From Equation 9.10,

$$\frac{J}{c} = \frac{\pi\,c^3}{2} = 0.816 \times 10^{-6}$$

Solving for c, one gets

$$c = \sqrt[3]{\frac{2(0.816)}{3.14}} \times 10^{-6} = 8 \times 10^{-3}\text{ m}$$

Therefore the diameter of the shaft must be at least

$$D = 2c = 16\text{ mm}$$

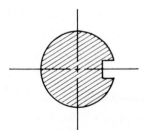

Figure 9.13   A keyway machined into a shaft.

Quite often there are two or more gears attached to a shaft. Most shafts are supported by bearings. Between bearings the shafts should be strong enough to resist deformations due to bending. Often the shaft diameter is changed between certain bearings or gears. In such instances the critical point or points where torsional stresses are the largest must be found by first constructing a *torque diagram*. One of these cases is illustrated by the following example.

Example 9.7

A drive arrangement is shown in Figure 9.14a. Determine the location and magnitude of the maximum torsional shear stress.

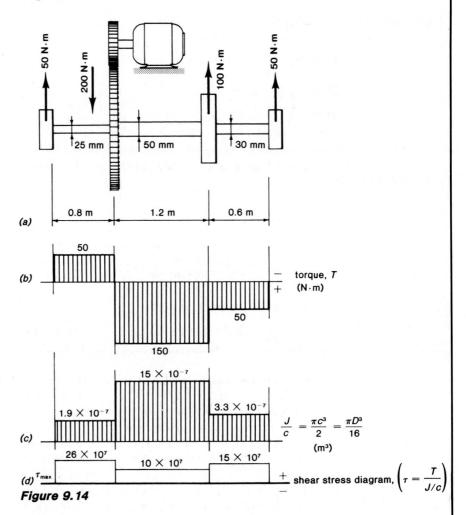

Figure 9.14

Solution:

First a torque diagram is drawn as in Figure 9.14b. The procedure is similar to the one used for drawing shear force diagrams. In this sketch, torques representing take-offs are considered negative and the driving torque by the electric motor is considered positive. At the ends of the shaft the value of the torque is zero. (Instead of torque, one may have given power and speed. In such cases Equation 9.13 or 9.14 may be used first.)

Figure 9.14c shows J/c values computed for the given shaft diameters. The values shown are in meters cubed.

The ratios of the graphed T and J/c values are shown. These represent Equation 9.12, giving torsional stresses.

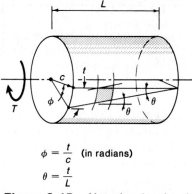

$$\phi = \frac{t}{c} \quad \text{(in radians)}$$

$$\theta = \frac{t}{L}$$

Figure 9.15 *Notation for the derivation of the angle of twist.*

9.4 The Angle of Twist

By virtue of Hooke's law transmission of a force or moment by a structural or machine element must always be associated with deformation. In the case of torsion, the twisting moment causes an angular deformation. This may be demonstrated by the distortion of a rectangular grid drawn on the surface of a cylindrical bar, as shown in Figure 9.10. Over the whole length of the twisted bar the total angular distortion results in the *angle of twist, ϕ*. The formula for the angle of twist may be derived without difficulty by geometric considerations. Figure 9.15 shows the notations used in this derivation. Expressing small angular deformations in radians, the angle of twist may be given as the ratio of the radius of the bar to the displacement of a point at the cylindrical surface caused by the twisting. If the latter is denoted by t, the twisting angle may be expressed by

(9.15)
$$\phi = \frac{t}{c}$$

Observing the distortion of a small rectangle drawn on the surface, like the element $ABCD$ in Figure 9.10, the angular distortion of the axial lines of the grid will be a constant throughout. This angle may be expressed in a manner similar to Equation 9.15 as

(9.16)
$$\theta = \frac{t}{L}$$

Solving for t and substituting Equation 9.16 into 9.15, one obtains

(9.17)
$$\phi = \frac{\theta L}{c}$$

By Equation 9.3, representing Hooke's law for torsion, the angular shear strain θ may be replaced by the shear stress τ, leading to

(9.18)
$$\phi = \frac{\tau_{max} L}{G c}$$

Example 9.8

The 3-foot length of polyethylene bar analyzed in Example 9.5 has a shear modulus, G, of 0.8×10^6 pounds per square inch. Using the data given in the cited example, determine the twist angle just before the bar fails.

Solution:
The bar's polar moment of inertia was calculated to be 25.13 in⁴; the twisting moment at failure as 16,334 pound inches. The length of the bar must be expressed in inches to conform with these values; the length is 36 inches. Substituting these terms into Equation 9.18, one obtains

$$\phi = \frac{16,334 \, (36)}{25.13 \, (0.8 \times 10^6)} = 0.029 \text{ rad}$$

Multiplying radians by $360/2\pi$, one gets the angle in degrees; hence,

$$\phi = 360 \, (0.029) / 2\pi = 1.66°$$

In turn, the shear stress may be replaced by the use of the torsion formula, Equation 9.12. This will result in

(9.19)
$$\phi = \frac{T \, L}{J \, G} \text{(radians)}$$

This is the *formula for the angle of twist.*

One of the most interesting applications of torsional theory is the design of *torsion bars* shown in Figure 9.16. They are applied in automobile suspensions and elsewhere where reliable spring action is desired. The rotation of the arm under the torque $P \times t$ is resisted by T in Equation 9.19. The angle of twist ϕ is directly proportional to T. The constant of proportionality, K, is more commonly expressed as its reciprocal:

(9.20)
$$\frac{1}{K} = \frac{J \cdot G}{L}$$

The dimension of $1/K$ is newton meters or pound inches. This term has a wider application in the solution of statically indeterminate structural problems involving pure torsion. In such cases, $1/K$ is referred to as *torsional stiffness factor.* One such example is Example 9.10.

Example 9.9

An aluminum bar 16 millimeters in diameter and 0.4 meter long is subjected to a torque of 35 newton meters. The shear modulus for aluminum is 28 gigapascals. Determine the angle of twist.

Solution:

First the polar moment of inertia is computed by Equation 9.10. It is

$$J = \pi (0.016)^4/32 = 6.43 \times 10^{-9} \, m^4$$

Now we are ready to substitute all required values into Equation 9.19:

$$\phi = \frac{35 (0.4)}{6.43 \times 10^{-9} (28 \times 10^9)} = 0.077 \text{ rad}$$

Converting into degrees, we get

$$\phi = 0.077 (360/2\pi) = 4.41°$$

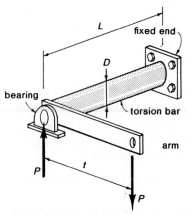

Figure 9.16 *Notation for the torsion bar concept.*

Example 9.10

Consider the problem shown in Figure 9.17, involving a combined bar made up of a round aluminum bar connected to a round magnesium alloy bar. There are three supports: fixed ends at the walls and a bearing (not shown in the figure) at the point where a torsional moment is applied by a cross bar. Determine the reactions.

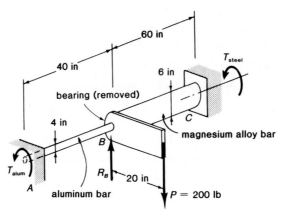

Figure 9.17 *Statically indeterminate problem involving torsion.*

Solution:

This problem cannot be solved by the application of the formulas of static equilibrium alone. Close observation of the structure reveals, however, that the bars are subjected to pure torsion. The force P applied at the crossbar is countered at the bearing by a force R that equals $-P$ and a torsional moment T that equals the product of the length of the crossbar and P, that is

$$T = 20 \text{ (in)} \times 200 \text{ (lb)} = 4{,}000 \text{ lb} \cdot \text{in}$$

With either of the torsion bars removed, the problem would become statically determinate. In that case, all the torsional moment T would be transferred to the remaining fixed end. Since the torsion bars are not loaded by a force, there will be no reaction forces at the walls.

Such statically indeterminate problems are solved by bringing in elastic considerations. As the two bars are rigidly connected at point B, the twist angle at B must be the same for both torsion bars. Both bars will contribute to the resisting of the external torsional moment. The contribution of each will depend on its torsional stiffness. The stiffer the bar, the larger proportion of T will it resist. (Two bars of equal torsional stiffness would split the external loading equally.) To learn how these two will share the load, we will have to compute their torsional stiffness coefficient.

First we find the shear modulus of the magnesium and aluminum from Table 9.1. These are

$$G_{Al} = 4 \times 10^6 \text{ psi}$$

$$G_{Mg} = 2.4 \times 10^6 \text{ psi}$$

Next the polar moment of inertia is computed for the two bars, using Equation 9.10:

$$\text{4-in aluminum bar, } J = \frac{\pi (4)^4}{32} = 25.1 \text{ in}^4$$

$$\text{6-in magnesium bar, } J = \frac{\pi (6)^4}{32} = 127.2 \text{ in}^4$$

Now we calculate the stiffness coefficients from Equation 9.20. These are

$$1/K_{Al} = \frac{25.1 \, (4 \times 10^6)}{60} = 1.67 \times 10^6 \text{ lb} \cdot \text{in}$$

$$1/K_{Mg} = \frac{127.2 \, (2.4 \times 10^6)}{40} = 7.69 \times 10^6 \text{ lb} \cdot \text{in}$$

These values indicate that the magnesium bar is almost five times stiffer than the aluminum bar. This means that the overwhelming portion of the external torque T will be carried by the magnesium bar. Let us now compute the exact figures. We have three equations. One, from statics, states that the sum of the moments carried by the torsion bars must equal the external moment T, that is

$$T = T_{Mg} + T_{Al} = 4{,}000 \text{ lb} \cdot \text{in}$$

The other two equations come from Hooke's law, that is, from the consideration of elastic properties. These are, by Equation 9.20,

$$\phi_B = T_{Mg} \cdot K_{Mg} = T_{Al} \cdot K_{Al}$$

From this it follows that

$$T_{Mg} = \left(\frac{K_{Al}}{K_{Mg}} \right) T_{Al} = \frac{1.67}{7.69} T_{Al} = 0.22 \, T_{Al}$$

This result can now be substituted into the static equilibrium formula:

$$4{,}000 = (0.22 + 1) \, T_{Al}$$

$$T_{Al} = 3{,}278 \text{ lb} \cdot \text{in}$$

$$T_{Mg} = 722 \text{ lb} \cdot \text{in}$$

Example 9.11

A spring is made of 0.5-inch steel wire consisting of 16 coils of 6-inch diameter. Determine the spring constant and the elongation of the spring under a 6-pound load.

Solution:

For steel the shear modulus, G, is 12×10^6 pounds per square inch. Substituting into Equation 9.23, one obtains

$$k = \frac{12 \times 10^6 (0.5)}{8 (16)} \left(\frac{0.5}{6} \right)^3 = 27 \, \text{lb/in}$$

To determine the elongation of the coil spring under a 6-pound load, one applies Equation 9.22.

$$6 \, (\text{lb}) = 27 \, (\text{lb/in}) \cdot t$$

From this we find the elongation as

$$t = \frac{6}{27} = 0.22 \, \text{in}$$

It is appropriate to mention here that *coil springs* are also classical examples of torsional loading. The derivation of formulas that describe their behavior is rather complex, so it will not be included, but the following relations are sufficient for approximate analysis.

Coil springs are loaded with concentrated forces acting through the center-line of the coil. Denoting the mean radius of the coil, taken through the centerline of the spring wire, as R_c, the torsional moment acting in the wire is

(9.21) $$T = P \cdot R_c$$

The stiffness of a coil spring is characterized by the *spring constant k* that is the ratio of the force P to the corresponding deflection t; that is,

(9.22) $$P = k \cdot t$$

An approximate formula for the spring constant is

(9.23) $$k = \frac{G \cdot d}{8 \, n} \left(\frac{d}{D} \right)^3$$

in which n is the number of coils, d is the diameter of the spring wire, and D is the mean diameter of the coil. Handbooks on machine design and brochures published by spring manufacturers provide additional information on this subject.

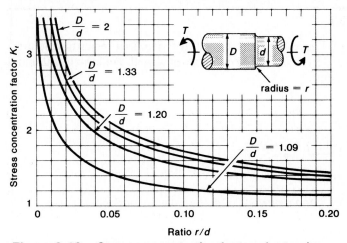

Figure 9.18   Stress concentration factors for torsion.

Source: Jacobsen, L. S. "Torsional-Stress Concentrations in Shafts of Circular Cross-Section and Variable Diameter." *Transactions of ASME* 47 (1926): 632.

9.5 Stress Concentration under Torsional Loading

In chapter 8 the subject of stress concentration in members subjected to axial stresses was presented. It was shown that a stress concentration may exist at any point within a member at which the cross section changes. In particular, the stress concentration at abrupt changes of cross section and near holes in a member were discussed.

The same type of stress concentration factor occurs in members subject to torsional loading. The problem of stress concentration in shafts is particularly severe at locations where the radius of the shaft changes, as shown in Figure 9.18. In the same manner as for axial stresses, a stress concentration factor K is defined such that the maximum shearing stress in the shaft τ_{max} is related to the maximum shearing stress in the shaft of smaller diameter (τ_s) by the relationship

$$\tau_{max} = K\,\tau_s \tag{9.24}$$

$$\tau_{max} = K\,\frac{T\left(\dfrac{d}{2}\right)}{J} \tag{9.25}$$

From the curves of the graph of Figure 9.18, the value of the stress concentration factor may be found, and then Equation 9.25 can be used to find the maximum shear stress in the shaft.

Example 9.12

The shaft of Figure 9.19 is subjected to a twisting moment of 16 newton meters at end A, as shown. The end B is fixed to a vertical wall. Find the maximum shear stress in the shaft.

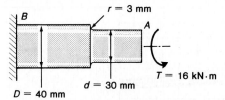

Figure 9.19 *Stress concentration in a shaft under torsional loading.*

Solution:

The polar moment of inertia of the smaller portion of the shaft is

$$J = \frac{\pi\,d^4}{32} = \frac{\pi\,(30)^4}{32}$$

$$= 7.952 \times 10^4 \text{ mm}^4$$
$$= 7.952 \times 10^{-8} \text{ m}^4$$

The maximum shear stress in the smaller shaft is therefore given by

$$\tau_s = \frac{T\left(\dfrac{d}{2}\right)}{J} = \frac{(16)\left(\dfrac{30}{2} \times 10^{-3}\right)}{7.952 \times 10^{-8}}$$

$$= 3.018 \times 10^6 \text{ Pa}$$

The stress concentration factor is found from Figure 9.18 by use of the following factors:

$$\frac{r}{d} = \frac{3}{30} = 0.10$$

$$\frac{D}{d} = \frac{40}{30} = 1.33$$

Thus K is found from the graph:

$$K = 1.7$$

Thus the maximum shear stress in the shaft is

$$\tau_{max} = K\,\tau_s$$
$$= 1.7 \times 3.018 \times 10^6$$
$$= 5.4 \times 10^6 \text{ Pa}$$
$$= 5.4 \text{ MPa}$$

Problems

9.1   What is the percentage difference in maximum torsional stress between a solid shaft 9 inches in diameter and a hollow shaft that has an external diameter of 9 inches and an internal diameter of 6 inches?

9.2   A composite beam made of steel and aluminum alloy is shown in Figure P 9.2. The 10-kilonewton load is acting at the center of the beam. For a section just to the right of support *A*, find the shearing stress at the intersection of steel and aluminum alloy.

9.3   For Problem 9.2, it is desired to put one rivet between the steel and the aluminum every 10 centimeters. How much shear force would each rivet take?

9.4   A 4-inch by 6-inch wood plank, shown in Figure P 9.4, is made of two pieces of wood by gluing the 45° intersurfaces. If the maximum shearing strength of the glued plank is 200 pounds per square inch, what is the maximum compression force *P* allowed?

9.5   See Problem 9.4: if the maximum shearing strength of the glued joint and the maximum compression strength of the wood are both 800 pounds per square inch, what would be the maximum force *P* allowed? Which material will fail first?

9.6   Compute the maximum shear stress for a solid carbon-steel shaft 20 millimeters in radius and subject to a torque of 300 newton meters.

9.7   The solid cylindrical shaft of variable size shown in Figure P 9.7 is acted upon by the torque indicated. What is the maximum torsional stress in the shaft, and between what two pulleys does it occur?

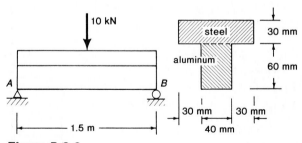

Figure P 9.2

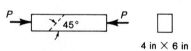

Figure P 9.4

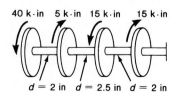

Figure P 9.7

9.8   What is the rate of torsional stress between the two cases shown in Figure P 9.8?

9.9   A 4-foot long pipe is end-loaded by concentrated torque of magnitude 1,500 pound feet. The allowable shear stress is 8,000 pounds per square inch and the allowable twist angle is 2.5°. From the table in Appendix A choose the lightest pipe that meets these requirements.

9.10   A solid cylindrical shaft of length 1.0 meter is used to transmit a torque of 200 newton meters. Given that the maximum shear stress is not to exceed 120 megapascals and the angle of twisting is not to exceed 3.0°, determine the required shaft diameter.

9.11   Calculate the location and magnitude of the maximum shear stress for the titanium alloy tube shown in Figure P 9.11. ($G = 6 \times 10^6$ psi; $R_o = 0.787$ in; $R_i = 0.394$ in.)

9.12   Design a hollow steel shaft to transmit 200 horsepower at 1.4 hertz without exceeding a shearing stress of 70 megapascals (outside diameter is 140 millimeters).

9.13   For the shaft given in Figure P 9.13, determine the required torque such that the shaft twists a total of 0.20°. ($G = 12 \times 10^6$ psi.)

9.14   A solid aluminum shaft 2.0 meters long and 70 millimeters in outside diameter is to be replaced by a tubular steel shaft of the same outside diameter and 1.5 meters long that carries the same torque and has the same angle of twist over the total length. What must the inner radius of the tubular steel shaft be? ($G_{st} = 80$ GPa; $G_{Al} = 26$ GPa.)

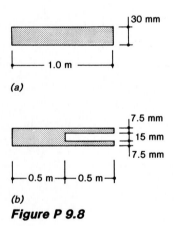

(a)

(b)

Figure P 9.8

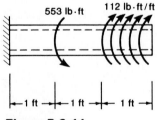

Figure P 9.11

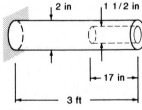

Figure P 9.13

9.15 Determine the required torque such that the shaft shown in Figure P 9.15 will twist 0.30°. ($G = 12 \times 10^6$ psi.)

9.16 For the shaft shown in Figure P 9.16, determine the angle of twist in degrees between the two ends. ($G = 84$ GPa.)

9.17 A tube of 50 millimeters outside diameter and 2 millimeters thickness is attached at the ends by means of rigid flanges to a solid shaft of 20 millimeters diameter as shown in Figure P 9.17. If both the tube and the shaft are made of the same linearly plastic material, determine maximum shear stress and angle of twisting. ($G = 84$ GPa.)

9.18 In Problem 9.17, what part of the applied torque T is carried by the tube?

9.19 For the shaft shown in Figure P 9.19, the length of the hollow part and the applied torque are changeable for x from 0 to 180. Find the twisting angle ϕ at the end in degrees for different given values of x. ($G = 84$ GPa; answers for each unit increment of x are available in the instructor's manual.)

9.20 A shaft with changing cross section is shown in Figure P 9.20; find the maximum shearing stress considering stress concentration.

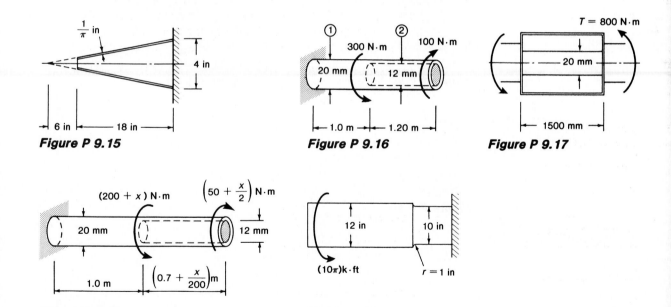

Figure P 9.15   Figure P 9.16   Figure P 9.17

Figure P 9.19   Figure P 9.20

In Problems 9.21 to 9.24 let $T^* = (\text{SS} + 50)^2 \, \text{N} \cdot \text{m}$ and $L^* = \sqrt{\text{SS} + 50}$ m, where SS are either the last two digits of your Social Security number or two digits assigned by the instructor.

9.21 A cantilever shaft, shown in Figure P 9.21, has the cross-sectional shape shown. If the material is steel ($G = 80$ GPa), find the maximum shear stress in the shaft and the angle of twist of the free end.

9.22 –
9.24 Resolve Problem 9.21 if the section is as shown in Figures P 9.22 to P 9.24.

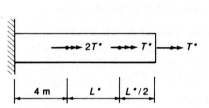

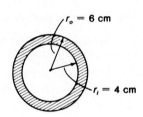

Figure P 9.21

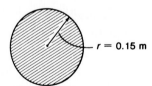

Figure P 9.22

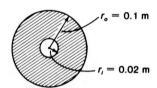

Figure P 9.23

Figure P 9.24

Bending Stresses

10

1. *To be able to find the flexural stress at any point on a section subject to a bending moment.*

2. *To learn to recognize potentially critical locations on a beam at which the stresses must be determined.*

3. *To know how to analyze reinforced concrete beams in which the stresses are within the elastic range.*

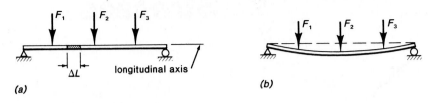

Figure 10.1 *Bending action of beams.*

10.1 Bending Theory

In Chapter 8 it was explained how axial forces cause axial stresses in a member. In the case of axial forces the force was applied parallel to the longitudinal axis of the member. It is the purpose of this chapter to investigate the stress distribution within the cross section of a member when the loads are applied transversely to the longitudinal axis of the member, as is shown in Figure 10.1a. These forces cause the beam to sag (Figure 10.1b) and with this deflection stresses are introduced into the member. Members with forces applied transversely to their longitudinal axis are called *beams.*

To determine the variation of stresses within a beam let us assume that we have a beam of rectangular cross-sectional area A and length ΔL as shown in Figure 10.2a. This beam has a vertical axis of symmetry and it is initially perfectly straight. In Chapter 6 the subject of bending moment diagrams was treated; here we will employ them to derive the value of internal bending moment at any point within a statically determinate beam. For the purposes of this derivation, a very short piece has been taken from a much longer beam such that the moment is the same on each end of the small piece of beam. The beam shown in Figure 10.2a is an enlarged view of the small piece of beam shown circled in Figure 10.1a. Figure 10.2b is an elevation view of the short beam shown with a moment M applied to each end. The longitudinal axis is defined as the x-direction and the perpendicular direction as the y-direction. As Figure 10.1b indicated, the beam tends to sag when subject to transverse loads, and thus the beam will deform as shown in Figure 10.2c.

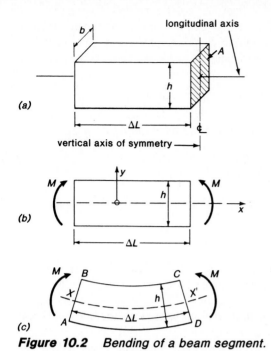

(a)

(b)

(c)

Figure 10.2 Bending of a beam segment.

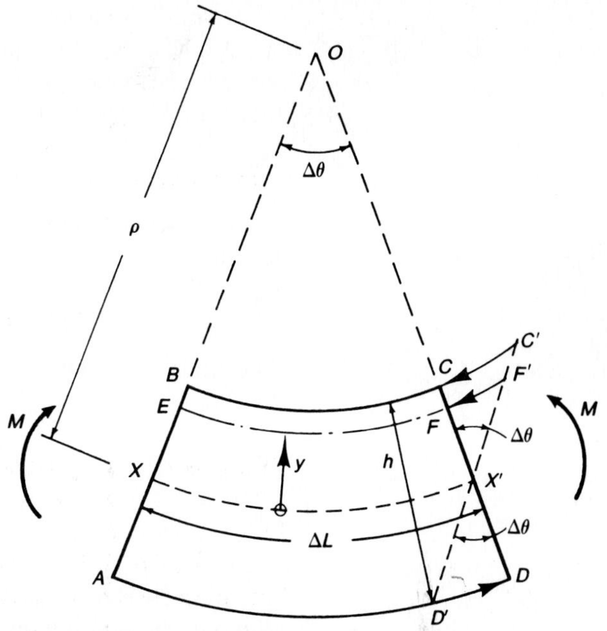

Figure 10.3 Strains in a beam segment.

The beam element shown in Figure 10.2c has been redrawn in Figure 10.3, in which the point O is the intersection of the lines AB and DC. The line $X-X'$ has remained the same length ΔL as the beam was bent. The line $X-X'$ is known as the *neutral axis* because it neither extends nor contracts as the beam is bending. It may be shown that the neutral axis represents a plane through the centroid of the section. (The finding of the centroid was described in Chapter 7.) In Figure 10.3 the line $C'-D'$ has been drawn parallel to the line $A-B$, indicating that the bottom of the beam, line $A-D$, has become longer. Since the the beam has become shorter, or has experienced a negative strain, it would be expected that this portion of the cross section would be subject to a compressive stress. For the same reason the bottom of the beam stretches and would therefore be subject to a tensile stress.

Consider the line $E-F$ in Figure 10.3, at a height y above the neutral axis $X-X'$. This line had an original length $E-F' = \Delta L$ before the beam was loaded. The line $E-F$ shortens by a distance $F-F'$ as the beam bends. The distance $F-F'$ is given by

(10.1)
$$F-F' = y \cdot \Delta\theta$$

Also the original length ΔL is related to the radius of curvature ρ by the relationship

(10.2)
$$\Delta L = \rho \cdot \Delta\theta$$

The strain in the line $E-F$ may now be found:

$$\epsilon_{E-F} = \frac{\text{change in length}}{\text{original length}} = \frac{F-F'}{E-F'} = \frac{F-F'}{X-X'} = \frac{-y \cdot \Delta\theta}{\rho \cdot \Delta\theta}$$

(10.3)
$$\epsilon_{E-F} = -\frac{y}{\rho}$$

The negative sign indicates that the line $E-F$ is shorter than the line $E-F'$ and thus a negative strain has occurred.

It is now assumed that the material remains in the elastic range of material behavior during the bending of the beam. If this is the case, then the stress in the beam at the location of the line $E-F$ may be computed using Equation 8.9. Thus

(10.4)
$$\sigma = E \cdot \epsilon = E\left(\frac{-y}{\rho}\right)$$

Notice in Equation 10.4 that the stress at any point on the cross section of the beam is linearly proportional to its distance y from the neutral axis. Using this it may be shown that the stress is given by

(10.5)
$$\sigma = \frac{M \cdot y}{I}$$

where M is the moment applied to the section (Figure 10.2b) and I is the moment of inertia of the section (see Chapter 7). Thus if the moment M is increased, the stresses will become larger. A higher moment of inertia leads to lower stresses.

The stresses are in the longitudinal or x-direction, and Figure 10.4 shows whether the stresses are tensile or compressive. The tensile or compressive nature of the bending stresses depends on the direction of the applied moment on the section. In Figure 10.4a a positive moment causes a compressive stress above the neutral axis and tensile stress at all locations below the neutral axis. Conversely, Figure 10.4b shows that a negative moment on the section causes compressive stresses below the neutral axis and tensile stresses above it.

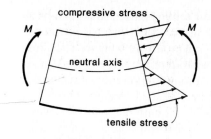

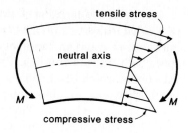

(a) positive moment   (b) negative moment

Figure 10.4   *Stresses due to positive and negative moments.*

Figure 10.4 indicates two other facts about bending stresses. First, since the neutral axis neither extends nor contracts during bending, the stress on the neutral axis will always be zero. Furthermore, the magnitudes of the bending stresses always increase as the distance from the neutral axis increases. Thus the maximum bending stress on a section will always be at that position on the section which is farthest from the neutral axis. Since it is very common for the maximum stress on the section to be required, in engineering calculations another property of a section called *section modulus* is often used. The section modulus is defined as

(10.6)
$$Z = \frac{I}{y_{max}}$$

where y_{max} is the maximum distance in the y-direction that any point on the section may have from the neutral axis. In most engineering texts y_{max} is denoted by c. Substituting Equation 10.6 into Equation 10.5 gives the expression for the maximum bending stress on the section.

(10.7)
$$\sigma_{max} = \frac{M}{Z} = \frac{M \cdot y_{max}}{I}$$

Since the moment of inertia, I, had units of (length)^4—e.g., in^4, m^4, or mm^4—the section modulus will have units of (length)^3—e.g., in^3, m^3, or mm^3. The values of section modulus Z for common engineering shapes are tabulated in Appendix A.

Careful examination of the properties of sections listed in Appendix A shows that many sections have two values of Z listed, one associated with bending about an x-axis and another with bending about a y-axis. Figure 10.5 illustrates both types of bending. The designer must identify correctly the axis or axes about which bending is taking place before attempting to calculate any bending stresses. In both the diagrams of Figure 10.5 the bending moment has been indicated using the familiar right-hand rule. These diagrams also show that the bending stress is constant for all points equidistant from the neutral axis.

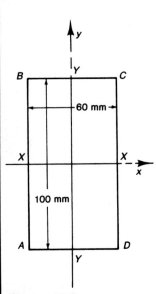

Figure 10.6
Cross section of beam in Example 10.1.

Example 10.1

A beam of rectangular cross section has the dimensions shown in Figure 10.6.

a. Find the maximum stress due to a moment of 2.0 kilonewton meters applied about the x-axis,

b. Find the maximum stress due to a moment of 5.0 kilonewton meters applied about the y-axis.

Solution:
Part a. Before the stresses may be found, the section modulus must be found. Throughout this example the calculations will be done in units of newtons and meters. From Chapter 7:

$$I_x = \frac{bd^3}{12} = \frac{(60 \times 10^{-3})(100 \times 10^{-3})^3}{12} = 5.0 \times 10^{-6} \text{ m}^4$$

Then, from Equation 10.6,

$$Z_x = \frac{I_x}{y_{max}} = \frac{5.0 \times 10^{-6}}{5.0 \times 10^{-2}} = 1.0 \times 10^{-4} \text{ m}^3$$

The distance y_{max} is the maximum distance that a point on the section may be from the $X-X$ axis. Then, using Equation 10.7,

$$\sigma_{max} = \frac{M_x}{Z_x} = \frac{2.0 \times 10^3}{1.0 \times 10^{-4}} = 2.0 \times 10^7 \text{ Pa} = 20.0 \text{ MPa}$$

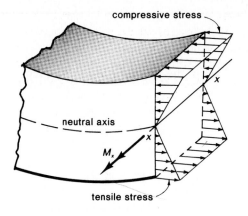

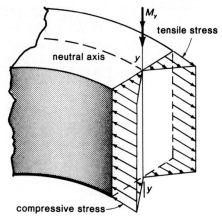

(a) bending about x-axis   **(b)** bending about y-axis

Figure 10.5 *Bending about perpendicular axes.*

Note that the section modulus relative to the *x*-axis must be used with the moment about the *x*-axis. The calculated value of maximum stress will occur along the edges *BC* and *AD* of the section.

Part b. A similar method is followed for finding the maximum stress caused by a moment applied about the *y*-axis. This time, however, the section modulus must be found relative to the $Y-Y$ axis. Again, from Chapter 7,

$$I_y = \frac{db^3}{12} = \frac{(100 \times 10^{-3})(60 \times 10^{-3})^3}{12} = 1.8 \times 10^{-6}\,\mathrm{m}^4$$

This is smaller than the value of I_x derived above. In derivation of the section modulus relative to the *y*-axis, Z_y, a different form of Equation 10.6 must be used, i.e.,

$$Z_y = \frac{I_y}{x_{max}} = \frac{1.8 \times 10^{-6}}{3.0 \times 10^{-2}} = 6.0 \times 10^{-5}\,\mathrm{m}^3$$

The distance x_{max} (30 mm $= 3.0 \times 10^{-2}$ m) is the maximum distance a point on the section may be from the $Y-Y$ axis. Now, using Equation 10.7,

$$\sigma_{max} = \frac{M_y}{Z_y} = \frac{5.0 \times 10^3}{6.0 \times 10^{-5}} = 8.33 \times 10^7\,\mathrm{Pa} = 83.3\,\mathrm{MPa}$$

The calculated value of maximum stress will occur along the edges *AB* and *CD* of the section.

Example 10.2

A wide-flange member W 18 × 46 is to be used as a beam.

a. If a moment of 50 kip feet is applied about the *x*-axis, find the maximum bending stress on the member.

b. If a moment of 15 kip feet is applied about the *y*-axis, find the maximum bending stress on the member.

Solution:

Part a. The applied moment $M_x = 50 \text{ kip} \cdot \text{ft} = 50 \times 10^3 \times 12 \text{ lb} \cdot \text{in} = 6.0 \times 10^5 \text{ lb} \cdot \text{in}$.

   Reference is now made to the tables of Appendix A, where the properties of a W 18 × 46 member are listed. From the table the section modulus of the member relative to the *x*-axis, Z_x, is found to be 78.8 in³. Now from Equation 10.7 the value of the maximum stress on the section due to the moment of 50 kip feet may be found:

$$\sigma_{max} = \frac{6.0 \times 10^5}{78.8} = 7.61 \times 10^3 \text{ psi} = 7.61 \text{ ksi}$$

This maximum stress is the value of the stress along the lines *AB* and *CD* on the member as shown in Figure 10.7, since all points on these lines are equidistant from the *x*-axis and they are at the greatest possible distance from the *x*-axis that any point on the section can be. The bending stress, whether tension or compression, is, of course, acting perpendicular to the cross section shown in Figure 10.7.

Part b. The applied moment M_y about the *y*-axis is first transformed to the chosen units:

$$M_y = 15 \text{ kip} \cdot \text{ft} = 15 \times 10^3 \times 12 \text{ lb} \cdot \text{in} = 1.8 \times 10^5 \text{ lb} \cdot \text{in}$$

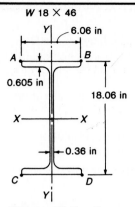

Figure 10.7 *Cross section of beam in Example 10.2.*

Using the tables of Appendix A again, the section modulus with respect to the y-axis, Z_y, is found to be 7.43 in³. From Equation 10.7 the value of the maximum stress on the section due to the moment of 15 kip feet may be found:

$$\sigma_{max} = \frac{1.8 \times 10^5}{7.43} = 2.42 \times 10^4 \, \text{psi} = 24.2 \, \text{ksi}$$

This maximum stress is the value of stress at the ends of the flanges—i.e., at points A, B, C, and D on the section—since these points are at the largest possible distances from the y-axis.

This example illustrates an important property of wide-flange sections. It is noted that the applied moment M_x is much larger than the applied moment M_y, and yet the maximum stress on the section due to M_x is much smaller than the maximum stress due to the moment M_y. Wide-flange members are very efficient in bending about their strong axis (the x-axis shown in the diagrams of Appendix A) but very inefficient in bending about their weak axis (the y-axis). Thus they need to be oriented so that the bending occurs about their strong axis wherever possible.

Example 10.3

A structural tee WT 18 × 150 is to be used as a beam. The beam is made of mild steel with a maximum allowable stress in bending of 22 kips per square inch.

a. Find the maximum moment that may be applied to the section about its x-axis without exceeding the maximum allowable bending stress.

b. Find the maximum moment that may be applied to the section about its y-axis without exceeding the maximum allowable bending stress.
    In both cases indicate on the section where the bending stress will be greatest.

Solution:

Part a. Since the maximum stress is now known, Equation 10.7 is now rearranged to give the maximum moment as the result. For the maximum moment about the x-axis, $M_{x\,max}$, the equation has the form:

$$M_{x\,max} = \sigma_{max} \cdot Z_x$$

From Appendix A the section modulus with respect to the x-axis, Z_x, for a WT 18 × 150 section is found to be 86.1 in³.

$$\begin{aligned} M_{x\,max} &= (22 \times 10^3) \cdot 86.1 \\ &= 1.89 \times 10^6 \, \text{lb} \cdot \text{in} \\ &= 1.89 \times 10^3 \, \text{kip} \cdot \text{in} \\ &= 158 \, \text{kip} \cdot \text{ft} \end{aligned}$$

The determination of where the maximum stress exists is often important. The maximum stress is at points that are the largest possible distance from the x-axis, and this may be at the top or at the bottom of the section, or both. In Appendix A the location of the x-axis is specified for this section as being a distance y below the top of the section, where y is 4.13 inches. The top of the section is therefore 4.13 inches from the x-axis. The bottom of the stem of the tee section is a distance $(d - y)$ from the x-axis. Since the table gives the distance d as 18.37 inches, the distance $(d - y)$ is given by

$$d - y = 18.37 - 4.13 = 14.24 \, \text{in}$$

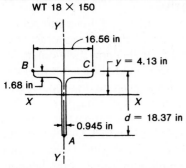

Figure 10.8 *Cross section of beam in Example 10.3.*

The bottom of the stem of the section is therefore farther from the x-axis of the section than is the top of the section, and the maximum stress due to the moment $M_{x\,max}$ will occur at the bottom of the stem of the tee—i.e., at the point A on the diagram of Figure 10.8.

Part b. An equation similar to that used above gives the largest permissible moment that may be applied to the section about the y-axis, $M_{y\,max}$, without exceeding the maximum allowable stress on the section:

$$M_{y\,max} = \sigma_{max} \cdot Z_y$$

From Appendix A the section modulus of the tee with respect to the y-axis is found to be 77.8 in³. Thus the largest permissible moment that may be applied is:

$$
\begin{aligned}
M_{y\,max} &= (22 \times 10^3) \cdot 77.8 \\
&= 1.71 \times 10^6 \, \text{lb} \cdot \text{in} \\
&= 1.71 \times 10^3 \, \text{kip} \cdot \text{in} \\
&= 143 \, \text{kip} \cdot \text{ft}
\end{aligned}
$$

In this instance it is no problem to find where the maximum stress will occur, since the ends of the flanges are the points farthest from the y-axis. Thus the maximum bending stress will occur at the points B and C on Figure 10.8. If the applied moment is 143 kip feet, then the value of the maximum stress will be 22 kips per square inch.

10.2 Critical Locations on Beams

The relationship between bending moment and section modulus or moment of inertia has just been discussed. However, in many cases the designer is faced rather with a situation in which he does not immediately know the moment for which the beam must be designed. Instead, the beam might be given with some known loads acting on it. From this the shear force diagram and the bending moment diagram may be drawn, using the principles of Chapters 5 and 6, respectively. It now remains to choose critical locations on the beam at which the stresses are likely to be the greatest. The stresses are the largest where the bending moment is the largest, and this is the indication of critical locations for beam design. It is likely that the maximum or minimum values of the bending moment diagram will be critical locations at which the bending stresses should be examined.

Example 10.4

The beam shown in Figure 10.9a is to be built using a structural tee section WT 18 × 150, oriented as shown in the accompanying section diagram. Find the maximum tensile and compressive stress in the beam and indicate where they occur.

Solution:

The first step in finding the maximum stresses is to determine the location or locations on the beam at which the moment is critical. This involves the equations of equilibrium to find the unknown support reactions, as described in Chapter 3. The shear force and bending moment diagrams may then be drawn, as discussed in Chapters 5 and 6. In this example the calculations are omitted, but the resulting shear force and bending moment diagrams are given in Figure 10.9b and c respectively. The deflection diagram shown in Figure 10.9d is derived by intuition only, and the magnitudes of the deflection at any point on the beam are unknown. However, the diagram has the form shown, indicating that the top of the beam will have fibers that are shortening along the whole beam length, whereas the bottom fibers will be lengthening. The top of the beam will thus be in compression and the bottom will be in tension, as indicated by the T and C on the deflection diagram. The moment diagram also helps in this determination, since the sign convention used always plots the diagram on the compression side of the beam.

The bending moment diagram is next examined, and it is found that the maximum moment in the beam has a value of 102 kip feet and it occurs at the point where the 20-kip load is applied. The maximum stresses will therefore occur at this location. Appendix A is now used to find the critical properties of the section so that Equation 10.7 may be used to find the maximum tensile and compressive stresses. The section is bending about its strong axis and so the moment of inertia required is that about the x-axis, i.e., $I_x = 1,230$ in^4.

Another problem arises from the necessity of knowing whether the maximum stresses due to bending are tensile stresses or compressive stresses. For unsymmetrical sections, the maximum tensile stress at a section may not be the same as the maximum compressive stress. To find the maximum stresses—and in deciding whether those stresses are tensile or compressive in nature—it is often convenient to draw a deflection diagram. (The subject of beam deflections is covered more thoroughly in Chapter 13, but some understanding of the shape of the deflected beam can assist the calculation of bending stresses.) The deflection diagram shows whether the top or bottom of the section will have the tensile stress in it. As Figure 10.3 showed, the fibers that lengthened due to the "sag" of the beam (the fiber AD') had a tensile stress in them, whereas the fibers that shortened due to the "sag" had compression stress in them (e.g., line BC'). Critical bending stresses may occur at points where stress concentration is present. The phenomenon of stress concentration in bending is similar to that in axial load, described in Section 8.10.

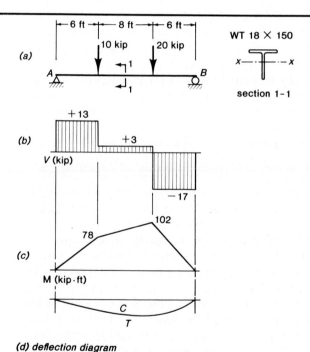

(d) deflection diagram

Figure 10.9 Beam of Example 10.4.

Example 10.4 *(Continued)*

Furthermore the tables give the location of the neutral axis as being a distance (y) of 4.13 inches below the top of the section, which has a total depth (d) of 18.37 inches. Since the maximum compressive stress will occur on the top of the section, its maximum value—at the top of the tee (all along the top of the flange) at the point of application of the 20-kip load—is

$$\sigma_{max\,comp} = \frac{M \cdot y}{I} = \frac{(102 \times 12) \times 4.13}{1,230} = 4.09 \text{ ksi}$$

Example 10.5

An overhanging beam is loaded as shown in Figure 10.10a. The beam is constructed by joining together two pieces of wood, each 150 by 50 millimeters in cross section, to form the tee section shown in Figure 10.10e. Find the maximum tensile and compressive stresses and the locations at which they occur.

Solution:
The analysis is commenced by drawing the shear force and bending moment diagrams as discussed in Chapters 5 and 6. These diagrams for this particular beam, with the loading condition given, are shown without the supporting calculations in Figures 10.10b and c respectively. In Figure 10.10c it is noted that the bending moment has two turning points—at B and C—and the stress will need to be examined at each of these locations. The deflection diagram of Figure 10.10d is now drawn. Again, the values of the deflection at particular points on the beam are not known, although it is known that the vertical deflection of the beam must be zero at the supports at B and D. The drawing of this diagram enables the designer to conclude that there will be tensile stress on the top of the beam at the support B and compression on the bottom of the beam at this point. However, at the point C—the other critical point noted on the bending moment diagram—the tensile stress will be on the bottom of the beam and the compressive stress on the top.

Before the stresses can be calculated, the location of the neutral axis and the moment of inertia of the section must be found. The method of finding these is discussed in Chapter 7 and will not be treated in detail here. However, the section may be divided into two rectangular pieces as shown in Figure 10.10e and a reference axis considered at the bottom of the section, as shown. Tables 10.1 and 10.2 are then used to find the required section properties.

In a similar diagram, the maximum tension stress must also occur at the point of maximum moment—i.e., at the point of application of the 20-kip load—but now it is on the bottom of the section (at the base of the web). The distance of this point from the neutral axis is given by

$$d - y = 18.37 - 4.13 = 14.24 \text{ in}$$

Thus the maximum tensile stress is

$$\sigma_{\text{max tens}} = \frac{M \cdot y}{I} = \frac{(102 \times 12) \times 14.24}{1,230} = 14.2 \text{ ksi}$$

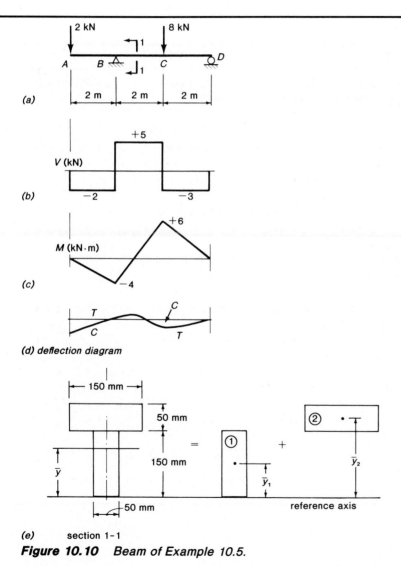

(a)

(b)

(c)

(d) deflection diagram

(e)   section 1-1

Figure 10.10   Beam of Example 10.5.

Example 10.5 *(Continued)*

Table 10.1

Section (i)	A_i(mm²)		$\bar{y}_i$(mm)	$A_i \cdot \bar{y}_i$ (mm³)
1	$150 \times 50 =$	7,500	75	562,500
2	$150 \times 50 =$	7,500	175	1,312,500
Total		15,000		1,875,000

$$\bar{y} = \frac{\Sigma A_i \cdot \bar{y}_i}{\Sigma A_i} = \frac{1,875,000}{15,000} = 125 \text{ mm}$$

Table 10.2

Section (i)	A_i (mm²)	$\bar{I}_{yi}$ (mm⁴)	$A_i \cdot (\bar{y} - \bar{y}_i)^2$ (mm⁴)	$\bar{I}_{yi} + A_i(\bar{y} - \bar{y}_i)^2$ (mm⁴)
1	7,500	$\dfrac{50 \times 150^3}{12}$ $= 1.406 \times 10^7$	$7,500 \cdot (125-75)^2$ $= 1.875 \times 10^7$	3.281×10^7
2	7,500	$\dfrac{150 \times 50^3}{12}$ $= 1.56 \times 10^6$	$7,500 \cdot (125-175)^2$ $= 1.875 \times 10^7$	2.031×10^7
Total				5.312×10^7

$$\bar{I}_y = \Sigma [\bar{I}_{yi} + A(\bar{y} - \bar{y}_i)^2]$$
$$= 5.312 \times 10^7 \text{ mm}^4$$
$$= 5.312 \times 10^{-5} \text{ m}^4$$

From the section properties thus determined, the stresses at any point on the section may now be calculated.

Consider first the point B on the beam, where the bending moment was found to be -4 kilonewton meters, which means there is tensile stress on the top and compression on the bottom at this location. The maximum values of the stresses are calculated using Equation 10.7, as follows:

$$\sigma_{\text{max tens}} = \frac{M \cdot y_{\text{max}}}{I} = \frac{(4 \times 10^3)(75 \times 10^{-3})}{5.312 \times 10^{-5}}$$
$$= 56.5 \times 10^5 \text{ Pa}$$
$$= 5.65 \text{ MPa}$$

$$\sigma_{\text{max comp}} = \frac{M \cdot y_{\text{max}}}{I} = \frac{(4 \times 10^3)(125 \times 10^{-3})}{5.312 \times 10^{-5}}$$
$$= 94.1 \times 10^5 \text{ Pa}$$
$$= 9.41 \text{ MPa}$$

Similar calculations are now carried out for the point C on the beam. Here, however, the bending moment is 6 kilonewton meters, and the maximum compressive stress is on the top of the beam, while the maximum tension is on the bottom:

$$\sigma_{\text{max tens}} = \frac{M \cdot y_{\text{max}}}{I} = \frac{(6 \times 10^3)(125 \times 10^{-3})}{5.312 \times 10^{-5}}$$
$$= 141.2 \times 10^5 \text{ Pa}$$
$$= 14.12 \text{ MPa}$$

$$\sigma_{\text{max comp}} = \frac{M \cdot y_{\text{max}}}{I} = \frac{(6 \times 10^3)(75 \times 10^{-3})}{5.312 \times 10^{-5}}$$
$$= 84.4 \times 10^5 \text{ Pa}$$
$$= 8.44 \text{ MPa}$$

Thus the maximum tensile stress in the beam is 14.12 megapascals, and it occurs on the bottom of the section at location C, the point of application of the 8-kilonewton force. The maximum compressive stress, however, is 9.41 megapascals, and it occurs on the bottom of the section at the support B of the beam.

Example 10.6

A beam loaded as shown in Figure 10.11a is to be built using a wide-flange section. The allowable stress in bending of the steel beam is 22 kips per square inch. Use the tables of Appendix A to select a suitable beam.

Solution:

The beam is first analyzed and the shear force and bending moment diagrams drawn. These diagrams are shown in Figures 10.11b and c, respectively. The maximum moment in the beam occurs at the point of application of the 12-kip load, and the maximum moment is 60 kip feet. Equation 10.7 is now rearranged to give the section modulus of the section required (Z_{req}) to sustain the given loading condition without exceeding the allowable bending stress. (Calculations are in kips and inch units.)

$$Z_{req} = \frac{M_{max}}{\sigma_{all}} = \frac{60 \times 12}{22} = 22.7 \text{ in}^3$$

Refer to Appendix A to find a section that has at least the required section modulus. The most economical section is the one that has a section modulus greater than or equal to the required modulus and is the lightest that meets the requirement. A search of Appendix A shows that these criteria are met by the wide-flange section W 12 × 22, which has a section modulus $Z_x = 25.4 \text{ in}^3$. This section is acceptable provided it is oriented such that it is bending about its strong axis.

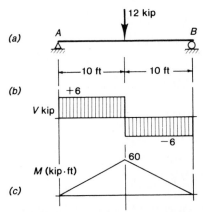

Figure 10.11 *Beam of Example 10.6.*

10.3 Reinforced Concrete Beams

It is common for buildings to be constructed from reinforced concrete. The members in such structures have cross sections that are essentially all concrete with a small proportion of the area taken up by steel bars. The problem in analyzing such beams is that the cross section of the beam contains different materials, which behave very differently from each other. The procedure to be considered here is applicable to sections made up of any two materials, although the reinforced concrete beam is the most common such structure.

Before analyzing a reinforced concrete beam, the behavior of both materials in the beam must be understood. Typical stress-strain diagrams for steel and concrete are shown in Figures 10.12a and b respectively. The dotted line in each diagram is the form of the true stress-strain diagram for each material, and the solid line is a simplified diagram ususally assumed for convenience of calculation. The stress-strain diagram for the steel is very similar to that shown in Figure 8.7, except that the strain-hardening portion has been omitted. The diagram has also been approximated as two straight lines in each of the tension and compression regions of material behavior. In each region the sloping straight-line portion represents the elastic region, while the horizontal portion represents its plastic region. Since the line assumed is straight and horizontal in this region, the material is said to be exhibiting *perfectly plastic* behavior. Note also that the behavior of steel may be assumed to be identical under both tensile and compressive stresses, as is shown by the fact that the diagram has point symmetry about the origin O.

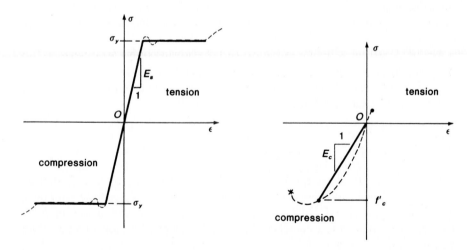

(a) *stress-strain diagram for steel*   (b) *stress-strain diagram for concrete*

Figure 10.12   *Stress-strain diagrams for steel and concrete.*

The modulus of elasticity of steel in the elastic range is given the symbol E_s to distinguish it from the modulus of elasticity of concrete E_c. As Figure 10.12b shows, however, the behavior of concrete is vastly different from that of steel. The most obvious difference is that the concrete has appreciable strength only under compressive stresses. In fact, concrete can sustain small tensile stresses, but this is usually ignored in engineering calculations. The concrete is thus assumed to be effective only while acted upon by compressive stresses. The diagram also shows that concrete is a brittle material; there is no significant portion of the diagram where the material may be said to be performing plastically. Indeed concrete has no readily definable yield stress: therefore some method, such as the 0.2 percent offset method, must be used to find an appropriate strength, labeled f_c'.

In analyzing reinforced beams, it is important to consider the different material behaviors involved, since Equation 10.5 implicitly assumes that the cross section is all of one material. It is usual, therefore, to convert one of the materials to an equivalent amount of the other material. It is usually easier to convert the steel area into an equivalent area of concrete. The transformed area should perform in a manner identical to the real section. Consider the beam having a cross section shown in Figure 10.13a subject to a bending moment M that places the top of the section in compression. The area of the steel reinforcing has a total value of A_y, and a neutral axis is assumed at a height of $\bar{y}$ above the level of the steel although at this point the distance $\bar{y}$ is unknown. Any material below the neutral axis is in tension and above the axis is in compression. Since it is assumed that concrete can support no tensile stresses, the material below the axis contributes nothing to the strength of the beam and so may be ignored. (This concrete does serve the essential purpose of keeping the section together and preventing the steel from falling away from the concrete.) With this in mind, a modified section is drawn in Figure 10.13b showing only those parts of the section that contribute to its load-carrying capacity.

The steel in the section is now transformed to an equivalent area of concrete. The transformed steel area, A_{st}, is given by

(10.8)
$$A_{st} = A_s \cdot \frac{E_s}{E_c}$$

The fraction E_s/E_c is the ratio of the modulus of elasticity of steel to that of concrete. This value is often called the *modular ratio* and is given the symbol n. The modulus of elasticity of steel is always much greater than that of concrete; the value of n is usually somewhere between 6 and 10. Equation 10.8 may thus be rewritten as

(10.9)
$$A_{st} = A_s \cdot n$$

where

(10.10)
$$n = \frac{E_s}{E_c}$$

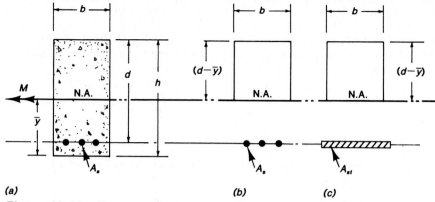

(a) **(b)** **(c)**

Figure 10.13 Transformation of a reinforced concrete beam to a beam of one material (N.A. = neutral axis).

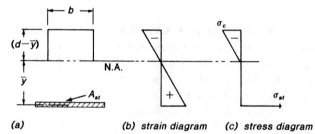

(a) **(b) strain diagram** **(c) stress diagram**

Figure 10.14 Strains and stresses at a section in a reinforced concrete beam (N.A. = neutral axis).

With the section now considered to be of one material the position of the neutral axis and the moment of inertia of the section may be found using the principles discussed in Chapter 7. A linear strain distribution is now assumed across the depth of the section as the beam bends. This distribution is shown in Figure 10.14b and is the same as that assumed in the bending stress distribution of Figure 10.3. As the diagram shows, the strain is negative at the top of the section and positive at the bottom. Thus when the strain is converted to stress using Equation 10.4, the compressive stress is greatest at the very top of the section. The tensile stress is, however, concentrated at the location of the steel reinforcing, as the concrete is unable to carry any tensile stress. The resulting stress diagram is shown in Figure 10.14c. To get this diagram, the strain diagram was multiplied by the modulus of elasticity of concrete, E_c. The stress σ_c is the maximum stress in the concrete, and this is usually one of the critical stress values on the section. It is often necessary to find the stress in the steel reinforcing. The stress σ_{st} is not the true steel stress but the stress in the transformed steel area, A_{st}. To find the true steel stress the value of σ_{st} must again be multiplied by the modular ratio, n, that is

(10.11)
$$\sigma_s = \sigma_{st} \cdot n$$

Example 10.7

A reinforced concrete beam having the cross section shown in Figure 10.15a is subject to a bending moment of 50 kilonewton meters. The steel reinforcing consists of four bars each of 12-millimeter diameter. The steel has a modulus of elasticity of 200 gigapascals and a yield stress of 250 megapascals. The concrete has a modulus of elasticity of 20 gigapascals and crushing strength, f_c', of 20 megapascals.

a. Find the maximum stress in the steel and in the concrete.

b. Are the stresses calculated in the elastic range of material behavior or is the analysis conducted not valid?

Solution:
The total area of steel in the section is required before any transformation may be made. Thus

$$A_s = 4 \left(\frac{\pi\, d^2}{4} \right) = 4 \left(\frac{\pi \cdot 12^2}{4} \right) = 453 \text{ mm}^2$$

The transformed steel area is calculated by multiplying the true steel area by the modular ratio, i.e.,

$$A_{st} = A_s \cdot n = 453 \left(\frac{200}{20} \right) = 4{,}530 \text{ mm}^2$$

Using the resulting concrete section shown in Figure 10.15b, the location of the neutral axis and the moment of inertia of the section may now be found using the principles of Chapter 7. The tabular calculations are shown in Table 10.3.

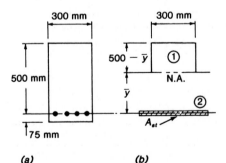

(a) (b)

Figure 10.15 Reinforced concrete beam of Example 10.7 (N.A. = neutral axis).

Table 10.3

Section (i)	A_i(mm²)	$\bar{y}_i$(mm)	$A_i \cdot \bar{y}_i$ (mm³)
1	$300(500 - \bar{y})$	$\bar{y} + \left(\dfrac{500 - \bar{y}}{2}\right)$	$150(500 - \bar{y})(500 + \bar{y})$
2	4,530	0	0
Total	$154{,}530 - 300\,\bar{y}$		$150(250{,}000 - \bar{y}^2)$

$$\bar{y} = \frac{\Sigma\, A_i \cdot \bar{y}_i}{\Sigma\, A_i} = \frac{150(250{,}000 - \bar{y}^2)}{154{,}530 - 300\bar{y}}$$

thus,

$$\bar{y} = 391.3 \text{ mm or } 638.9 \text{ mm}$$

Of these two solutions to the quadratic equation in $\bar{y}$, only the smaller answer is of any use as the larger solution represents a neutral axis above the top of the beam, which is impossible. Thus the location of the neutral axis is given by $\bar{y} = 391.3$ mm $= 0.391$ m. The depth of the concrete compression block is given by $500 - \bar{y} = 108.7$ mm $= 0.1087$ m. The moment of inertia is found from Table 10.4.

Table 10.4

Section (i)	A_i(mm²)	$\bar{I}_{yi}$(mm⁴)	$A_i(\bar{y}-\bar{y}_i)^2$ (mm⁴)	$\bar{I}_{yi}+A_i(\bar{y}-\bar{y}_i)^2$ (mm⁴)
1	300×108.7 $= 32{,}610$	$\dfrac{300 \times 108.7^3}{12}$ $= 3.211 \times 10^7$	$32{,}610\left(\dfrac{108.7}{2}\right)^2$ $= 9.633 \times 10^7$	1.284×10^8
2	4,530	0	$4{,}530 \times 391.3^2$ $= 6.936 \times 10^8$	6.936×10^8
Total				$8.22 \times 10^8\,$mm⁴

Example 10.7 (Continued)

$$\bar{I}_y = 8.22 \times 10^8 \, \text{mm}^4$$
$$= 8.22 \times 10^{-4} \, \text{m}^4$$

Part a. The stresses on the section may now be found using Equation 10.5. The maximum stress in the concrete (a compressive stress), σ_c, is given by

$$\sigma_c = \frac{M \cdot y}{I} = \frac{(50 \times 10^3) \times (0.1087)}{8.22 \times 10^{-4}} = 6.60 \times 10^6 \, \text{Pa} = 6.6 \, \text{MPa}$$

In a similar manner, the maximum stress in the transformed steel section is given by

$$\sigma_{st} = \frac{My}{I} = \frac{(50 \times 10^3) \times (0.3913)}{8.22 \times 10^{-4}} = 2.38 \times 10^7 \, \text{Pa} = 23.8 \, \text{MPa}$$

Example 10.8

The concrete beam shown in Figure 10.16*a* is reinforced by four steel bars, each having an area of 1.0 square inch. The modulus of elasticity of the steel is 30,000 kips per square inch and that of the concrete is 3,000 kips per square inch. If the allowable stress in compression of the concrete is 1,100 pounds (1.10 kips) per square inch and the allowable stress in tension of the steel is 18.0 kips per square inch, find the maximum moment that may be applied to the section.

Solution:
The total area of steel in the section is $A_s = 4 \cdot 1.0 = 4.0 \, \text{in}^2$. Thus the transformed steel area is given by

$$A_{st} = A_s \cdot n = 4.0 \left(\frac{30,000}{3,000} \right) = 40.0 \, \text{in}^2$$

The transformed section is shown in Figure 10.16*b*. Using this diagram the centroid of the section may be found by means of Table 10.5.

To obtain the true steel stress, the value of σ_{st} will need to be multiplied by the modular ratio; i.e.,

$$\sigma_s = \sigma_{st} \cdot n = 23.8 \times \left(\frac{200}{20} \right) = 238 \text{ MPa}$$

Part b. Since Equation 10.5 was based on the assumption that the material was performing elastically, it is important to determine whether this was indeed the case. The maximum compressive stress in the concrete was calculated to be 6.6 megapascals, while the specified crushing strength of the concrete is 20 megapascals, and thus the assumption of elastic material behavior is satisfied by the concrete. The maximum stress in the steel is found to be 238 megapascals, while the specified yield strength of the material is 250 megapascals. The assumption of elastic behavior of the steel is thus valid, but the steel is very close to its yield point. Since both materials are still behaving elastically, the analysis conducted is valid.

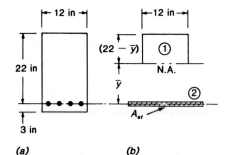

(a)   (b)

Figure 10.16 *Reinforced concrete beam of Example 10.8 (N.A. = neutral axis).*

Table 10.5

Section (i)	A_i(in²)	$\bar{y}_i$(in)	$A_i \cdot \bar{y}_i$ (in³)
1	$12(22 - \bar{y})$	$\bar{y} + \left(\dfrac{22 - \bar{y}}{2} \right)$	$6(22 - \bar{y})(22 + \bar{y})$
2	40	0	0
Total	$304 - 12\,\bar{y}$		$6(484 - \bar{y}^2)$

Example 10.8 (Continued)

Now

$$\bar{y} = \frac{\Sigma\, A_i \cdot \bar{y}_i}{\Sigma\, A_i} = \frac{6(484 - \bar{y}^2)}{304 - 12\bar{y}}$$

Solving this quadratic in $\bar{y}$ gives

$$\bar{y} = 12.77 \text{ in or } 37.89 \text{ in}$$

The solution $\bar{y} = 37.89$ inches is impossible, since it would place the neutral axis above the top of the section. Therefore,

$$\bar{y} = 12.77 \text{ in}$$

The depth of the concrete stress block is then $22 - \bar{y} = 9.23$ inches. The moment of inertia of the section is then found from Table 10.6.

Table 10.6

Section (*i*)	A_i(in²)	$\bar{I}_{yi}$	$A_i(\bar{y}_i - \bar{y})^2$	$\bar{I}_{yi} + A_i(\bar{y}_i - \bar{y})^2$
1	12×9.23	$\dfrac{12 \times 9.23^3}{12}$	$110.8 \times \left(\dfrac{9.23}{2}\right)^2$	3,146.1
	= 110.8	= 786.3	= 2,359.8	
2	40	0	$40 \times (12.77)^2$ = 6,522.9	6,522.9
Total				9,669 in⁴

$$\bar{I}_y = 9{,}669 \text{ in}^4$$

At this stage the value of the maximum moment that may be applied to the section without exceeding the given allowable stresses is not known, but this moment may be taken as the unknown quantity in Equation 10.7.

First find the moment M_c that would cause the maximum stress in the concrete portion of the beam to equal the maximum allowable stress. This maximum stress occurs at the top of the beam.

$$\sigma_c = \frac{M_c \cdot y}{I}$$

i.e.,

$$1.10 = \frac{M_c \cdot 9.23}{9{,}669}$$

Thus

$$M_c = 1{,}152 \text{ kip} \cdot \text{in}$$

To find the moment M_s that would cause the maximum permissible stress in the steel reinforcing bars, the transformed steel stress must be used as an intermediary. Thus

$$\sigma_{st} = \frac{M_s \cdot y}{I} = \frac{M_s \times 12.77}{9{,}669} \text{ ksi}$$

But

$$\sigma_s = \sigma_{st} \cdot n$$

i.e.,

$$18.0 = \frac{M_s \times 12.77}{9{,}669} \times \left(\frac{30{,}000}{3{,}000} \right)$$

Solving this gives

$$M_s = 1{,}363 \text{ kip} \cdot \text{in}$$

Thus the maximum moment that may be applied to the section without exceeding the allowable stresses is 1,152 kip inches. At this moment the maximum stress in the concrete is 1,100 pounds per square inch, and it occurs at the top of the section.

10.1–

10.10 For the given beams with their different loadings shown in Figures P 10.1 to P 10.10, determine the critical bending moments and their locations.

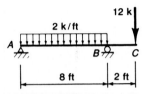

Figure P 10.1

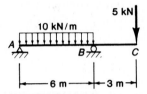

Figure P 10.2

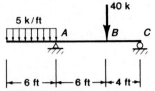

Figure P 10.3

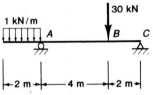

Figure P 10.4

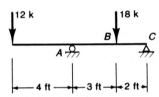

Figure P 10.5

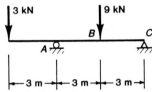

Figure P 10.6

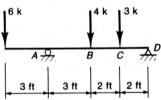

Figure P 10.7

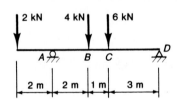

Figure P 10.8

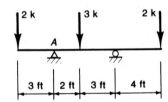

Figure P 10.9

10.11 A 4- by 6-inch wooden cantilever beam carries a 10-kip concentrated
       force at the end, as shown in Figure P 10.11. Determine the
       maximum bending stresses at points *A* and *B*. Neglect the weight of
       the beam.

10.12 From Figure P 10.12, determine the maximum bending stresses at
       points *A* and *B* for the W 18 × 50 steel beam with the 5-kilonewton
       concentrated force. Neglect the weight of the beam.

10.13 A simply supported W 16 × 36 steel beam is shown in Figure
       P 10.13. Find the maximum bending stress at point *A* caused by the
       weight of the beam only.

10.14 Determine the maximum bending stress at point *A* for the
       W 24 × 100 steel beam with the given load as shown in
       Figure P 10.14. Neglect the weight of the beam.

10.15 Determine the maximum bending stresses at points *A* and *B* for the
       W 27 × 94 steel beam shown in Figure P 10.15. Neglect the weight.

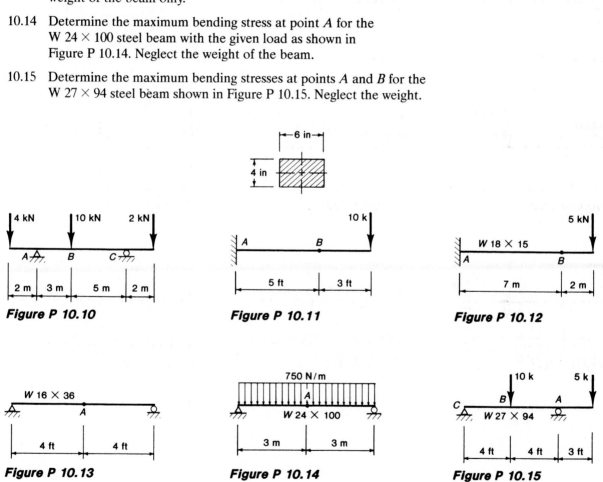

Figure P 10.10

Figure P 10.11

Figure P 10.12

Figure P 10.13

Figure P 10.14

Figure P 10.15

10.16 Determine the maximum bending stresses at points A and B for the wooden beam with the triangular cross section shown in Figure P 10.16. Neglect the weight of the beam.

10.17–
10.21 For the steel beams shown in Figures P 10.17 to P 10.21, determine the maximum bending stress and its location respectively. Neglect the weight of the beam. ($E = 30 \times 10^6$ psi or 207 GPa.)

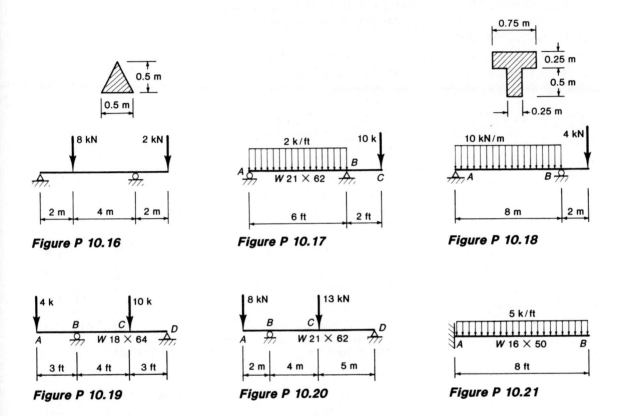

Figure P 10.16

Figure P 10.17

Figure P 10.18

Figure P 10.19

Figure P 10.20

Figure P 10.21

10.22 Determine the maximum bending stress and its location for the cylindrical beam shown in Figure P 10.22. Neglect the weight of the beam.

10.23 The reinforced concrete beam shown in Figure P 10.23 is subjected to a positive bending moment of 12,000 pound-feet. Determine the maximum tensile stress in steel. Assume $n = 16$.

10.24 Determine the moment-carrying capacity of a reinforced concrete beam with 160 by 220 millimeter rectangular section and two 100 square millimeter steel bars 20 millimeters from the bottom, as shown in Figure P 10.24. Assume:

Allowable stresses in concrete: Compression = 20 MPa
                                    Tension   = 0 Pa
Allowable stresses in steel: Compression = 150 MPa
                                      Tension   = 150 MPa
Young's modulus: E concrete   = 20 GPa
                                      E steel   = 207 GPa

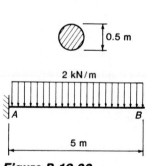

Figure P 10.22

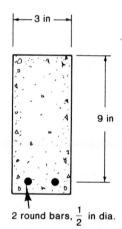

2 round bars, $\frac{1}{2}$ in dia.

Figure P 10.23

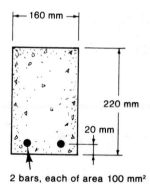

2 bars, each of area 100 mm²

Figure P 10.24

10.25 The cross section of a reinforced concrete beam is given in Figure P 10.25. Determine the moment capacity. Assume $n = 12$.

Allowable stresses in concrete: Compression = 1,000 psi
Tension   = 0 psi

Allowable stresses in steel: Compression = 20,000 psi
Tension   = 20,000 psi

10.26 Determine the moment-carrying capacity of a reinforced concrete beam with the section shown in Figure P 10.26. Assume $n = 12$.

Allowable stresses in concrete: Compression = 15 MPa
Tension   = 0 Pa

Allowable stresses in steel: Compression = 140 MPa
Tension   = 140 MPa

10.27 Determine the moment capacity for the beam with a T-section shown in Figure P 10.27. Assume $n = 10$. The allowable stresses for concrete and steel are 1,600 and 20,000 pounds per square inch, respectively.

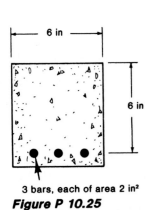

3 bars, each of area 2 in²

Figure P 10.25

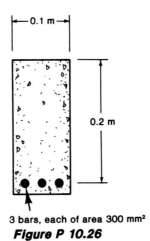

3 bars, each of area 300 mm²

Figure P 10.26

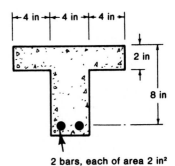

2 bars, each of area 2 in²

Figure P 10.27

10.28 Determine the moment-carrying capacity for the beam with a
       T-section shown in Figure P 10.28. Assume $n = 12$. The allowable
       stresses for the concrete and steel are 14 and 140 megapascals,
       respectively.

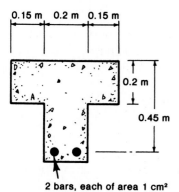

2 bars, each of area 1 cm²

Figure P 10.28

Shear in Bending

11

1. *To understand the concept of shear stresses in structural and machine elements subjected to bending.*

2. *To learn how to compute the distribution of shear stresses acting in the axial direction and over the cross sections of beams.*

Objectives

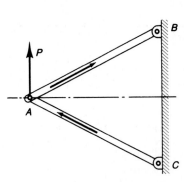

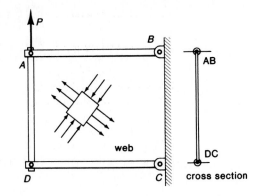

(a) V frame under shear force P

(b) rectangular rigid hinged frame with an elastic web

Figure 11.1 Transfer of shear force in hinged frames with or without a web.

11.1 The Transfer of Shear in a Beam

The magnitude of internal shear forces acting in the cross sections of beams perpendicular to their axis can be determined according to the techniques introduced in Chapter 5. The concept of shear stresses was introduced in Chapter 9. While it is true that the shear force divided by the area on which it acts results in a shear stress, it must be understood that this results in the value of the *average shear* only. As we will see later on in this chapter, shear due to bending forces is almost never distributed uniformly across the cross section of the beam. For the purposes of design and analysis of mechanical and structural components, one needs to know the *distribution of shear stresses* over the cross section. The development of the method of calculating such shear stress distributions is the primary aim of this chapter. It is based on two major concepts already introduced: the bending stress formula (Equation 10.5) and the basic conclusions of Section 9.1 concerning the nature of shear.

To gain understanding of the transfer of shear forces across a beam subjected to bending moments, let us first consider the forces created in a V-frame by the action of a force P, as shown in Figure 11.1a. This transverse force will result in a compressive force in member AB and a tensile force in member AC. Generally, this example suggests that transverse forces are transferred in bending by forces that are acting in the axial direction. As the next step in this analysis, consider an alternate structure consisting of a rectangular frame—hinged at points A, B, C, and D—surrounding an elastic plate called the web, shown in Figure 11.1b. Because of the hinges the frame itself could not resist the force P without the web filling the space between its members. In other words, the transverse force is transferred to the wall through the web. This transfer is accomplished through compressive and tensile stresses as shown in one particu-

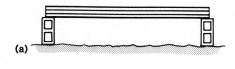

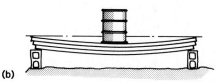

Figure 11.2 *An intuitive approach to understand the existence of axially oriented shear forces in a beam under bending.*

lar rectangular element shown in a typical location within the web. The stress distribution on this rectangular element would result in an elongation in one direction and a compression in a perpendicular direction. Replacing the rectangular element with a circular one would show that the tensile and compressive stresses would distort the circle into an ellipse. On Figure 9.4 this loading condition was already considered. There it was found that the deformation can also be considered to be caused by shear stresses acting in directions diagonal to the normal stresses. In the example shown in Figure 11.1*b*, these shear stresses act in directions parallel and perpendicular to the force P, as long as the frame is considered rigid and the web is connected firmly to it all along its edges. The shear thus generated would be pure shear. The compressive and tensile stresses acting on the rectangular element shown would then be principal stresses, acting at 45° to the direction of P. Thus the internal shear forces that were studied in Chapter 5 create shear stresses not only in the cross sections perpendicular to the axis of the beam but also in planes parallel to the axis of the beam.

An identical conclusion may be reached intuitively by considering the deformation of a stack of unconnected boards placed across two blocks, as shown in Figure 11.2. Without any loading on it the ends of the boards are aligned, as shown in Figure 11.2*a*. When a barrel is placed in the middle of the span, as shown in Figure 11.2*b*, the boards will bend. The ends of the boards now will move apart. The differential movement means that the boards are sliding relative to one another, and this is, in effect, a shearing movement. If nails are driven through the board before the barrel is placed on it, the sliding could not occur. The nails then would be subject to shearing forces in the planes between the boards. Therefore, the transverse force (the weight of the barrel) generates shear forces parallel to the axis of the beam.

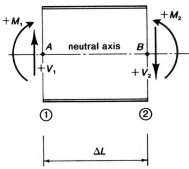

Figure 11.3 *Static equilibrium of moments on a free body of a beam between two cuts at an elementary ΔL distance apart.*

11.2 Derivation of the Bending Shear Formula

As stated earlier, the internal shear force in a beam subjected to bending shall be derived with the aid of the formula for bending stresses due to bending moments. Up to this point we have been referring to the internal shear force V as the cause of the axially oriented shear forces in the beam. Now we will find a functional relationship between bending moments and the internal shear force. In Figure 6.8 such a relationship was already established for a number of specific cases. To obtain a generalized relationship we will consider the static equilibrium of moments on a small free body of a beam subjected on both ends to bending moments and shear forces. To do this consider the small element of a beam shown in Figure 11.3. Writing the formula for the equilibrium of moments we have

(11.1) $$\Sigma M \, \zeta^+ = M_1 - M_2 + V_2 \, (\Delta L) = 0$$

Rearranging to express the shear at point B, one obtains

(11.2) $$V_2 = \frac{M_2 - M_1}{\Delta L}$$

where the distance ΔL is a very small quantity. If the external loading over ΔL is negligible, then $V_2 = V_1 = V$.

The fundamental conclusion one may draw from Equation 11.2 is that there is no internal shear force in the beam when M_1 equals M_2, that is, when there is no change in the bending moment along the beam. Conversely, the formula states that *the rate of change of the bending moment* over the distance ΔL equals the *shear force V.*

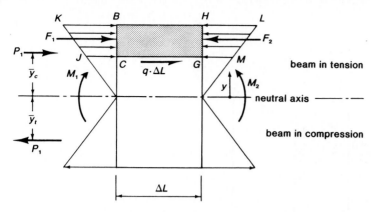

Figure 11.4 *Equilibrium of forces on the portion of the beam above the CG axial cut, subjected to bending stresses at the ends and an internal shear force on the CG plane.*

Next divide the free body of length ΔL by cutting it along an arbitrarily selected plane parallel to the axis of the beam, as shown in Figure 11.4. The horizontal cut from point C to G will result in a new free body, shown by the shaded area $BCGH$. This free body is subjected to three separate forces: F_1, the resultant of the bending stresses on the left, indicated by the stress distribution shown by the trapezoid $KBCJ$; F_2, the corresponding bending force from the right, which is the resultant of the stresses of the area $HLMG$ acting on the surface on the right side; and also the shear force acting along the new cut between points C and G, denoted by $q \cdot \Delta L$. Writing the formula for static equilibrium of all horizontal forces results in

(11.3) $$\Sigma F \overset{+}{\rightarrow} = F_1 - F_2 + q \cdot \Delta L = 0$$

Rearranging to express the horizontal shear q, we get

(11.4) $$q = \frac{F_2 - F_1}{\Delta L}$$

The dimension of this shear is pounds per foot or newtons per meter, not that of a true force; therefore it is called *shear flow.*

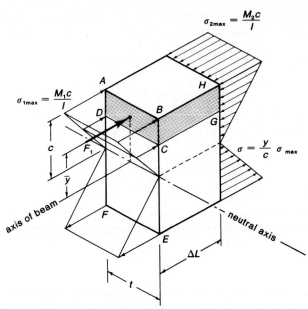

Figure 11.5 *Notations for the derivation of the shear force formula (shear force sought acts in the plane DCG).*

The third step in our derivation is the determination of F_1 and F_2 in terms of bending stresses due to the bending moments at the two ends. Looking at Figure 11.5 we may realize that F_1 is the resultant of the bending stresses acting on the area $ABCD$. According to Equation 10.5, the bending stress at any distance y above the neutral axis is

(11.5)
$$\sigma = \frac{My}{I}$$

Breaking up the area $ABCD$ into elementary small slices parallel with the line CD and denoting them by ΔA, the force F_1 at the left end can thus be written as

(11.6)
$$F_1 = \left(\frac{M_1}{I}\right) \Sigma \, y \, \Delta A$$

The term in the summation sign can be recognized as the partial moment Q of the area $ABCD$ introduced in Section 7.5. Accordingly the force F_1 can be written as

$$(11.7) \qquad F_1 = \frac{M_1 Q}{I}$$

in which I is the *moment of inertia for the whole area ABEF.* A similar equation may be written for the force F_2 acting from the right.

Substituting these expressions for F_1 and F_2 into Equation 11.4, we obtain

$$(11.8) \qquad q = \frac{Q}{I}\left(\frac{M_2 - M_1}{\Delta L}\right)$$

The term in parentheses may be replaced by the internal shear force V according to Equation 11.2; hence the formula for the shear flow takes the form

$$(11.9) \qquad q = \frac{VQ}{I}$$

Recall that in this equation Q is the first moment of the area above the arbitrary cut and away from the neutral axis of the beam, and that I is the moment of inertia of the whole cross-sectional area of the beam. When shear stress is desired rather than shear flow, Equation 11.9 should be divided by the width of the beam at the cut, t, that is,

$$(11.10) \qquad \tau = \frac{q}{t} = \frac{VQ}{It}$$

In Section 9.1 it was shown that the orientation of this shear stress is not only axial but transverse as well. This fact then gives us an indication of the distribution of the shear stresses in the cross section of the beam perpendicular to its axis.

Example 11.1

An I-beam is made by welding three steel plates together as shown in Figure 11.6. Determine the required strength of each weld if the shear force acting on the beam is 113 kilonewtons.

Solution:

First we determine the moment of inertia I for the whole composite beam. From Figure 7.9 and Equation 7.18, the moment of inertia about the neutral axis will be

$$I = I_x + A \, d_y^2$$

$$\frac{20 \, (500)^3}{12} + 2 \left[\frac{250 \, (20)^3}{12} + 250 \, (20) \, 260^2 \right] = 884 \times 10^6 \text{ mm}^4$$

The partial moment of the upper flange may be computed by Equation 7.22. Hence

$$Q_A = A_p \cdot \bar{y}_A$$

$$Q = 250 \, (20) \, 260 = 1.3 \times 10^6 \text{ mm}^3$$

Substituting into Equation 11.9, one obtains

$$q = \frac{113 \, (1.3 \times 10^6)}{884 \times 10^6} = 2.8 \text{ kN/mm}$$

Since the structure is symmetrical about the neutral axis, the same shear flow must be transferred from the lower flange also. Since there are two welds holding each flange, for the design of each weld only one half of the shear flow will be carried by each weld. The required weld strength therefore is $q/2$, or 1.4 kilonewtons per millimeter.

Example 11.2

Two identical wide-flanged beams are placed upon each other to form a composite beam shown in Figure 11.7. The beams are to be connected by rivets placed symmetrically on both sides. The distance between the rivets is 6 inches along the axis of the beam. If the shear force carried by the beam is 70 kips, what should be the size of the rivets used? The allowable shear stress in the rivets is 20,000 pounds per square inch.

Solution:

Appendix A contains the information necessary for the computation of all geometric properties. From the appropriate table the following data may be found for one W 10×19 beam; area $A = 5.62$ in²; depth $d = 10.24$ in; moment of inertia $I_x = 96.3$ in⁴ around the neutral axis of each beam. For the composite

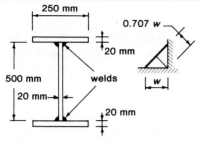

Figure 11.6 *Determination of welding strength requirement.*

In the specification of a weld the width (the "leg") of the seam is quoted. This is indicated in the small sketch in Figure 11.6 as w. However, the weld will fail over its weakest section, called the weld throat, oriented diagonally to the width. The allowable shearing stress of the weld must refer to this section. The strength of such a *fillet weld* would then be equal to the width of the throat times the allowable shear strength of the weld. The latter is usually taken as 30 percent of the electrode's tensile strength. For instance an E70 electrode has a tensile strength of 483 megapascals. Hence its allowable shear strength is 0.3 (483) or 145 megapascals. The required width for each weld could then be computed from

$$q = 0.707 \; w \; \tau_{weld} = 0.707 \; (w)145 = 0.102 \; (w) \; kN/mm$$

Since our requirement is 1.4 kN/mm, the necessary weld width using E70 electrodes will be

$$w = \frac{1.4}{0.102} = 13.7 \; mm, \; say \; 15 \; mm$$

for all four welds shown in Figure 11.6.

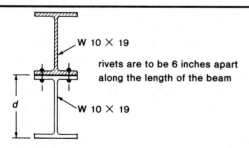

W 10 × 19

rivets are to be 6 inches apart
along the length of the beam

W 10 × 19

d

Data from Appendix A:

$A = 5.62 \; in^2$
$d = 10.24 \; in$
$I_x = 96.3 \; in^4$

Figure 11.7
*Sizing of rivets connecting two
identical wide-flange beams.*

Example 11.2 *(Continued)*

beam the common neutral axis is located along the plane where the beams are in contact. Accordingly the moment of inertia for the composite section should be computed by the use of Equation 7.18. Hence the moment of inertia will be

$$I = 2 (I_x + A (d/2)^2) = 2 (96.3 + 5.62 (10.24/2)^2) = 487.2 \text{ in}^4$$

The value of Q is based on only one of the beams since the rivets are to prevent the two from relative movement. Therefore the partial moment of the area is

$$Q = A (d/2) = 5.62 (10.24/2) = 28.8 \text{ in}^3$$

Substituting into Equation 11.8, one obtains

$$q = \frac{VQ}{I} = \frac{70,000 (28.8)}{487.2} = 4,138 \text{ lb/in}$$

As there are two rows of rivets, one row will take half of this shear flow, that is,

$$q/2 = 4,138/2 = 2,069 \text{ lb/in}$$

11.3 Shear Stress Distribution in Beams

By Equation 11.10 the magnitude of shear stresses can be computed for any location over the cross section of beams. In the previous examples, shear flow or shear stresses in the direction of a beam's axis were sought. As it was mentioned earlier, shear stresses act in both axial and transverse directions. In this section our attention is directed to transverse shear. By inspecting Equation 11.10 several important conclusions may be drawn. Both the internal shear force V and the moment of inertia I are invariant. In the determination of the shear stress distribution only Q and t change with y, the distance from the neutral axis.

At the extreme fibers, on the top and bottom of the beam, the shear stress *must* be zero. With y at its extreme value there is no area remaining above it, and hence Q is zero. Another way to explain this is to recall Figure 9.2 showing the two pairs of shear stresses acting on an elementary cube. If one of the surfaces is a free one, with no material contact, there can be no shear on that plane. One of the shearing couples is then zero. Consequently the other one must be zero as well, from Equation 9.2. Accordingly the shear stress distribution graph begins and ends with zero value on the top and bottom of a beam, regardless of the beam's geometrical configuration.

The rivets are to be placed 6 inches apart. This means that each rivet will have to carry 6 inches times 2,069 pounds per inch or

$$\text{shear force on each rivet} = 6 \, (2{,}069) = 12{,}414 \text{ lb}$$

Bolts and rivets are commonly specified by standards of the American Society for Testing Materials (ASTM). Manuals of the American Institute for Steel Construction (AISC) contain tables on the sizes and bearing capacity of bolts and rivets. However, in this example we shall use the direct method. Assuming that the rivets have an allowable shear strength of 20,000 pounds per square inch, the rivet diameter may be computed from

$$\tau_{all} = \frac{P}{\left(\dfrac{\pi D^2}{4} \right)}$$

$$D^2 = \frac{4 \, P}{\pi \, \tau_{all}} = \frac{(4) \, 12{,}414}{(3.14) \, 20{,}000} = 0.79 \, in^2$$

From this D equals 0.89 inch, so 1-inch diameter rivets are used. This is quite a large rivet; therefore it may be advisable to reduce the distance between the rivets and use correspondingly smaller sizes.

The shear stress also varies inversely with the width of the beam. For instance, in a wide-flanged section the shear stress in the flanges is very small when compared to the shear stress in the web. Most beam failures due to transverse shear will therefore occur in the web.

The location of the maximum shear appears on the neutral axis. This is where the magnitude of Q is the greatest. Only unusually shaped beams are exempt from this rule. In such sections the thickness of the web is enlarged at the neutral axis to counteract the effect of the large Q.

Depending on the shape of a beam, the distribution of shear stresses in its cross section vary. As an example let us consider the distribution in a rectangular beam, as follows.

Example 11.3

Determine the shape of the shear stress distribution in the rectangular beam shown in Figure 11.8.

Solution:

Using the notations shown in the corresponding drawing, the value of Q may be written as

$$Q = t(c-y)\left(\frac{c+y}{2}\right) = \frac{t}{2}(c^2 - y^2)$$

Substituting into Equation 11.10 to find the shear stress, we obtain

$$\tau = \frac{V(c^2 - y^2)}{2I}$$

indicating parabolic distribution of the shear stress. Substituting

$$y = 0, c = \frac{h}{2}, I = \frac{th^3}{12}$$

we obtain the maximum value for the shear stress at the neutral axis as

$$\tau_{max} = \frac{3}{2}\frac{V}{th} = \frac{3}{2}\tau_{av}$$

As the foregoing example indicates, the maximum shear in a *rectangular* cross section, in terms of the average shear, V/A, is

(11.11)
$$\tau_{max} = \frac{3}{2}\tau_{av}$$

Similar equations may be derived for *cylindrical bars,* for which the result is

(11.12)
$$\tau_{max} = \frac{4}{3}\tau_{av}$$

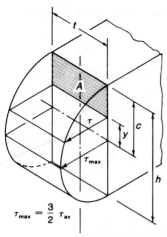

$$\tau_{max} = \frac{3}{2}\,\tau_{av}$$

Figure 11.8   *Shear stress distribution in the cross section of a rectangular beam.*

For *pipes and tubes* the corresponding result is

(11.13)
$$\tau_{max} = 2\,\tau_{av}$$

Care should be taken with use of Equation 11.13 as it is only approximate, and its use should be restricted to thin tubes. For structural shapes and others the results are more complicated, and they may be obtained by the direct application of Equation 11.10 for the neutral axis.

Example 11.4

Construct the shear stress distribution graph for the composite beam considered in Example 11.1.

Solution:
Figure 11.6 shows the dimensions of the beam. The moment of inertia, I, has been determined to be 211×10^6 mm^4. The shear force, V, is 113 kN. The width t equals 250 millimeters for the flanges and 20 millimeters for the web.

At the extreme fibers the shear stress is zero. As the beam is composed of rectangular sections, the shear stress distribution will be made up of parabolic components. On the inside edges of the two flanges as well as at the top and bottom of the web the magnitude of y is 250 mm. Q for these locations was computed to be 1.3×10^6 mm^3. From this it follows that the shear stress along the inner edges of the flanges will be

$$\tau = \frac{113\,(1.3 \times 10^6)}{211 \times 10^6\,(250)} = 0.0028 \text{ kN/mm}^2 = 2.8 \text{ MPa}$$

On the top of the webs only the width t is different, hence

$$\tau = \frac{113\,(1.3 \times 10^6)}{211 \times 10^6\,(20)} = 0.0348 \text{ kN/mm}^2 = 34.8 \text{ MPa}$$

From the latter value, the shear stress increases parabolically toward the neutral axis, where it reaches its maximum value. The change results from the increase of Q. As the flange's contribution to Q was already computed, the additional amount comes from the upper part of the web. This equals

$$Q_{web} = 250\,(20)\,125 = 6.87 \times 10^6 \text{ mm}^3$$

11.4 Approximation for Wide Flange Shapes

In Section 11.3 the general expression for the shear stress in a member due to bending action was given as Equation 11.10. That expression was generally applicable for all cross sections. However, consider now the special case of a wide-flange section (or I-section) commonly used in engineering technology. Such a section is shown in Figure 11.10a. If a shear force V is applied to the section, then the shear stress distribution through the depth of the member may be shown to have the shape shown in Figure 11.10b. It is noted that the shear stress in the flanges of the section is very small, and that the largest shear stress occurs at mid-height of the section (i.e., at C).

In many applications it is necessary to find the *maximum* shear stress in the section. It is rather difficult to apply Equation 11.10 at point C, and for this reason an acceptable approximation to the maximum shear stress may be written as

(11.14)
$$\tau_{max} \cong \frac{V}{d \cdot t_w}$$

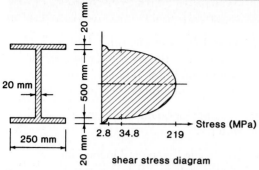

Figure 11.9 *Shear stress distribution in a wide-flanged beam cross section (not to scale).*

The sum of Q's to be considered for the neutral axis is

$$Q_{flange} + Q_{web} = (1.3 + 6.87) \times 10^6 = 8.17 \times 10^6 \text{ mm}^3$$

With this result the maximum shear stress at the neutral axis is

$$\tau = \frac{113 (8.17 \times 10^6)}{211 \times 10^6 (20)} = 0.219 \text{ kN/mm}^2 = 219 \text{ MPa}$$

The resulting shear stress distribution is shown in Figure 11.9. The increase of the shear stress in the welded connections was neglected in these computations. From the table of material properties the reader may note that the latter value considerably exceeds the allowable shear stresses: hence a redesign of the beam is necessary.

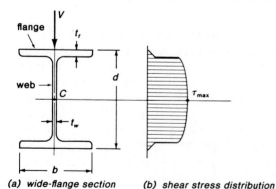

(a) wide-flange section   (b) shear stress distribution

Figure 11.10 *Shear stress distribution in wide-flanged sections.*

where d is the depth of the member, and t_w is the thickness of the web. (Many values of d and t_w are given in Appendix A.)

Although only an approximation, the value of maximum shear stress obtained from Equation 11.14 is usually acceptable for design of wide-flange sections.

11.1 A solid cylindrical member 3 feet long is simply supported as shown in Figure P 11.1. The diameter of the circular cross section is 2 inches. With the given loading condition, find the locations and magnitudes of the maximum and minimum shearing stresses.

11.2 For the 50- by 80-millimeter rectangular beam shown in Figure P 11.2, rework Problem 11.1.

11.3 A hollow cylindrical beam with inside diameter 2 inches and outside diameter 3 inches is shown in Figure P 11.3. For the given loading condition, find the shearing stresses at points 1 and 2 as indicated.

11.4 Rework Problem 11.3 for the triangular section and loading condition shown in Figure P 11.4.

11.5 –
11.6 Determine the location and magnitude of maximum shearing stress for the sections and loading conditions shown in Figures P 11.5 and P 11.6, respectively.

11.7 A W 14 × 87 and a W 14 × 34 are welded together at top and bottom flanges, as shown in Figure P 11.7. Find the shearing stresses at the levels indicated 1, 2, and 3.

11.8 Find the maximum shearing stress on the beam shown in Figure P 11.8. Where does it occur?

11.9 If a shearing force V of 100 kips acts on the section shown in Figure P 11.9, find the shearing stress at levels 1 and 2. (Assume no torsional action.)

11.10 If a shearing force V of 100 kilonewtons acts on the section shown in Figure P 11.10, find the shearing stress at levels 1 and 2. (Assume no torsional action.)

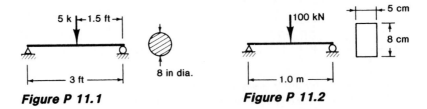

Figure P 11.1   **Figure P 11.2**

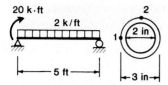

Figure P 11.3

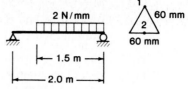

Figure P 11.4

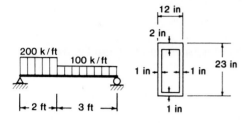

Figure P 11.5

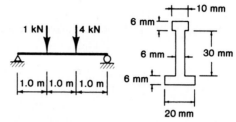

Figure P 11.6

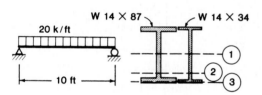

Figure P 11.7

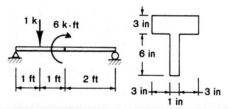

Figure P 11.8

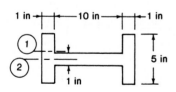

Figure P 11.9

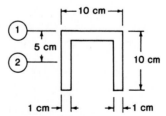

Figure P 11.10

11.11 With L^* and P^* as defined in Problem 7.19, find the shear stress at locations A and B on the section at the point $C-C$ of the beam given in Figure P 11.11.

11.12 –
11.15 Repeat Problem 11.11 using the cross sections shown in Figures P 11.12 to P 11.15.

11.16 With L^* and P^* defined as in Problem 7.19, find the shear stress at locations A and B on the section at the point $C-C$ of the beam given in Figure P 11.16.

11.17–
11.21 Repeat Problem 11.16 using the cross sections shown in Figures P 11.17 to P 11.21.

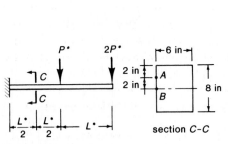

Figure P 11.11

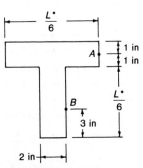

Figure P 11.12

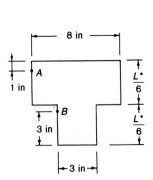

Figure P 11.13

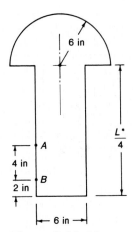

Figure P 11.14

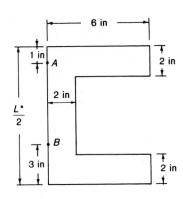

Figure P 11.15

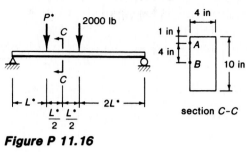

Figure P 11.16

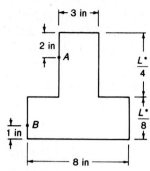

Figure P 11.17

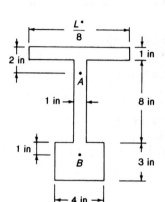

Figure P 11.18

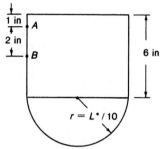

Figure P 11.19

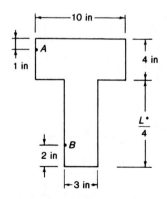

Figure P 11.20

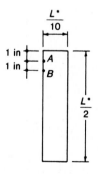

Figure P 11.21

Combined Stresses

12

Objectives

1. To be able to evaluate the combined effects of axial and bending stresses.

2. To know how to determine the stresses in the walls of thin-walled pressure vessels.

3. To be able to find the state of plane stress on any plane using either the equations for transformation of plane stress or Mohr's circle construction.

4. To learn how to find the principal stresses acting on an element and the critical orientation of the element.

5. To be able to assess the combined effects of normal stresses and shear stresses.

12.1 Introduction

Considerable attention has now been directed towards finding the stresses occurring in members subjected to various modes of loading. Axial stresses or normal stresses arising from axial forces (Chapter 8) and bending moments (Chapter 10) have been investigated, as well as shear stresses arising from twisting (Chapter 9) and shear due to bending (Chapter 11). In practice these stresses rarely occur by themselves, but rather one action usually occurs with one of the others. For example, at most locations on a beam normal stresses usually are introduced by the shear forces on the beam. It is the purpose of this chapter to investigate how the various stresses influence one another, and, in particular, find the critical situation that a designer should consider when designing a part in which more than one stress acts.

Several situations in which stresses may be readily combined are discussed in this chapter. These include addition of bending and axial stresses, as well as the special cases encountered in considering thin-walled pressure vessels.

Example 12.1

The beam *ABC* is loaded as shown in Figure 12.1*a* and has a rectangular cross section of 2 by 3 inches, as shown. Show the variation of normal stress across the section of the beam at the point *B*.

Solution:

As usual the first step in solving for the forces in a beam is the finding of the support reactions. These are shown on the free body diagram of Figure 12.1*b*. With this information the shear force diagram, the bending moment diagram, and the axial force diagram may be drawn as shown in Figures 12.1*c, d,* and *e* respectively. At point *B* in the beam, therefore, the axial force is +6 kips and the bending moment is 9 kip inches. The area *A* of the section and the moment of inertia *I* of the section are given by

$$A = b \cdot h = 2 \times 3 = 6.0 \text{ in}^2$$

$$I = \frac{bh^3}{12} = \frac{2 \times 3^3}{12} = 4.5 \text{ in}^4$$

Thus the uniform axial stress over the section at *B* due to the axial force is given by

$$\sigma = \frac{P}{A} = \frac{6}{6} = 1.0 \text{ ksi}$$

Two methods of combining stresses are then presented—an analytical method in which two equations are used and a graphical technique. These methods each have advantages, but unfortunately they often are used with different sign conventions, and so care must be exercised in using either approach.

12.2 Combining Axial and Bending Stresses

The first example involves the summation of normal stresses. Chapter 8 demonstrated that axial forces produce normal stresses that are uniformly distributed across the cross-sectional area. In Chapter 10, it was demonstrated that bending moments also produce normal stresses at a cross section, but that these stresses vary in both magnitude and sign across the section. In the addition of these, some care must be taken.

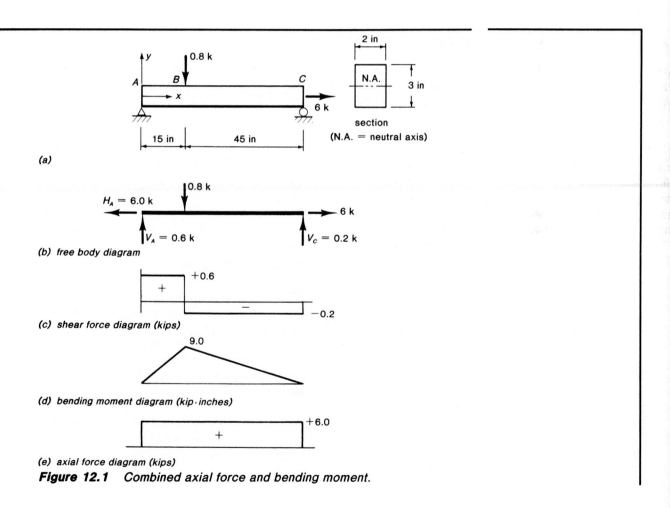

(a)

(b) free body diagram

(c) shear force diagram (kips)

(d) bending moment diagram (kip·inches)

(e) axial force diagram (kips)

Figure 12.1   Combined axial force and bending moment.

Example 12.1 *(Continued)*

This is shown in diagrammatic form in Figure 12.1*f*.

The bending stresses vary linearly across the section with compression stresses above the neutral axis (NA) and tension stresses below it. The normal stresses due to bending are given by

$$\sigma = \frac{My}{I}$$

Thus the maximum bending stress at the point *B* (tension on the bottom and compression on the top) is

$$\sigma_{max} = \frac{My}{I} = \frac{9.0 \times 1.5}{4.5} = 3.0 \text{ ksi}$$

The bending stresses therefore vary over the section as shown in Figure 12.1*g*.

Since the stresses due to bending moments and axial forces are both normal stresses, they can be added directly:

$$\sigma_{total} = \sigma_{axial} + \sigma_{bending}$$

Thus, the resulting stresses at top and bottom of the section are

$$\sigma_{top, \, final} = \sigma_{top, \, axial} + \sigma_{top, \, bending}$$

12.3 Thin-Walled Pressure Vessels

Another situation involving stresses in more than one direction is that of pressure vessels or boilers. Here, however, as we shall see, the problem involves recognizing that normal stresses are simultaneously occurring in two perpendicular directions.

Basically, a vessel is considered to be thin-walled if the stresses in it do not vary through the thickness of the wall. In practice, many pressure vessels obey this approximation with sufficient accuracy. It is a good rule of thumb that a vessel may be considered to be thin-walled if its radius is at least ten times its wall thickness.

There are two types of thin-walled pressure vessels to be considered. The first of these, the cylindrical pressure vessel, has two different stresses acting in perpendicular directions. The second, the spherical pressure vessel, has only one value of normal stress in the wall.

i.e.,

$$\sigma_{top} = +1.0 - 3.0 = -2.0 \text{ ksi (i.e., compression)}$$

and

$$\sigma_{bottom, final} = \sigma_{bottom, axial} + \sigma_{bottom, bending}$$

i.e.,

$$\sigma_{bottom} = +1.0 + 3.0 = 4.0 \text{ ksi}$$

The variation of normal stresses across the depth of the section resulting from these two actions is shown in Figure 12.1h. Note that the final stress at the centroid of the section (i.e., at mid-height) is not now zero. The stresses are, of course, constant across the width of the section at any height y above or below the neutral axis.

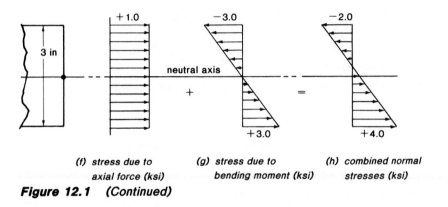

(f) stress due to axial force (ksi)

(g) stress due to bending moment (ksi)

(h) combined normal stresses (ksi)

Figure 12.1 (Continued)

Cylindrical Pressure Vessels

A typical cylindrical pressure vessel is shown in Figure 12.2. The vessel consists of a long cylinder with capped ends; it has an internal pressure p. In order to find the stresses produced in the wall of the cylinder by the pressure p, two free bodies are cut from the cylinder, as shown in Figure 12.3. At this point, it is important to realize that the stresses in the wall of the cylinder are produced solely by the internal pressure, and not by any bending of the vessel between its supports.

If the longitudinal stress within the wall of the vessel is defined as σ_L and it is assumed that this value is constant through the thickness of the cylinder wall, then the free body of Figure 12.3a may be used to find the value of the stress. Recall from physics that pressure acts in all directions. For this particular

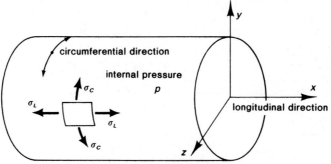

Figure 12.2 *A thin-walled cylindrical pressure vessel.*

application the only unbalanced forces that the pressure causes on the free body act on the closed end of the cylinder. The value of the force that this pressure exerts on the end is given by

(12.1)
$$F_P = p \cdot A_{\text{end}} = p\,(\pi R_i^2)$$

The longitudinal stress is distributed over the exposed area of the wall, and the net force represented by this stress, F_L, is

(12.2)
$$F_L = \sigma_L \cdot A_{\text{wall}} = \sigma_L\,(\pi R_o^2 - \pi R_i^2)$$

where

$$R_o = \text{external radius of cylinder}$$
$$R_i = \text{internal radius of cylinder}$$

The cylinder must be in equilibrium in the longitudinal direction. Thus

(12.3)
$$\Sigma F_x = 0 \overset{+}{\rightarrow}$$
$$F_L - F_p = 0$$

Substituting Equations 12.1 and 12.2 into 12.3:

(12.4)
$$\sigma_L\,(\pi R_o^2 - \pi R_i^2) = p\,\pi R_i^2$$

i.e.,

(12.5)
$$\sigma_L\,(R_o + R_i)(R_o - R_i) = p\,R_i^2$$

Now the wall thickness, t, is

(12.6)
$$t = R_o - R_i$$

if also the vessel is thin-walled,

$$R_o \simeq R_i$$

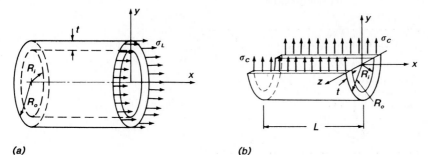

(a)   **(b)**

Figure 12.3   *Free bodies cut from a thin-walled cylindrical pressure vessel.*

and thus we can define an "average" radius R, such that

(12.7)
$$R_o \simeq R_i = R$$

and so

(12.8)
$$R_i + R_o = 2R$$

Substitution into Equation 12.5 gives

$$\sigma_L \cdot 2R\,t = p\,R^2$$

i.e.,

(12.9)
$$\sigma_L = \frac{pR}{2t}$$

This is the expression for longitudinal stress in a thin-walled pressure vessel.

Now define the stress in the circumferential direction within the vessel wall as σ_C. The value of σ_C may be found with the aid of the free body diagram in Figure 12.3b. To get this free body, a ring of length L was cut from the cylinder and was then cut in half along a diameter. The stresses on the ends of the element have been omitted for clarity. The net force in the negative y-direction on this free body is

(12.10)
$$F_\rho = p \cdot A = p \cdot (2R_i \cdot L)$$

The force F_C representing the summation of the circumferential stresses is given by

(12.11)
$$F_C = \sigma_C \cdot A = \sigma_C (2L \cdot t)$$

Now resolving forces in the y-direction:

(12.12)
$$\Sigma F_y = 0 \uparrow +$$

$$F_C - F_\rho = 0$$

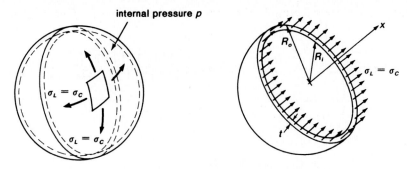

(a)   **(b)** *free body of half a sphere*

Figure 12.4   *Thin-walled spherical pressure vessel.*

Substituting Equations 12.10 and 12.11 into 12.12 gives

$$\sigma_C(2Lt) = p \cdot 2\,R_i L \simeq 2p\,RL$$

thus,

(12.13)
$$\sigma_C = \frac{pR}{t}$$

This is the expression for the circumferential stress in a thin-walled cylindrical pressure vessel. It is noted that the circumferential stress is twice the longitudinal stress.

Spherical Pressure Vessels

A typical thin-walled spherical pressure vessel is shown in Figure 12.4*a*. Such a situation may be approximated by a satellite orbiting the earth with a vacuum outside the sphere and some finite pressure inside. A pressurized sports ball may also behave somewhat like a spherical pressure vessel. In Figure 12.4*b* a free body diagram is shown of half of a thin-walled spherical pressure vessel. It does not matter which orientation is chosen when the free body is cut from the vessel, and thus the longitudinal stress in the wall of the vessel has the same magnitude as the circumferential stress, i.e.,

(12.14)
$$\sigma_L = \sigma_C$$

The net pressure force acting in the negative x-direction on the free body is

$$(12.15) \qquad F_p = p \cdot A = p \cdot (\pi R_i^2)$$

The net longitudinal (or circumferential) stress acting on the wall section of the free body is

$$(12.16) \qquad F_s = \sigma_L \cdot A = \sigma_L (\pi R_o^2 - \pi R_i^2)$$

From the equilibrium condition,

$$(12.17) \qquad \Sigma F_x = 0 \overset{+}{\longrightarrow}$$

$$F_s - F_p = 0$$

substituting Equations 12.15 and 12.16 into 12.17

$$(12.18) \qquad \sigma_L \cdot (\pi R_o^2 - \pi R_i^2) - p \pi \cdot R_i^2 = 0$$

$$\sigma_L (R_o + R_i)(R_o - R_i) = p R_i^2$$

Using the thin-wall approximation again,

$$(12.19) \qquad \sigma_L = \sigma_C = \frac{pR}{2t}$$

Note that the same stress will be acting in any two perpendicular directions within the walls of a thin-walled spherical pressure vessel.

A note of caution is appropriate at this point. While it is true that the theory developed for this discussion is valid in some situations, it is not universally true for all pressure vessels. The stresses in the walls of real pressure vessels are very easily altered from the values predicted by this theory. For example, most pressure vessels have at least one opening in them, such as an inlet valve and/or a safety valve. The presence of such alterations to the vessel may have a large effect on the stresses in the walls of the vessel, particularly in the region of the opening, in much the same manner as the stress concentrations discussed in Chapter 8. Furthermore, in some pressure vessels the internal pressure may be large and relatively thick walls may be needed. In such cases the stresses may not be uniform throughout the thickness of the wall, and prediction of stresses is more complex.

Example 12.2

A steel boiler, 4 feet in diameter and 10 feet long, is maintained at an internal pressure of 200 pounds per square inch. If the boiler is made from 3/4-inch plate, find the longitudinal and circumferential stresses acting in the walls of the boiler. Draw the resulting stresses on a stress element.

Solution:

A diagram of the boiler is shown in Figure 12.5a. From Equations 12.9 and 12.13, the longitudinal and circumferential stresses in the wall of the boiler are found to be:

$$\sigma_L = \frac{pR}{2t} = \frac{200 \times 24}{2 \times 0.75} = 3,200 \text{ psi}$$

and

$$\sigma_C = \frac{pR}{t} = \frac{200 \times 24}{0.75} = 6,400 \text{ psi}$$

In the above calculations the outside radius of the boiler was taken to be the radius R. A stress element showing these stresses is given as Figure 12.5b.

Example 12.3

A satellite orbiting the earth is maintained at an internal pressure of 30 kilopascals. If the satellite is spherical, has an external diameter of 1.2 meters, and is made of aluminum with thickness 5 millimeters, find the stress acting in the satellite. Draw these stresses on a stress element.

Solution:

Before any calculations are made, the units are standardized in newtons and meters. Thus,

$$R = 0.6 \text{ m}$$
$$t = 5 \text{ mm} = 0.005 \text{ m}$$
$$\text{and } p = 30 \text{ kPa} = 30 \times 10^3 \text{ Pa}$$

Equation 12.19 is now used to find the normal stress in any direction in the wall of the satellite:

$$\sigma_L = \sigma_C = \frac{pR}{2t} = \frac{30 \times 10^3 \times 0.6}{2 \times 0.005}$$

$$= 18 \times 10^5 \text{ Pa}$$
$$= 1.8 \text{ MPa}$$

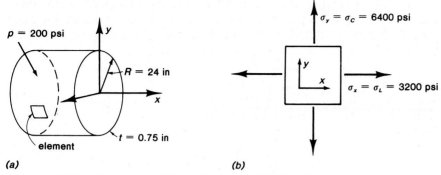

(a)

(b)

Figure 12.5 *Stresses in thin-walled cylindrical pressure vessel.*

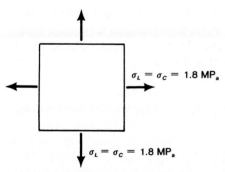

Figure 12.6 *Stress element for Example 12.3.*

The stresses are shown on the stress element of Figure 12.6, although the element may rotate in any direction and still have the same stresses acting on its faces.

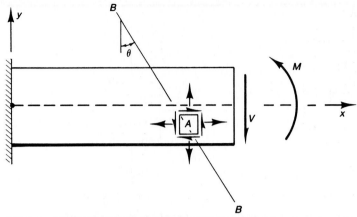

Figure 12.7 *Stress transformation example.*

12.4 Transformation of Plane Stresses

The derivation and discussion of this chapter will be confined to situations in which the stresses are essentially two-dimensional in nature. In other words, only *plane stress* situations are to be considered. Consider the cantilever beam of Figure 12.7, subject to a moment M and a shear force V, as shown. The bending moment will introduce normal stresses into the beam while the shear force will introduce shear stresses. At some point A in the beam, the stresses can be represented pictorially as stresses on a small element as shown. This small element is enlarged in Figure 12.8a, where the various stresses that may act on the element are shown. (In this case the stress σ_y is probably zero, but it is included here to allow a derivation of a general case.) Particular note should be taken of the directions in which the various stresses are acting. The directions indicated will be the positive stress directions in this derivation. (It is common to take normal stresses as being positive when they are tensile stresses, as is shown here, but the sign convention for shear stresses is somewhat more arbitrary.)

The values of the stresses may be found from the applied loadings (and previous chapters have described how this is done) when the small element is oriented with sides parallel to the x- and y-axes. With this known, it is appropriate to ask: What stresses are acting on another plane (labeled $B-B$ in Figure 12.7) which makes an angle θ with the vertical, as indicated? To answer this question, the small rectangular element of Figure 12.8a is cut into a triangular shape as shown in Figure 12.8b, where the hypotenuse of the right-angled triangle is parallel to the plane on which the stresses are desired. If the directions normal and tangential to this new plane are defined as the n- and t-directions, respectively, then the normal stress on this new plane may be referred to as σ_n and the shear stress as τ_t. If these stresses σ_n and τ_t may be found, then the question asked has been answered. Note the directions assumed for the unknown stresses—they have been shown in their positive sense in Figure 12.8b.

In order to consider equilibrium of the small triangular element, it is necessary to convert the stresses on each face to equivalent forces. It is remembered that the small triangular element is really a small triangular wedge-

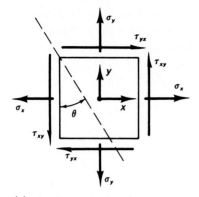

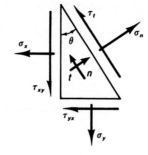

(a) stresses on rectangular element

(b) stresses on triangular element

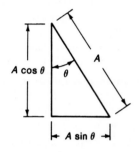

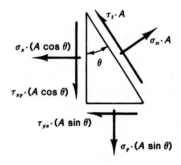

(c) sides of triangular element

(d) forces on triangular element

Figure 12.8 *Transformation of stresses on an element.*

shaped piece of the beam of Figure 12.7. Thus each side of the triangle represents a certain area. If the area of the hypotenuse of the triangle of Figure 12.8b is taken to be A, then the areas represented by the other two sides of the triangle as shown in Figure 12.8c. The forces represented by the stresses on each face may now be found by multiplying each stress by the appropriate cross-sectional area on which it is acting. The forces on the wedge are shown in Figure 12.8d. Since the stresses have now been transformed to forces, the equations of equilibrium may now be applied to the wedge, giving the equations

(12.20)
$$\Sigma F_x = 0 \overset{+}{\rightarrow}$$

$$(\sigma_n \cdot A) \cos \theta - (\tau_t \cdot A) \sin \theta - \sigma_x \cdot (A \cos \theta) - \tau_{xy} \cdot (A \sin \theta) = 0$$

and

(12.21)
$$\Sigma F_y = 0 \uparrow +$$

$$(\sigma_n \cdot A) \sin \theta + (\tau_t \cdot A) \cos \theta - \sigma_y \cdot (A \sin \theta) - \tau_{xy} \cdot (A \cos \theta) = 0$$

Using the trigonometric identities for double angles, these equations may be rewritten to give the unknown stresses σ_n and τ_t in terms of the other stresses, which are already known. The resulting equations are

(12.22)
$$\sigma_n = \frac{(\sigma_x + \sigma_y)}{2} + \frac{(\sigma_x - \sigma_y)}{2}\cos 2\theta + \tau_{xy}\sin 2\theta$$

and

(12.23)
$$\tau_t = -\frac{(\sigma_x - \sigma_y)}{2}\sin 2\theta + \tau_{xy}\cos 2\theta$$

Example 12.4

The state of stress at a point in a body is represented by the stresses on the small element shown in Figure 12.9a. Find the stresses acting on a plane making an angle of $60°$ with the vertical as shown.

Solution:
Using Equations 12.22 and 12.23, the normal and shear stresses acting on the required plane may be found:

$$\sigma_n = \frac{\sigma_x + \sigma_y}{2} + \frac{(\sigma_x - \sigma_y)}{2}\cos 2\theta + \tau_{xy}\sin 2\theta$$

$$= \frac{6 + 2}{2} + \frac{(6 - 2)}{2}\cos 120° + 4 \cdot \sin 120°$$

$$= 6.46 \text{ MPa}$$

and

$$\tau_t = -\frac{(\sigma_x - \sigma_y)}{2}\sin 2\theta + \tau_{xy}\cos 2\theta$$

$$= -\frac{(6 - 2)}{2}\sin 120° + 4\cos 120°$$

$$= -3.73 \text{ MPa}$$

The positive value of σ_n indicates that it is a tensile stress acting in the direction assumed in Figure 12.8. The negative value of τ_t indicates that it acts in a direction opposite to that assumed in Figure 12.8d. The results are shown on the wedge of Figure 12.9b.

These expressions are the general equations for transformation of plane stresses. The only caution that is necessary in applying these equations is with the sign convention, which must have the positive directions of the stresses in directions shown in Figure 12.8b. The angle θ is positive only when it is a counterclockwise rotation from the vertical as shown in Figure 12.8b.

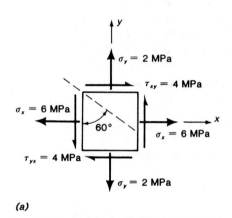

(a)

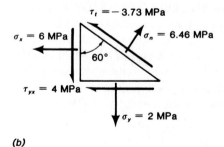

(b)

Figure 12.9 Stress at a point for Example 12.4.

Example 12.5

The stresses on the small element of Figure 12.10a represent the state of stress at some point in a body. Find the state of stress on a plane that has an angle of 30° to the vertical as shown.

Solution:

Before proceeding with the substitution of the known stresses into the stress transformation equations, several facts must be noted. Attention must be drawn to the sign of the stresses. In particular, σ_y is a compressive stress and thus will have a negative sign. Furthermore, examination of Figure 12.10a shows that the shear stresses act in the opposite direction to that assumed in deriving the equations. The shear stresses will thus be negative also. The angle between the vertical and the required plane involves a *counterclockwise* rotation from the vertical of 30°, whereas the derivation assumed a clockwise rotation. The rotation may be considered as a positive rotation of 330° or as a negative rotation of −30°. For convenience the rotation will be taken as −30°.

With this preparation, substitution may be made in Equations 12.22 and 12.23 to find the required normal and shear stresses:

$$\sigma_n = \frac{\sigma_x + \sigma_y}{2} + \frac{(\sigma_x - \sigma_y)}{2} \cos 2\theta + \tau_{xy} \sin 2\theta$$

$$= \frac{18 + (-6)}{2} + \frac{[18 - (-6)]}{2} \cos(-60°) + (-12) \sin(-60°)$$

$$= 22.39 \text{ ksi}$$

12.5 Critical Values of Plane Stress Transformation

Equations 12.22 and 12.23 are general expressions of normal stress and shear stress on an arbitrary plane rotated at an angle θ to the vertical. There are, however, certain critical values of the angle θ that produce important results. For example, there are two angles θ at which there are no shear stresses acting on the plane, only a normal stress. The stresses on the planes on which this is true are called the *principal stresses,* which are designated σ_1 and σ_2:

(12.24) $$\sigma_1 = \sigma_{n\,max} = \frac{\sigma_x + \sigma_y}{2} + \sqrt{\left(\frac{\sigma_x - \sigma_y}{2}\right)^2 + \tau_{xy}^2}$$

(12.25) $$\sigma_2 = \sigma_{n\,min} = \frac{\sigma_x + \sigma_y}{2} - \sqrt{\left(\frac{\sigma_x - \sigma_y}{2}\right)^2 + \tau_{xy}^2}$$

σ_1 and σ_2 may also be shown to be the largest and smallest normal stresses, respectively, that may exist on any plane found by rotating the wedge through

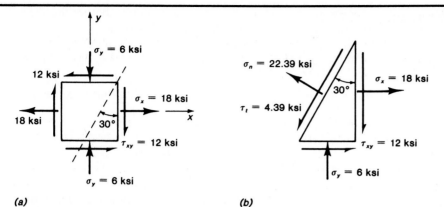

(a)   **(b)**

Figure 12.10   *Stress at a point for Example 12.5.*

and

$$\tau_t = -\frac{(\sigma_x - \sigma_y)}{2} \sin 2\theta + \tau_{xy} \cos 2\theta$$

$$= -\frac{[18 - (-6)]}{2} \sin(-60°) + (-12)\cos(-60°)$$

$$= 4.39 \text{ ksi}$$

The results are indicated on the wedge of Figure 12.10*b*.

any angle θ. The planes on which σ_1 and σ_2 act may be shown to be always at right angles to one another, and their inclinations can be found from the equation

(12.26)
$$\tan 2\theta_1 = \frac{\tau_{xy}}{\left(\dfrac{\sigma_x - \sigma_y}{2}\right)}$$

where θ_1 is the angle shown in Figure 12.11. Since σ_1 is larger than σ_2, σ_1 is often called the *major principal stress*, and σ_2 the *minor principal stress*.

Another case to be considered is that of maximum and minimum shear stresses. The magnitudes of the maximum and minimum shear stresses may be shown to be the same, but they differ in sign. The value of the maximum or minimum shear stress is given by

(12.27)
$$\tau_{\text{max or min}} = \pm\sqrt{\left(\frac{\sigma_x - \sigma_y}{2}\right)^2 + \tau_{xy}^2}$$

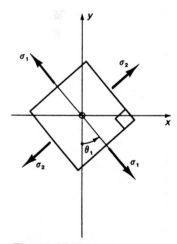

Figure 12.11
Principal stresses and directions.

At this value of shear stress, the normal stress is the same whether the shear stress is positive or negative. The appropriate normal stress is

(12.28) $$\sigma_n \text{ at } \tau_{max/min} = \frac{\sigma_x + \sigma_y}{2}$$

The angle at which this situation occurs is referred to as θ_2 (indicated on Figure 12.12) and has the value

(12.29) $$\tan 2\theta_2 = -\frac{\left(\dfrac{\sigma_x - \sigma_y}{2}\right)}{\tau_{xy}}$$

In a similar manner to the derivation of principal stresses it may be shown that the planes of maximum and minimum shear stress are always at right angles to one another. Comparison of Equations 12.26 and 12.29 shows that the tangent of $2\theta_1$ is the negative of the inverse of the tangent of $2\theta_2$. It is a trigonometric identity that this occurs only when the two angles differ by $90°$. Thus, $2\theta_1$ differs from $2\theta_2$ by $90°$, and the difference between θ_1 and θ_2 is always $45°$.

In many instances the values of principal stresses and/or maximum shear stress determine whether a material will behave safely at a given loading.

Example 12.6

The state of plane stress at a point in a body is shown on the small element of Figure 12.13a. Using the equations for transformation of plane stress, (a) calculate the principal stresses; (b) calculate the maximum shear stress and associated normal stress; (c) show the stresses from parts (a) and (b) on properly oriented elements.

Solution:
From the diagram on which the stresses are given,

$$\sigma_x = 40 \text{ ksi}$$
$$\sigma_y = 0$$
$$\tau_{xy} = -10 \text{ ksi}$$

Part a. Using Equation 12.24, the major (or larger) principal stress may be found:

$$\sigma_1 = \frac{\sigma_x + \sigma_y}{2} + \sqrt{\left(\frac{\sigma_x - \sigma_y}{2}\right)^2 + \tau_{xy}^2}$$

$$= \frac{40 + 0}{2} \qquad \sqrt{\left(\frac{40 - 0}{2}\right)^2 + (-10)^2}$$

$$= 20.0 + 22.36$$
$$= 42.36 \text{ ksi}$$

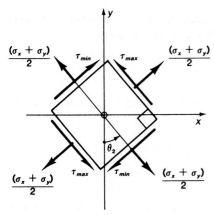

Figure 12.12 *Maximum and minimum shear stress.*

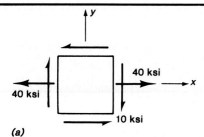

(a)

Figure 12.13 *Plane stress elements for Example 12.6.*

Similarly, using Equation 12.25, the minor (or smaller) principal stress is given by:

$$\sigma_2 = \frac{\sigma_x - \sigma_y}{2} - \sqrt{\left(\frac{\sigma_x - \sigma_y}{2}\right)^2 + \tau_{xy}^2}$$

$$= 20.0 - 22.36$$

$$= -2.36 \text{ ksi}$$

(The negative sign indicates that this stress is compressive.)

Part b. Equation 12.27 for maximum shear stress is

$$\tau_{max} = + \sqrt{\left(\frac{\sigma_x - \sigma_y}{2}\right)^2 + \tau_{xy}^2}$$

Example 12.6 *(Continued)*

It is noted that this is part of the equations for σ_1 and σ_2. Thus

$$\tau_{max} = 22.36 \text{ ksi}$$

The associated normal stress is given by Equation 12.28:

$$\sigma_n = \frac{\sigma_x + \sigma_y}{2} = \frac{40 + 0}{2} = 20.0 \text{ ksi}$$

Part c. Having found the principal stresses in *part a,* the angle at which they occur is found by Equation 12.26.

$$\tan 2\theta_1 = \frac{\tau_{xy}}{\left(\dfrac{\sigma_x - \sigma_y}{2}\right)} = \frac{-10.0}{\left(\dfrac{40 - 0}{2}\right)} = -0.5$$

thus,

$$2\theta_1 = -26.6°$$

and

$$\theta_1 = -13.3°$$

It is now known that a counterclockwise rotation of 13.3° yields a plane on which one of the principal stresses acts, but it is not known whether the major or minor principal stress acts on this plane. To determine this, use Equation 12.22, substituting $\theta = -13.3°$:

$$\sigma_n = \frac{(\sigma_x + \sigma_y)}{2} + \frac{(\sigma_x - \sigma_y)}{2} \cos 2\theta + \tau_{xy} \sin 2\theta$$

$$\sigma_{(-13.3°)} = \frac{40 + 0}{2} + \left(\frac{40 - 0}{2}\right) \cdot \cos(-26.6°) + (-10) \cdot \sin(-26.6°)$$

$$= 20.0 + 17.88 + 4.48$$
$$= 42.36 \text{ ksi}$$
$$= \sigma_1$$

Thus the element is properly oriented as shown in Figure 12.13*b*. Note that the minor principal stress acts on a plane at right angles to the plane of the major principal stress.

In *part b* the maximum shear stress was found. Thus, from Equation 12.27,

$$\tau_{max} = 22.36 \text{ ksi}$$

$$\tau_{min} = -22.36 \text{ ksi}$$

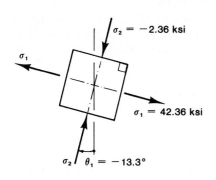

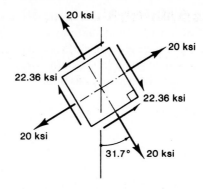

(b) principal stresses

(c) maximum and minimum shear stress

Figure 12.13 (Continued)

The associated normal stresses are

$$\sigma_n = 20.0 \text{ ksi}$$

since σ_n is positive it is a tensile stress. The orientation of this is given by Equation 12.29:

$$\tan 2\theta_2 = -\frac{\left(\dfrac{\sigma_x - \sigma_y}{2}\right)}{\tau_{xy}} = -\frac{\dfrac{40 - 0}{2}}{(-10)} = 2.0$$

thus,

$$2\theta_2 = 63.4°$$
$$\theta_2 = 31.7°$$

To discover whether τ_{max} or τ_{min} acts on this plane, substitution is made into Equation 12.23:

$$\tau_t = -\left(\frac{\sigma_x - \sigma_y}{2}\right) \sin 2\theta + \tau_{xy} \cos 2\theta$$

$$\tau_{31.7°} = -\left(\frac{40 - 0}{2}\right) \cdot \sin(63.4°) + (-10) \cdot \cos(63.4°)$$

$$= -22.36 \text{ ksi}$$
$$= \tau_{min}$$

Thus the stresses on the oriented element are as shown in Figure 12.13c.

It is noted that the difference between θ_1 and θ_2 is 45°, as pointed out in Section 12.5.

12.6 Mohr's Circle of Plane Stress

In Section 12.4 the equations for the normal stress and shear stress on an arbitrary plane were derived (Equations 12.22 and 12.23). Since these expressions are based on given values of the stresses on the x- and y-planes (σ_x, σ_y, and τ_{xy}), let us now define the constants a, b, and c as

$$(12.30) \qquad a = \frac{\sigma_x + \sigma_y}{2}$$

$$(12.31) \qquad b = \tau_{xy}$$

$$(12.32) \qquad c = \frac{\sigma_x - \sigma_y}{2}$$

Using these, Equation 12.22 may be rewritten as

$$\sigma_n = a + c \cdot \cos 2\theta + b \cdot \sin 2\theta$$

or

$$(12.33) \qquad \sigma_n - a = c \cdot \cos 2\theta + b \cdot \sin 2\theta$$

Similarly, Equation 12.23 may be rewritten as

$$(12.34) \qquad \tau_t = -c \cdot \sin 2\theta + b \cdot \cos 2\theta$$

Squaring Equations 12.33 and 12.34 gives

$$(12.35) \qquad (\sigma_n - a)^2 = c^2 \cdot \cos^2 2\theta + 2bc \cdot \sin 2\theta \cdot \cos 2\theta + b^2 \cdot \sin^2 2\theta$$

and

$$(12.36) \qquad \tau_t^2 = c^2 \cdot \sin^2 2\theta - 2bc \cdot \sin 2\theta \cdot \cos 2\theta + b^2 \cdot \cos^2 2\theta$$

Adding Equations 12.35 and 12.36 and using common trigonometric identities,

$$(12.37) \qquad (\sigma_n - a)^2 + \tau_t^2 = c^2 + b^2$$

In the Cartesian Coordinate System of Figure 12.14, the equation of the circle centered at the point (m,n) with radius r is

$$(12.38) \qquad (x - m)^2 + (y - n)^2 = r^2$$

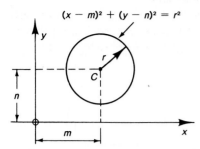

Figure 12.14   *General equation of a circle.*

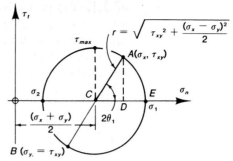

Figure 12.15   *Mohr's circle of plane stress.*

Comparison of Equations 12.37 and 12.38 shows that they are very similar in form. In fact, they are made identical by the following substitutions:

a. the x-axis becomes the σ_n-axis,

b. the y-axis becomes the τ_t-axis,

c. $m = a = \dfrac{\sigma_x - \sigma_y}{2}$

d. $n = 0$

e. $r^2 = c^2 + b^2 = \tau_{xy}^2 + \left(\dfrac{\sigma_x - \sigma_y}{2}\right)^2$

Thus the equations representing the stresses on a plane may be represented by a circle known as Mohr's circle of plane stress, which is shown as a diagram in Figure 12.15. Also shown on this diagram are the *major principal stress* σ_1, the *minor principal stress* σ_2, and the *maximum shear stress* τ_{max}. The angle $2\theta_1$ shown on the diagram is the angle between the x-direction plane and the plane on which the maximum principal stress acts. This is the same angle that occurred in Equation 12.26.

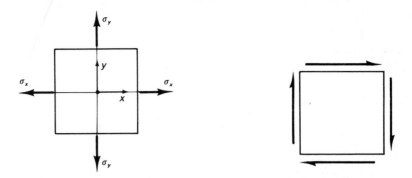

(a) positive normal stresses   (b) positive shear stresses

Figure 12.16   *Sign convention for Mohr's circle.*

Some confusion may easily arise with the sign conventions necessary in the use of Mohr's circle, and these are shown in Figure 12.16. There is no problem with the direct stresses, as these follow the usual convention of positive stresses being tensile stresses, as indicated in Figure 12.16*a*. The convention for shear stresses, however, is a little more complex. In order for Mohr's circle to be valid it is necessary to use the convention given in Figure 12.16*b*. It is noted that all the shear stresses given tend to cause the element to rotate clockwise, in contrast to the convention of shear stresses for the transformation equations illustrated in Figure 12.8*a*. Thus on any element the shear stress on two faces will be positive and on the other two faces will be negative. This is a characteristic of Mohr's circle, and the stresses on an element will always be represented by points like *A* and *B* on Figure 12.15. The point *A* in Figure 12.15 represents the state of stress on either the face *A* or the face *A'* in Figure 12.17. Similarly, the point *B* in Figure 12.15 represents the state of stress on either the face *B* or the face *B'* in Figure 12.17.

The drawing of Mohr's circle of plane stress may be summarized in the following steps:

1. Draw the stresses on an element (similar to Figure 12.17);
2. Locate the center of Mohr's circle, which is given by

(12.39)
$$OC = \frac{\sigma_x + \sigma_y}{2}$$

3. Draw Mohr's circle with the radius given by

(12.40)
$$r = \sqrt{\tau_{xy}^2 + \left(\frac{\sigma_x - \sigma_y}{2}\right)^2}$$

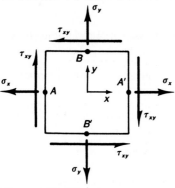

Figure 12.17 Stress at a point.

Thus the circle may be drawn, and the points representing the state of stress on the given element may easily be located (the points A and B on Figure 12.15).

The most common uses for Mohr's circle in this form is the finding of the principal stresses, the maximum shear stress, and the angle θ_1, which represents the rotation from the vertical to the plane on which the major principal stress σ_1 acts. This angle is given by

$$\tan 2\theta_1 = \frac{AD}{CD}$$

$$= \frac{\tau_{xy}}{\sigma_x - \frac{(\sigma_x + \sigma_y)}{2}}$$

i.e.,

(12.41) $$\tan 2\theta_1 = \frac{\tau_{xy}}{\left(\dfrac{\sigma_x - \sigma_y}{2}\right)}$$

It is noted that Equation 12.41 is identical to Equation 12.26.

Example 12.7

The state of stress at a point is represented by the stresses on the element shown in Figure 12.18a.

Part a. Draw Mohr's circle of plane stress for this stress situation and label on it the stresses on faces A and B of the element.

Part b. Use this diagram to find the major and minor principal stresses, and show them on a properly oriented element.

Part c. Give the value of the maximum shear stress.

Solution:
Part a. From Figure 12.18a we find the stresses on faces A and B.
On face A,

$$\sigma_x = 800 \text{ kPa}$$

$$\tau_{xy} = -900 \text{ kPa}$$

On face B,

$$\sigma_y = -1{,}200 \text{ kPa (compression)}$$

$$\tau_{xy} = 900 \text{ kPa}$$

Now, from Equations 12.39 and 12.40, the center of the circle and its radius are found:

$$OC = \frac{\sigma_x + \sigma_y}{2} = \frac{800 + (-1{,}200)}{2} = -200 \text{ kPa}$$

$$r = \sqrt{\tau_{xy}^{\,2} + \left(\frac{\sigma_x - \sigma_y}{2}\right)^2} = \sqrt{900^2 + \left[\frac{800 - (-1{,}200)}{2}\right]^2}$$

$$= 1{,}345 \text{ kPa}$$

Thus the Mohr's circle of Figure 12.18b is drawn with the stresses on faces A and B as shown.

Part b. From Mohr's circle, the major and minor stress values are found to be

$$\sigma_1 = OC + r = -200 + 1{,}345 = 1{,}145 \text{ kPa}$$
$$\sigma_2 = OC - r = -200 - 1{,}345 = -1{,}545 \text{ kPa}$$

(Note the major principal stress is always the stress at the maximum positive or minimum negative intercept of the circle with the σ-axis.)

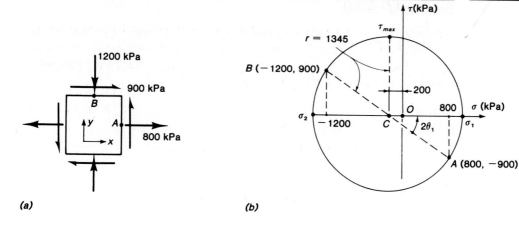

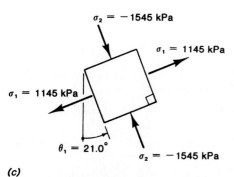

(c)

Figure 12.18 Mohr's circle for Example 12.4.

The orientation of the element when the principal stresses are acting is given by

$$\tan 2\theta_1 = \frac{900}{(800 + 200)} = 0.90$$

$$2\theta_1 = 41.99°$$

$$\theta_1 = 21.0°$$

It is noted that a counterclockwise rotation is required from the line CA to the line $C–\sigma_1$, and thus the element is oriented as shown in Figure 12.18c.

Part c. From Mohr's circle it follows that

$$\tau_{max} = r = 1,345 \text{ kPa}$$

Example 12.8

The state of stress at a point is shown in Figure 12.19a. Draw Mohr's circle for this case. Label and find the maximum shear stress, as well as the major and minor principal stresses. Draw a properly oriented element, showing the major and minor principal stresses on it.

Solution:
From Figure 12.19a we find the stresses on the faces A and B.
On face A:

$$\sigma_x = -15 \text{ ksi (compression)}$$

$$\tau_{xy} = 15 \text{ ksi}$$

On face B:

$$\sigma_y = 25 \text{ ksi}$$

$$\tau_{xy} = -15 \text{ ksi}$$

Using Equations 12.39 and 12.40 the center and radius of Mohr's circle are found:

$$OC = \frac{\sigma_x + \sigma_y}{2} = \frac{-15 + 25}{2} = 5 \text{ ksi}$$

$$r = \sqrt{\tau_{xy}^2 + \left(\frac{\sigma_x - \sigma_y}{2}\right)^2} = \sqrt{15^2 + \left(\frac{-15 - 25}{2}\right)^2} = 25 \text{ ksi}$$

From this, Mohr's circle is drawn as shown in Figure 12.19b. The stresses on the faces A and B are marked on the circle, and the following critical values are determined:

$$\tau_{max} = r = 25 \text{ ksi}$$

$$\sigma_1 = OC + r = 5 + 25 = 30 \text{ ksi}$$

$$\sigma_2 = OC - r = 5 - 25 = -20 \text{ ksi}$$

It is noted that there is a clockwise rotation $2\theta_1$ marked on Mohr's circle from the plane on which σ_1 acts. Thus,

$$\tan \alpha = \frac{AD}{DC} = \frac{15}{15 + 5} = 0.75$$

i.e.,

$$\alpha = 36.87°$$

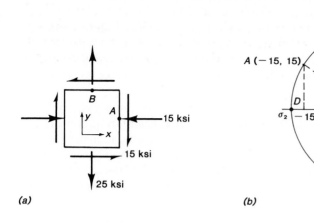

(a)

(b)

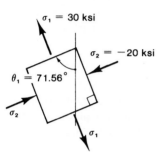

(c)

Figure 12.19 *Mohr's circle for Example 12.5.*

Thus,

$$2\theta_1 = 180^\circ - \alpha = 180^\circ - 36.87^\circ$$
$$= 143.13^\circ$$
$$\theta_1 = 71.56^\circ$$

As mentioned, this is a clockwise rotation, and so the principal stresses are indicated on an element shown in Figure 12.19c.

12.7 Combining Normal and Shear Stresses

In Sections 4 and 6 of this chapter, two methods of finding the state of stress on an arbitrary plane were presented, using either the equations for transformation of plane stress or the Mohr's circle construction. In each of these cases the analysis was commenced with a small square element on which some combination of normal stresses and shear stresses was acting. In this section we consider just how such combinations of stress may be generated.

The combined loading of a shaft is one of the most common cases in which the combined effects of shear stresses and normal stresses must be considered. In Chapter 9 it was shown that a shaft acted upon by a twisting moment produces shear stresses internal to the shaft. If such a shaft is subjected to a bending moment as well as a twisting moment (and this is almost always the case in shafts), then a section within the shaft will have both shear stress and normal stress acting on it. Furthermore, since both the normal stress due to bending and the shear stress due to twisting vary across a cross section, the combination of stress considered is valid for one position on a section. One of the reasons that it is often necessary to consider the combined state of stress on a shaft

Example 12.9

A shaft BA shown in Figure 12.20a is subject to a vertical force of 1.0 kilonewton, as indicated. The shaft has a 50-millimeter diameter and is rigidly fixed to a vertical wall at B. Find the stresses acting on a small rectangular element shown at A.

Solution:

The 1.0-kilonewton force may be replaced by a force of 1.0 kilonewton and a twisting moment, T, of $1.0 \times 0.7 = 0.7$ kilonewton meter acting at C, as is shown in the free body diagram of Figure 12.20b. The reactions at the support at B are also indicated on this diagram. Using the free body diagram, the bending moment diagram and the twisting moment diagram may be drawn as shown in Figures 12.20c and d, respectively. Thus, the twisting moment at A is 0.7 kilonewton meter and the bending moment at A is −0.6 kilonewton meter (i.e., the bending moment introduces tensile stresses at the element location). The moment of inertia, I, and the polar moment of inertia, J, are now required:

$$I = \frac{\pi r^4}{4} = \frac{\pi (25)^4}{4} = 3.068 \times 10^5 \, \text{mm}^4$$
$$= 3.068 \times 10^{-7} \, \text{m}^4$$

$$J = \frac{\pi r^4}{2} = \frac{\pi (25)^4}{2} = 6.136 \times 10^5 \, \text{mm}^4$$
$$= 6.136 \times 10^{-7} \, \text{m}^4$$

At the element, the shear stress is given by

$$\tau = \frac{Tr}{J} = \frac{700 \times (0.025)}{6.136 \times 10^{-7}} = 2.85 \times 10^7 \, \text{Pa}$$
$$= 28.5 \, \text{MPa}$$

is that there is often a point or points on a cross section at which the bending stress and the shear stress are greatest.

There are also a number of other instances where the combination of normal and shear stresses may be considered. These would include a shaft that is being acted upon by an axial force and a twisting moment simultaneously. Such a situation might occur if the operator of an auger was pushing on the connecting rod as well as trying to turn the auger.

Another common case is that of a beam, which typically has both shear force and bending moment present. In this problem the combination of shear and normal stresses is usually not critical. From Chapter 10 we remember that the bending stresses are largest at the greatest distance from the neutral axis and zero on the neutral axis. Chapter 11, on the other hand, shows that the shear stress due to the shear force is greatest on the neutral axis, and zero at the maximum distance from the neutral axis. Thus the conditions of maximum normal stress and maximum shear stress do not occur simultaneously, and thus the combined effect is usually not critical.

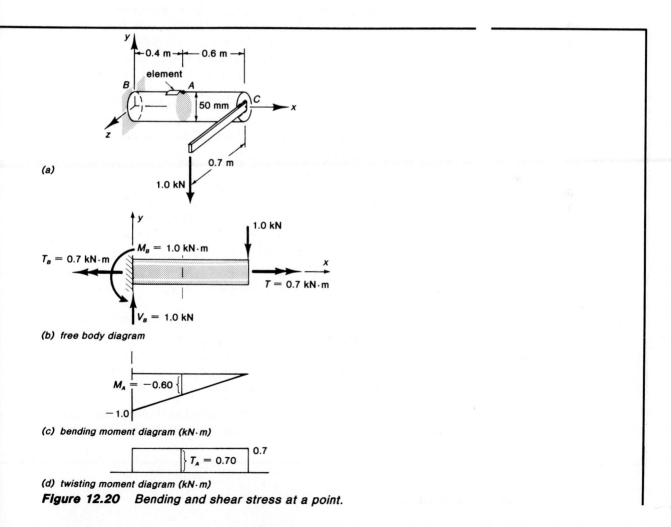

(a)

(b) free body diagram

(c) bending moment diagram (kN·m)

(d) twisting moment diagram (kN·m)

Figure 12.20   *Bending and shear stress at a point.*

Example 12.9 *(Continued)*

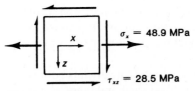

$\sigma_x = 48.9$ MPa

$\tau_{xz} = 28.5$ MPa   *(e) stresses on element at A (MPa)*

Figure 12.20   (Continued)

Example 12.10

A man is attempting to drill a post-hole using an auger, as shown in Figure 12.21a. In order to drill the hole, the man placed one hand on the handle at each of the points B and C. He attempts to turn the auger by applying a horizontal force of 30 pounds at each of B and C, as shown, as well as leaning on the handle so as to apply vertical forces also of 20 pounds at each of B and C. If the auger blades at E are stuck firmly in the soil, find the stresses at the point A on the shaft DAE. The shaft is of $1\frac{1}{2}$ inches diameter.

Solution:
If the shaft is cut at the point A and a free body diagram is drawn of the portion of the auger above the cut, then the forces acting at A may be found. Such a diagram is drawn in Figure 12.21b. The axial force N and the twisting moment T and A are found using the equilibrium equations:

$$\Sigma F_y = 0 \downarrow +$$
$$20 + 20 + N = 0$$
$$N = -40\,\text{lb (compression)}$$

$$\Sigma M_x = 0$$
$$30 \cdot (18) - T = 0$$
$$T = 540\,\text{lb} \cdot \text{in}$$

The area A and the polar moment of inertia J of the shaft are now found:

$$A = \pi r^2 = \quad (0.75)^2 \quad = 1.767\,\text{in}^2$$
$$J = \frac{\pi r^4}{2} = \frac{\pi (0.75)^4}{2} = 0.494\,\text{in}^4$$

Clearly, this problem once again involves normal and shear stresses, since axial forces produce normal stress and twisting moments produce shear stress.

Similarly, the tensile bending stress at the element is

$$\sigma = \frac{My}{I} = \frac{600 \times (0.025)}{3.068 \times 10^{-7}} = 4.89 \times 10^7 \text{ Pa}$$
$$= 48.9 \text{ MPa}$$

Thus the stresses on the element, as seen by an observer looking down on the element from above, are shown in Figure 12.20e. The Mohr's circle construction could now be used to find the principal stresses. (Note that the stress on this element is not the maximum stress occurring in the shaft.)

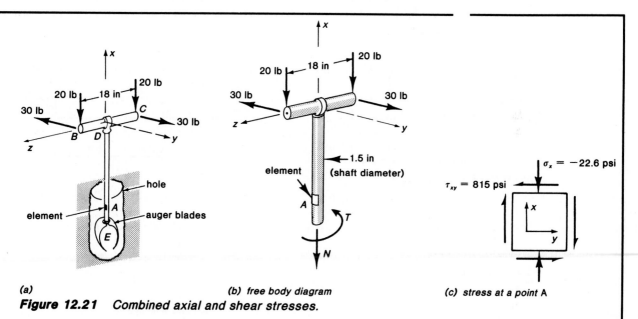

(a)

(b) free body diagram

(c) stress at a point A

Figure 12.21 *Combined axial and shear stresses.*

The values of each of these stresses may be found using the usual techniques:

$$\sigma = \frac{P}{A} = \frac{-40}{1.767} = -22.6 \text{ psi}$$

$$\tau_{max} = \frac{Tr}{J} = \frac{540 \cdot (0.75)}{0.494} = 815 \text{ psi}$$

The stresses on the element at A are shown on the diagram of Figure 12.21c. It is clear from this diagram that the magnitude of the shear stress is much greater than the magnitude of the normal stress due to the axial force. Therefore the stress due to the axial force is rarely a consideration when the shear is critical, but it should be noted that a higher compression stress will result, as well as a relatively high tension, due to the effect of the shear stresses.

12.1–
12.4   For the beams shown in Figures P 12.1 to P 12.4, determine the
        maximum tensile and compressive stress developed. Indicate the
        location at which they occur.

12.5   Determine the maximum shear stress at the wall on Figure P 12.5
        due to the 40-kip load acting in the negative *y*-direction.

12.6   Determine the normal and shear stresses at points *A*, *B*, *C*, and *D*
        on the bar shown in Figure P 12.6. The diameter of the bar is
        80 millimeters.

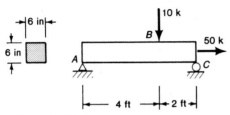

Figure P 12.1

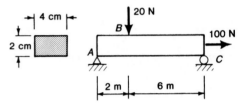

Figure P 12.2

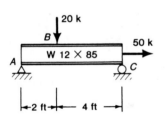

Figure P 12.3

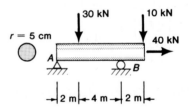

Figure P 12.4

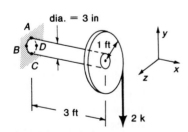

Figure P 12.5

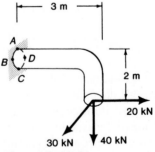

Figure P 12.6

12.7   A crank consists of a solid shaft of 10-inch diameter, as shown in Figure P 12.7. Determine the direct stress in the z-direction and the shear stresses at point A.

12.8   For the member shown in Figure P 12.8, determine the normal stress and shear stress at point A. The diameter of the member is 2 centimeters.

12.9   If the cylindrical pipe of Figure P 12.9, with an outer diameter of 10 inches and an inner diameter of 9.5 inches, is subjected to an internal pressure of 1,000 pounds per square inch, determine the longitudinal and circumferential stresses on the outer surface of the pipe.

12.10 Rework Problem 12.9 for the thin-walled cylindrical pipe shown in Figure P 12.10.

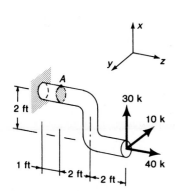

Figure P 12.7

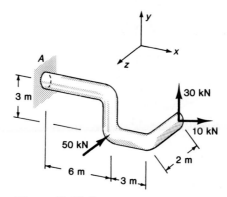

Figure P 12.8

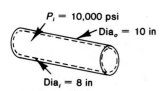

Figure P 12.9

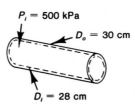

Figure P 12.10

12.11 Determine the maximum internal pressure, P, for the seamless steel sphere with cross section shown in Figure P 12.11. Assume the ultimate stress of the steel is 40,000 pounds per square inch.

12.12 A piece of 300-millimeter diameter pipe with a 5-millimeter thick wall was closed off at the ends, as shown in Figure P 12.12. This assembly was then placed into a testing machine and subjected simultaneously to an axial pull, P, and an internal pressure of 2.5 megapascals. If the distance between the gage points A and B is observed to increase from 220 to 220.05 millimeters, determine the magnitude of the applied force, P. ($E = 207$ GPa; $v = 0.20$.)

12.13 A closed thin-walled cylinder, shown in Figure P 12.13, is fixed at one end to a vertical wall. If the cylinder has an external diameter of 12 inches and a wall thickness of 0.15 inch, find the direct stress in the x-direction at point A, due to the internal pressure, P, and the external applied load.

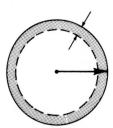

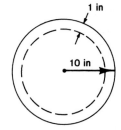

Figure P 12.11

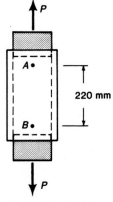

Figure P 12.12

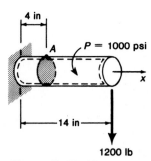

Figure P 12.13

12.14–

12.16 For the infinitesimal elements shown in Figures P 12.14,
P 12.15, and P 12.16, find the normal and shear stresses acting on
the indicated planes. Use the equations for transformation of plane
stress.

12.17–

12.22 For the elements given in Figures P 12.17 through P 12.22:
   a. Determine the principal stresses and show their sense on a
      properly oriented element.
   b. Determine the maximum shearing stress with associated normal
      stresses and show the results on a properly oriented element.

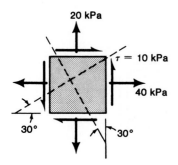

Figure P 12.14

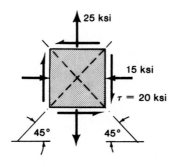

Figure P 12.15

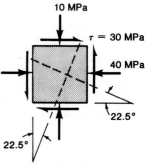

Figure P 12.16

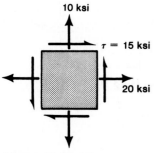

Figure P 12.17

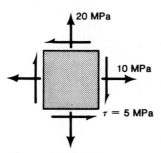

Figure P 12.18

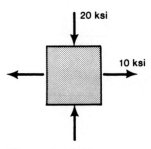

Figure P 12.19

12.23 The state of stress for a wooden beam is shown on the infinitesimal element in Figure P 12.23. The direction of the grain in the wood makes a $+45°$ angle with the x-axis. The allowable shearing stress parallel to the grain is 200 pounds per square inch for this wood. Is the state of stress permissible? Verify your answer by calculations.

12.24 The state of stress for a wooden beam is shown on the infinitesimal element in Figure P 12.24. The direction of the grain in the wood makes a $-60°$ angle with the x-axis. The allowable shearing stress parallel to the grain is 10 kilopascals for this wood. Is the state of stress permissible? Verify your answer by calculations.

12.25 Using P^* as defined in Problem 7.19, find the maximum direct stress occurring within the walls of the closed cylinder shown in Figure P 12.25, with an internal pressure of 300 pounds per square inch in the cylinder.

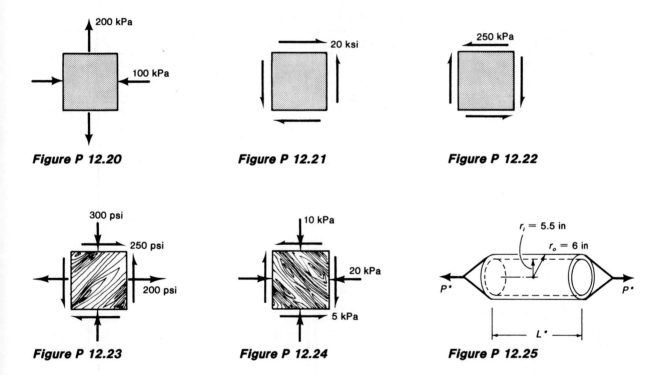

Figure P 12.20

Figure P 12.21

Figure P 12.22

Figure P 12.23

Figure P 12.24

Figure P 12.25

Deflection of Beams

13

1. *To know how to apply deflection formulas and the technique of superposition.*

2. *To review the basic principles of geometry pertaining ot the deflection, slope, and curvature of a beam.*

3. *To understand the relationship between bending moment and deformation of a beam.*

4. *To know how deflection formulas are developed.*

Objectives

13.1 Deflection Formulas

The determination of deflections of beams subjected to bending is an important problem of mechanics and is directly applicable to beam design. Design specifications require not only that maximum stresses not exceed certain allowable limits, but also that beams must not deflect excessively. Usually the *allowable deflection* of a beam is specified in terms of a percentage of its length. For instance the *Handbook of the American Institute of Steel Construction* specifies that the maximum allowable deflection under live load is 1/360 of the span. For structures where deformations may be critical, the maximum allowable deflection is specified directly. In some cases the designer needs to know not only the maximum deflection of a beam under a certain loading but also the deflection and slope of the beam at any point along the beam's axis. Computation of such quantities is usually made easy by tabulated solutions for simple cases of loading. One is included in Appendix B. The table shows equations for 38 cases of beam

Example 13.1

A 9-foot-long cantilever beam carries at its center a 4,000-pound load. The beam is a W 8 × 31 size steel beam. Determine the deflection and the slope of the beam at its end. Neglect the weight of the beam.

Solution:

Figure 13.1 shows the arrangement. We shall denote the end of the cantilever as point A, the point of load application at the middle as point B, and the fixed end as point C. The deflection y_A at point A is desired as well as the slope θ at that point.

The solution is found in the deflection table as Case 2. The concentrated load is denoted in this table as W, the length of the beam as l. To place the load at the center, the distance of the load to the fixed end is $a = b = 4.5$ feet. Although by intuition we may realize that the deflection at A is the maximum deflection, y_{max}, for which a formula is given in the table, for practice we may use the first formula, giving the deflection at any point between A and B. Substituting our values, with all lengths converted into inches, we get

$$y_A = -\frac{1}{6} \frac{W}{EI} (-a^3 + 3\,a^2\,l - 3\,a^2\,x)$$

$$= -\frac{1}{6} \frac{4,000}{EI} [-54^3 + 3\,(54)^2\,108 - 3\,(54)^2\,0]$$

The last term, x, in the equation is zero since the origin of the coordinate system shown in the sketch in the table is located at point A. Before we solve our equation we will need the modulus of elasticity E and the moment of inertia I for the beam. For the former, we may use 30×10^6 pounds per square inch. For

loadings, including both statically determinate and statically indeterminate beams. Concentrated and distributed loadings are shown. Similar deflection tables may be found in most handbooks of mechanical and civil engineering as well as in design codes for steel, concrete, and timber. The cases of loading included are selected so that a designer will find a solution directly for most loading situations that come up in everyday practice.

When a combination of loads is placed on a beam, a solution cannot be found directly in deflection tables. Such is the case, for example, when there are two or more concentrated forces on a beam, or when concentrated and distributed forces appear jointly. Such problems may be solved by using the *principle of superposition*. This allows the separation of the loadings into two or more simple cases for which tabular solutions exist, solving each of these individually, and adding all of these solutions at the end.

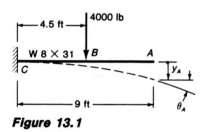

Figure 13.1

the latter, the steel table at the end of the book has to be used. In the table for wide flange (W) sections we find that $I = 110$ in⁴. Substituting these values and solving the formula leads to

$$y_A = -0.16 \text{ in}$$

The negative sign comes from the coordinate system adopted in the deflection table, pointing upward for positive deflections.

To obtain the slope at point A, the table gives the following formula

$$\theta = +\frac{1}{6}\frac{Wl^2}{EI} \quad (\text{from } A \text{ to } B)$$

As there is no load between A and B, bending does not occur between these points; hence the slope is constant in this portion of the beam. Substituting our values we obtain

$$\theta = +\frac{1}{6}\frac{4,000\,(108)^2}{EI}$$

$$\theta = 2.3 \times 10^{-3} \text{ rad}$$

Example 13.2

Solve Example 13.1, including the weight of the beam in the computations.

Solution:

The weight of the W 8×31 beam is 31 pounds per foot or 2.58 pounds per inch. The total weight of the 9-foot-long piece is $9 \times 31 = 279$ pounds, which is only about 7 percent of the concentrated load carried by it. For this reason the consideration of the beam's own weight would often be neglected in practice.

The case of a cantilever beam loaded with a uniformly distributed force is Case 3 in our deflection table. The maximum deflection at point A is given as

$$y_A = \frac{1}{8} \frac{W}{E} \frac{l^3}{I}$$

in which

$$W = w \cdot l = 2.58 \text{ (lb/in)} \times 108 \text{ (in)} = 279 \text{ lb}$$

Substituting all known values into the deflection equation, our result is

$$y_A = -0.013 \text{ in}$$

The calculation of beam deformations has important application in the analysis of statically indeterminate beams. To understand the derivation of the techniques for analyzing indeterminate structures, one should know how deflection formulas originated. The remainder of this chapter addresses this question.

13.2 Geometric Fundamentals

Many different methods are used in the analysis of beam deformations. They are all based on the same simple geometric relationships between the radius and angle of curvature, tangent, arc length, and deflection of a line. In this section these fundamental concepts will be reviewed.

Let us consider a smooth curving line connecting points A, B, C, and D, as shown in Figure 13.3. Let us assume that the portions of this line between each adjacent pair of these points may be well approximated by circular arcs. The centers of each of these arcs are located at O_1, O_2, and O_3. The radii of curvature and central angles are denoted by ρ_1, ρ_2, and ρ_3 and ϕ_1, ϕ_2, and ϕ_3, respectively. Perpendicular to each radius one may draw the corresponding tangent. Observing the drawing, one may recognize that the change of slope between adjacent points is equal to the central angle of the connecting arc. It is also evident from the drawing that the sum, $\Sigma \phi$, of all tangential slope changes from point A through D equals the angle between the lines EF and EG.

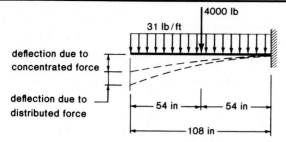

Figure 13.2 *Superposition of deflections due to concentrated force and distributed force acting on a cantilever beam.*

As shown in Figure 13.2, the deflections due to the concentrated force and to the distributed force can be simply added together. The result of Example 13.1 was $y_A = -0.16$ inch. Adding this to the last result of -0.013 inch, the combined deflection will be

$$y_A = -0.173 \text{ in}$$

The slope of the beam at its end may be considered in a similar manner.

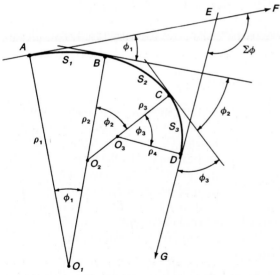

Figure 13.3 *Geometric relationships between the parameters of a curving line replaced by elemental circular arcs.*

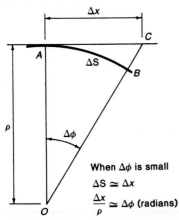

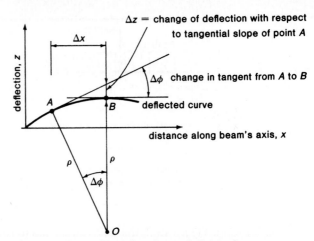

When $\Delta\phi$ is small

$\Delta S \simeq \Delta x$

$\dfrac{\Delta x}{\rho} \simeq \Delta\phi$ (radians)

Figure 13.4 *The fundamental assumption in deflection geometry: For a small incremental change of the central angle, arc length ΔS may be replaced by Δx.*

Figure 13.5 *Notation for a deflected curve.*

The central angle of a circular arc may be measured in degrees or in radians. One radian is defined as the angle at which the arc length equals the radius. A complete circle is then $360°$, or 2π radians, and the

(13.1) $$\text{angle in radians} = \frac{2\pi}{360} \text{ (angle in degrees)}$$

On this basis the relationship between arc length S, radius of curvature ρ, and central angle ϕ in Figure 13.1 is

(13.2) $$\phi \text{ (in radians)} = \frac{S}{\rho}$$

When the curved line is replaced by elementary small arcs, the central angle of the corresponding circles will be small as well. However, the radius of curvature ρ will not diminish along with the arc length. On the other hand, as the arc length is taken to be a small quantity it may be replaced by the tangential length Δx, as shown in Figure 13.4. The assumption of $\Delta S = \Delta x$ is a reasonable one in practical mechanics, since the deformation of beams subjected to bending is usually very small. Thus the radius of curvature ρ is very large. The error introduced by taking ρ to be the distance between points OC rather than OB, as shown in Figure 13.4, is negligible. With this assumption the relationship described by Equation 13.2 may be rewritten as

(13.3) $$\Delta\phi = \frac{\Delta x}{\rho}$$

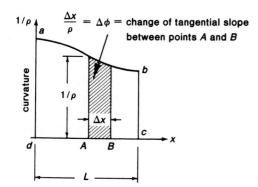

$\frac{\Delta x}{\rho} = \Delta\phi$ = change of tangential slope between points A and B

Figure 13.6
Plot of curvature
of a beam subjected to bending.

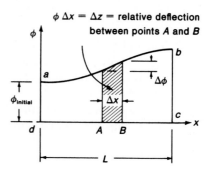

$\phi \Delta x = \Delta z$ = relative deflection between points A and B

Figure 13.7
Tangential slope versus distance,
plotted by the use of Equation 13.5.

Using these notations, the change of slope in a deflected curve shown in Figure 13.5 may be written as

(13.4)
$$\frac{\Delta\phi}{\Delta x} = \frac{1}{\rho} = \text{curvature}$$

Along a deflected beam the magnitude of curvature may be plotted as shown in Figure 13.6. For each small increment Δx along the axis of the beam, the shaded area equals the rate of change of the tangential slope, according to Equation 13.6. The total change of slope from one end of the beam to the other, that is, from $x = 0$ to $x = L$, is the total area enclosed by points $a, b, c,$ and d shown in the figure. This is an important result of these considerations. For later use we may rephrase it as follows: *The change of slope of a line between two points equals the area under the plot of curvature (l/ρ) verus distance (x) between the two points in question.* Putting it into mathematical format, this statement may be written as

(13.5)
$$\Delta\phi_{AB} = \sum_{A}^{B}\left(\frac{1}{\rho}\right)\Delta x = \text{area under the } \frac{1}{\rho} \text{ diagram}$$

between A and B

Since by Equation 13.5 the slope of a deflected line may be determined at any point, it may be computed and plotted for the whole line, as shown in Figure 13.7. In this plot the initial tangent (at $x = 0$) was assumed to have a known value. This is a necessary requirement for plotting the results of Equation 13.5, since that equation provides only the change of the tangential slopes between two points along x.

Once the slope is defined for all points of the line, we now can find the deflection itself. *Slope is defined as the rate of change of deflection.* In the form of an equation this means

(13.6)
$$\tan \phi = \frac{\Delta z}{\Delta x} \cong \phi$$

The right side of this equation is valid only for very small angles. This requirement is fully satisfied under the assumption made in mechanics that deflections are small. Therefore, as Δz is very much smaller than Δx, the tangent of ϕ may be replaced by ϕ itself.

Rearranging Equation 13.6 to express the relative deflection Δz between two points, e.g., A and B in Figure 13.5, one obtains

(13.7)
$$\Delta z = \phi \Delta x$$

From Figure 13.7 it is clear that this equation represents the area under the ϕ curve. The relative deflection between points A and B is shown as the shaded area in the plot. Of course the total deflection from $x = 0$ to $x = L$ is the area under the whole ϕ curve encompassed by points a, b, c, and d. This is another important result. Generally it may be restated as follows: *The change in deflection of a line between two points equals the area under the plot of tangential slope (ϕ) versus distance (x) between the two points in question.*

The change of deflection is taken with respect to the slope of the tangential line at the point from which the relative deflection is measured, as shown in Figure 13.5.

When the relative deflection between two points (such as points A and B in Figures 13.5, 13.6, and 13.7) is desired and only the curvature diagram $1/\rho$ is given initially, one must go through two successive summations of areas under the $1/\rho$ and ϕ curves. As outlined in Equation 7.4, this double summation of the two dependent functions may be regarded as the first moment of the $1/\rho$ function. The moment arm in this case is the distance between the centroid under the $1/\rho$ function between points A and B and the point A whose tangent is used as a base line for the relative deflection. Figure 13.8 illustrates this concept. Using the notations shown in this drawing, the relative deflection Δz between points A and B is given by

$$\Delta z = \bar{x} \cdot (\text{area under } 1/\rho \text{ diagram between } A \text{ and } B)$$
$$\Delta z = \bar{x} \cdot \Delta \phi_{AB}$$

where, in the second equation, $\Delta \phi_{AB}$ is given by Equation 13.5. Of course, $\Delta \phi_{AB}$ is the difference between the tangents at A and B. Note that as one draws the tangents to the deflected shape at points A and B, their point of intersection is at the same location as the centroid of the area under the $1/\rho$ curve between points A and B.

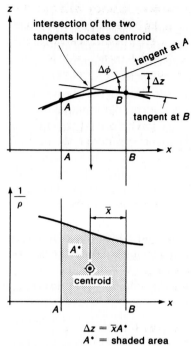

Figure 13.8
Determination of the
relative deflection of point B with
respect to the tangent at point A.

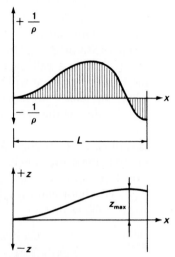

Figure 13.9
Deflection curve (z versus x)
constructed on the basis of curvature
data (1/ρ versus x).

With Equation 13.8 providing the relative deflection of a deflected shape between any two points, one may compute and plot the deflection for all points between $x = 0$ and $x = L$. As the equation results in relative deflections only, it is necessary to locate at least one point of the line as an initial location, say $z = 0$ at $x = 0$. With this the complete curve of deformation may be plotted with respect to the curvature $1/\rho$, as shown in Figure 13.9.

13.3 Elastic Considerations

The purpose of this section is to find an expression for the curvature of a deflected beam in terms of the bending moment and the properties of a beam influencing its stiffness. In order to do this, let us consider the effect of a bending moment on a small part of length Δx from a beam as shown in Figure 13.10. The geometry as well as the elastic modulus of the beam is known. Within the elastic range, the deformation of the beam will follow Hooke's law. As the deformation is small, one can state that $\Delta S = \Delta x$. The central angle of the deformed element $\Delta\phi$ equals $\Delta S/\rho$ or $\Delta x/\rho$. On the left side of the element considered, the beam is rigidly fixed to a wall and will not undergo any deformation. On the right side, the bending moment will cause compressive stresses on the lower part of the beam and tensile stresses on its upper part. There will be no stress, hence no deformation, at the neutral axis. Following Hooke's law, the compressive strain caused by the bending is described by the triangle ABD. The tensile strain is described by the triangle BCE. From geometric considerations one will recognize that the central angle $\Delta\phi$ at point O is equal to the angle ABD at the neutral axis. Both, of course, are then equal to Equation 13.3.

The deformation of the extreme fiber AD at the compressive side is u_{max}. It also follows that

$$(13.9) \qquad u_{max} = c\,\Delta\phi$$

As this is the total deformation over the length Δx, the maximum compressive strain may be written by Hooke's law as

$$(13.10) \qquad \sigma_{max} = E\,\epsilon_{max} = \frac{u_{max}E}{\Delta x} = \frac{c\,\Delta\phi\,E}{\Delta x}$$

Substituting Equation 10.5 in place of the maximum compressive stress σ_{max}, and by Equation 13.3, $1/\rho$ in place of $\Delta\phi/\Delta x$, and rearranging the result, we find for the curvature of a beam subjected to bending moments is

$$(13.11) \qquad \frac{1}{\rho} = \frac{M}{E\,I}$$

The right side of this equation is referred to as *reduced bending moment*. If the loading on a beam is known, the distribution of bending moments over it can be determined according to the methods introduced in Chapter 6. If the material and the geometry of the beam are known also, the diagram showing this reduced bending moment may be constructed. According to Equation 13.11, this diagram then depicts the curvature of the deflected beam. As described in the previous section, this is all the information one needs to determine the deformation of a beam subjected to bending.

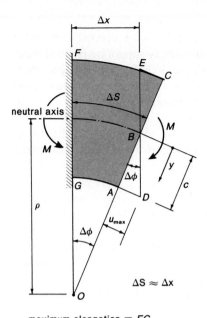

$$\Delta S \approx \Delta x$$

maximum elongation $= EC$

maximum tensile strain $= \dfrac{EC}{EF}$

maximum compression $= AD$

maximum compressive strain $= \dfrac{AD}{GA}$

Figure 13.10   *Notation for the elastic considerations of beam deformations.*

Example 13.3

Construct the reduced moment diagram for the simply-supported beam shown in Figure 13.11. The elastic modulus for the steel is 30×10^6 pounds per square inch. The beam consists of two parts, as shown. Determine the area of the reduced moment diagram and the location of its centroid.

Solution:

First the reactions at the two supports are computed by writing equilibrium formulas for moments. This leads to

$$R_1 = 1.714 \text{ kips}$$

$$R_2 = 1.286 \text{ kips}$$

The moment diagram will be of triangular shape. Its maximum at the point of application of the 3-kip load is

$$M_{max} = 6 \ (12) \ 1.714 = 123 \text{ kip} \cdot \text{in}$$

The moment of inertia for part A is

$$I_A = \frac{5 \ (4)^3}{12} = 26.6 \text{ in}^4$$

For part B,

$$I_B = \frac{5(6.45)^3}{12} - I_A$$

$$I_B = 85.2 \text{ in}^4$$

The product of E and I will then be

$$EI_A = 0.8 \times 10^6 \text{ kip} \cdot \text{in}^2$$

$$EI_B = 2.56 \times 10^6 \text{ kip} \cdot \text{in}^2$$

The $\frac{M}{EI}$ terms may be computed by dividing the maximum moment first by EI_A, resulting in 154×10^{-6} in^{-1}, then by EI_B, resulting in 48×10^{-6} in^{-1}. These results are plotted in Figure 13.11. The area of the reduced moment diagram is

$$A^* = 6(12) \ 154 \ (10^{-6}) \ 0.5 + 8 \ (12) \ 48 \ (10^{-6}) \ 0.5$$
$$= 5.544 \ (10^{-3}) + 2.3 \ (10^{-3})$$
$$= 7.84 \times 10^{-3} \text{ in/in, a dimensionless number}$$

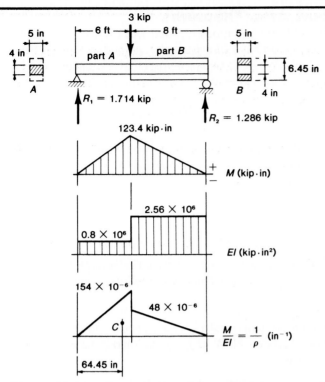

Figure 13.11 *Construction of the reduced moment diagram.*

Measured from the left support, the centroid of the left triangle under part A in the $1/\rho$ diagram is located at $(2/3)\, 6\,(12) = 48$ inches. The centroid of the right triangle is at $6\,(12) + (1/3)\, 8\,(12) = 104$ inches. The centroid of the whole $1/\rho$ diagram is then

$$\bar{x} = \frac{48\,(5.544) + 104\,(2.3)}{7.84} = 64.45\,\text{in}$$

measured from the left support.

13.4 The Moment Area Theorems

Substituting Equation 13.11 into Equations 13.5 and 13.8 results in two important formulas that serve as the basis of a powerful method of solving beam deflection problems. They give rise to two principles that are called *moment-area theorems.* The first one is written as

(13.12)
$$\Delta\phi_{AB} = \sum_{A}^{B} \frac{M\Delta x}{EI} = A^*$$

stating that *the change of slope in radians between two points A and B on a deflected elastic beam is equal to the corresponding area A* under the reduced moment diagram between the two points.*

Example 13.4

Determine the tangents of the deflected beam analyzed in Example 13.3 at the two supports and the deflection of the point where the load is applied.

Solution:

Let us denote the two unknown tangential angles as ϕ_1 and ϕ_2, referring to the left and the right supports, respectively. With this notation the change of slope between the two supports is

$$\Delta\phi = \phi_2 + \phi_1$$

According to the first moment area theorem, this change of slope equals the area of the reduced moment diagram between the supports. From Example 13.3, this equals

$$A^* = 7.84 \times 10^{-3} = \Delta\phi \text{ (in radians)}$$

According to the third moment area theorem, the tangential lines at the two supports must intersect under the location of the centroid of the reduced moment area. This was computed to be 64.45 inches to the right from the left support, as shown in Figure 13.12. Denoting the distance between the horizontal line and the intersection of the two tangential lines by r, the relative magnitude of ϕ_1 and ϕ_2 can be derived to be

$$r = 64.45 \, \phi_1 = (168 - 64.45) \, \phi_2$$

From this it follows that

$$\phi_1 = \frac{168 - 64.45}{64.45}\phi_2 = 1.6 \, \phi_2$$

Hence

$$(1.6 + 1) \, \phi_2 = 7.84 \times 10^{-3} \text{ rad}$$

The second one is written

(13.13)
$$\Delta z = \bar{x} \cdot A^*$$

stating that *the deflection of point* B *measured from the tangent at point* A *is equal to the moment of the corresponding area* A* *under the reduced moment diagram about point* B. (The centroidal distance $\bar{x}$ is measured from point B.)

As a consequence of the geometric properties, an additional third theorem may also be introduced by stating that *the tangents of a deflected elastic beam at points* A *and* B *intersect at a point that coincides with the location of the centroid of* A*.

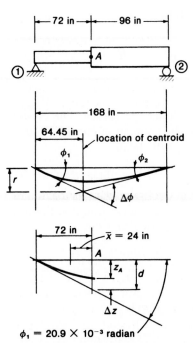

$\phi_1 = 20.9 \times 10^{-3}$ radian

Figure 13.12 *Application of the moment area theorems.*

as a consequence,

$$\phi_2 = 3.0 \times 10^{-3} \text{ rad}$$

and

$$\phi_1 = 1.6 \, (3.0 \times 10^{-3}) = 4.8 \times 10^{-3} \text{ radians}$$

Example 13.4 *(Continued)*

The determination of the deflection of point A, where the two beams are connected, requires the use of the second moment area theorem. It must be remembered that this theorem allows the determination of a relative deflection measured from the tangent line of another point. In this case the tangent at the left support, ϕ_1, will be used as a base. Equation 13.13 will then give Δz, as shown in Figure 13.12. As a matter of practicality we seek to know the deflection from the horizontal line, which is $z_A = d - \Delta z$. The distance d can be obtained by writing

$$\frac{d}{72} = \phi_1 = 4.8 \times 10^{-3}$$

from which

$$d = 72 \, (4.8 \times 10^{-3}) = 0.34 \text{ in}$$

Example 13.5

The cantilever beam shown in Figure 13.13 is loaded as shown. The elastic modulus E of the beam is 12 gigapascals, and the moment of inertia is 110×10^6 mm⁴. Determine the deflection at the end of the beam.

Solution:

The moment diagram is triangular, as shown in the figure, with its maximum value at the wall, M_{max}, being 5×10^6 newton millimeters. Note that there is no moment between points B and C. The product of E and I is

$$EI = 12 \times 10^3 \, (\text{N/mm}^2) \times 110 \times 10^6 \, (\text{mm}^4)$$
$$= 132 \times 10^{10} \, (\text{N} \cdot \text{mm}^2)$$

The maximum value of the reduced moment diagram is

$$\left(\frac{1}{\rho}\right)_{max} = \left(\frac{M}{EI}\right)_{max} = \frac{5 \times 10^6}{132 \times 10^{10}} = 3.79 \times 10^{-6} \, \text{mm}^{-1}$$

The slope of the beam at the wall remains zero. Therefore the change of the slope between points A and B is found using the first moment area theorem:

$$A^* = \frac{1}{2}(2{,}500) \, 3.79 \times 10^{-6} = 0.0047 \text{ rad} = \phi_B$$

The relative deflection Δz is computed from Equation 13.13, in which the partial moment area between the left support and point A was computed in Example 13.3 as

$$A^* = 5.544 \times 10^{-3}$$

Combining this value with that of $\bar{x}$ (24 inches, measured from point A to the centroid) in Equation 13.13, one obtains

$$\Delta z = 24 \ (5.544 \times 10^{-3}) = 0.133 \text{ in}$$

The actual deflection will then be

$$z_A = d - \Delta z = 0.34 - 0.133 = 0.21 \text{ in}$$

By similar computations, the deflection of any other point of the span may be determined. Calculating the deflections at a reasonable number of locations along the span, a deflection curve may be constructed for the beam.

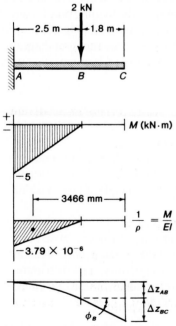

Figure 13.13 *Deflection computation using the moment area theorems.*

Example 13.5 *(Continued)*

As there is no curvature between points B and C, the same slope will remain throughout over that portion of the beam. Because of this initial (at point B) slope the deflection at point C will be composed of two parts: one, the deflection of point B due to the bending moment causing curvature between points A and B; the other, the deflection caused by the slope at point B.

The deflection at point B is to be determined by the second moment area theorem. The centroid of the triangle representing the reduced moment area lies two-thirds of the way along the 2,500-millimeter span. The moment of the area is therefore

$$\Delta z_{CA} = \left(\frac{2}{3} \cdot 2{,}500 + 1{,}800 \right) 0.0047 = 3{,}466 \, (0.0047) = 16.29 \, \text{mm}$$

In the case of complex loading conditions, when several forces or distributed forces are placed on a structure, the analysis of the moment diagram gets more difficult. In Chapter 6 we have learned that such problems can be simplified by using the principle of superposition. In this technique, the loading may be separated into its simpler elements. After determining the moment diagram for the parts, the results are added together to obtain the composite moment diagram. In deflection analysis the same principle applies. Reduced moment diagrams

Example 13.6

A two-supported beam is loaded as shown in Figure 13.14. Determine the deflection at the end of the overhang. The elastic modulus is 35 megapascals, and the moment of inertia is $200 \times 10^6 \, \text{mm}^4$. Apply the principle of superposition.

Solution:

For the concentrated force alone, the moment diagram is composed of a triangle with its peak at point B. The magnitude of this moment equals 1.2 (m) $\times$ 1.1 (kN) $= 1.32 \, \text{kN} \cdot \text{m} = M_B{}'$ and is a negative quantity. The distributed force between A and B results in a parabolic moment distribution. The maximum moment will be at mid-span. Its magnitude M''_{center}, is

$$\frac{0.8 \, (4)^2}{8} = 1.6 \, \text{kN} \cdot \text{m}$$

a positive quantity.

The second part of the deflection of the end of the beam may also be computed by the aid of the angle ϕ_B:

$$\frac{\Delta z_{BC}}{1,800} = \phi_B = 0.0047 \text{ rad}$$

therefore,

$$\Delta z_{BC} = 1,800 \, (0.0047) = 8.46 \text{ mm}$$

The deflection of point B will then be

$$z_B = 16.29 - 8.46 = 7.83 \text{ mm}$$

for elements of the composite loading have simpler geometries; hence their areas and the location of their centroids are less difficult to determine. Once these terms are known, angles of rotation and deflections can be superimposed. The sum of the results is the same as if the whole effect of the loading had been analyzed in one single step. Of course, in this process one has to be very careful in maintaining the sign convention. Drawing individual deflection diagrams for each of the separate loadings aids in avoiding errors. The table in Appendix B includes many deflection formulas applied to common loading conditions.

Before the second moment area theorem is applied to obtain the relative deflection of the end of the overhang, the tangential slope of the beam at point B must be determined. This angle is the sum of two angles, one due to the distributed load, the other due to the concentrated one. To obtain ϕ'', the first moment area theorem is to be applied to the distributed loading. Equation 13.12 leads to

$$A^{*\prime\prime} = \frac{2}{3} \, (4,000) \, 1.6 \, \frac{10^6}{35 \times 10^3 \, (200 \times 10^6)} = 0.61 \times 10^{-3} \text{ rad}$$

As the loading is symmetrical in the span AB, the slope at point B is one-half of the value of $A^{*\prime\prime}$. Hence

$$\phi'' = 0.305 \times 10^{-3} \text{ rad}$$

Example 13.6 *(Continued)*

To obtain the angle of rotation at point B due to the concentrated force the reduced moment area between points A and B must first be determined. This is

$$A^{*\prime} = \frac{1}{2} (4,000)\, 1.32\, (10^6)\, \frac{1}{35 \times 10^3\, (200 \times 10^6)}$$

$$= 3.77 \times 10^{-4}\ \text{rad}$$

This is the change of the slope between points A and B. As the moment distribution is not symmetrical between these points, the angle at B will be different from the angle at A. Example 13.4 shows the method of solution of one such problem; using the technique applied there the location of the centroid is determined. It is $(2/3)\, 4,000$ millimeters to the right of point A. Consequently, $\phi_B{}' = 2\, \phi_A$ and since $\phi_A{}' + \phi_B{}' = 3.77 \times 10^{-4}$, it follows that

$$\phi' = 2.5 \times 10^{-4}\ \text{rad}$$

Observing the two sketches that show the expected deformation of the beam for the two loadings, it is obvious that ϕ' rotates clockwise and ϕ'' rotates counterclockwise. The sum of these rotations is

$$\phi_B = (2.5 \times 10^{-4}) - (3.05 \times 10^{-4}) = -0.55 \times 10^{-4}\ \text{rad}$$

i.e., counterclockwise. Finally, the deflection of point C will be composed of two terms. One is due to the counterclockwise rotation of point B. The other is due to the downward relative deflection of point C from point B caused by the reduced moment diagram or curvature of the beam between points B and C. The first term equals

$$z_1 = -\phi_B (1,200) = -0.55 \times 10^{-4}\, (1,200)$$

$$= -0.066\ \text{mm (upward)}$$

The second, by the application of the second moment area theorem,

$$\Delta z = \frac{1}{2}(1,200)\, 1,320,000 \left(\frac{2}{3} \right) 1,200 \left[\frac{1}{35 \times 10^3\, (200 \times 10^6)} \right]$$

$$= 0.09\ \text{mm (downward, measured from the tangent line)}$$

The final result is

$$z_1 - \Delta z = z_C\ 0.09 - 0.066 = 0.024\ \text{mm (downward)}$$

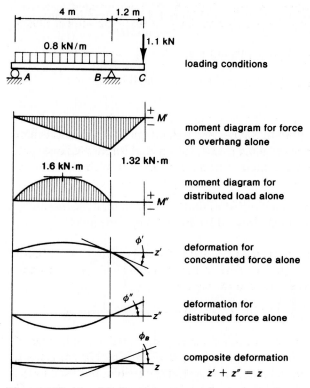

loading conditions

moment diagram for force
on overhang alone

moment diagram for
distributed load alone

deformation for
concentrated force alone

deformation for
distributed force alone

composite deformation

$z' + z'' = z$

Figure 13.14 *Method of superposition in deflection analysis.*

13.1   For the cantilever beam shown in Figure P 13.1, determine the slopes and deflections at B and C. (EI = constant.)

13.2   Repeat Problem 13.1 with Figure P 13.2.

13.3   Verify your answer for Problem 13.1 by using the table in Appendix B.

13.4–

13.6   Use the table to find the slopes and deflections at A and B for the structures shown in Figures P 13.4, P 13.5, and P 13.6. Assume EI = constant. (Use superposition method.)

13.7–

13.8   Use the moment area method to find the maximum deflection for the beams shown in Figures P 13.7 and P 13.8, respectively.

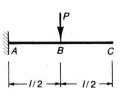

Figure P 13.1

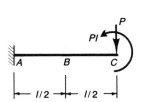

Figure P 13.2

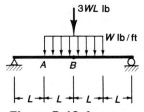

Figure P 13.4

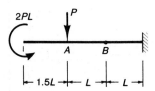

Figure P 13.5

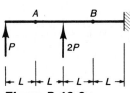

Figure P 13.6

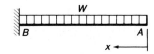

Figure P 13.7

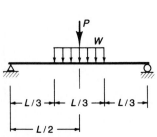

Figure P 13.8

13.9–
13.17 By using the moment area method, find the deflection and slope of the elastic curve at point A for each beam and loading condition as shown in Figures P 13.9 to P 13.17.

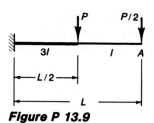

Figure P 13.9

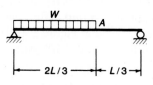

Figure P 13.10

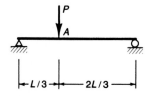

Figure P 13.11

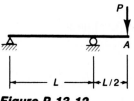

Figure P 13.12

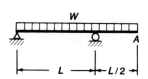

Figure P 13.13

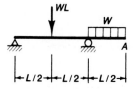

Figure P 13.14

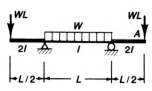

Figure P 13.15

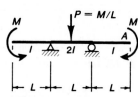

Figure P 13.16

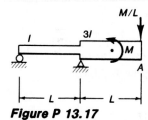

Figure P 13.17

13.18 For the cantilever beam with different cross sections shown in Figure P 13.18, the length of the stiffer part is changeable for x from 0 to 180 inches. Find the angle of rotation and deflection at A for different given values of x. (Answers for each unit increment of x are available in instructor's manual.) ($E = 29 \times 10^6$ psi.)

13.19 With P^* and L^* as defined in Problem 7.19, use the moment area techniques to find deflection and rotation at the point A on the beam shown in Figure P 13.19, if the beam has the constant section shown. ($E = 30 \times 10^6$ psi.)

13.20–
13.22 Repeat Problem 13.19 with the sections shown in Figures P 13.20 to P 13.22.

13.23 With P^* and L^* as defined in Problem 7.19, use moment area techniques to determine the deflection and rotation at the point A on the beam shown in Figure P 13.23 if the beam has the constant section shown. ($E = 25 \times 10^6$ psi.)

13.24–
13.26 Repeat Problem 13.23 with the sections shown in Figures P 13.24 to P 13.26.

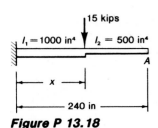

Figure P 13.18

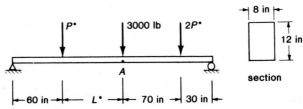

Figure P 13.19

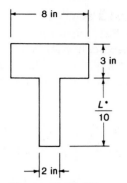

Figure P 13.20

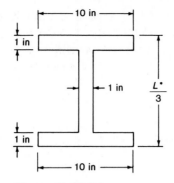

Figure P 13.21

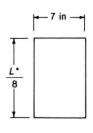

Figure P 13.22

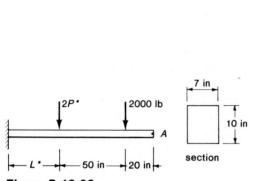

Figure P 13.23

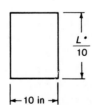

section

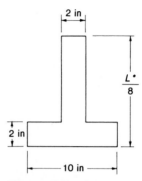

Figure P 13.24

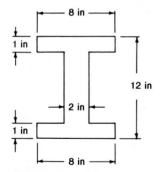

Figure P 13.25

Figure P 13.26

Statically Indeterminate Beams

14

1. To review the fundamental concepts of static indeterminacy.

2. To learn how statically indeterminate beam problems are solved by techniques of superposition using the moment area method.

3. To develop the capability of handling indeterminate beam problems through simultaneous solution of three-moment equations.

4. To understand how indeterminate structures are solved by the iterative technique of moment distribution.

Objectives

14.1 Indeterminate Structures

When all reaction forces and moments supporting a structure or a machine component can be determined by using only the equilibrium equations, the situation is known as *statically determinate*. When the static equilibrium equations introduced in Chapter 3 are insufficient to obtain a solution of a problem, it is called *statically indeterminate*. For instance, a simply-supported beam with one pinned and one roller support has three unknown reactions. The three unknowns require three independent equations for complete solution. An equation stating the equilibrium of forces and one stating the equilibrium of moments will be insufficient for solving the problem. The same holds for the case of a cantilever beam. However, if a cantilever beam is also supported by a roller at its end, as shown in Figure 14.1, the static equilibrium equations are insufficient to obtain a solution. As there are four unknowns, M, R_{Ax}, R_{Ay}, and R_B, Equations 3.2 and 3.3 need to be supplemented with yet another equation.

Problems similar to this have already been considered in this book. For example, in Example 8.7, the reactions created by the thermal expansion of a beam between two rigid walls required the use of the formula for thermal expansion (Equation 8.20) in order to find the solution. Similarly, in Example 9.8, the torsional moments acting at the ends of two different circular bars twisted in the middle were determined by writing the equation for the twist angle (Equation 9.19) for both bars in addition to the moment equilibrium formula. Both of these examples represented *statically indeterminate* problems. Generalizing on the basis of the examples quoted above, we may state that *statically indeterminate problems may be solved by making use of equations defining appropriate elastic deformations in addition to the equations of static equilibrium.*

The number of such elastic deformation equations that may be required to supplement the static equilibrium formulas for a particular problem is expressed by the term *degree of redundancy.* For example, in the case of the structure shown in Figure 14.1, the problem would become a statically determinate one either by removing the roller support from point B or by making the fixed end at A into a pinned support. By removing any *one* of these *constraints* the need for a supplementary elastic formula would be eliminated. Problems like this are said to have *one degree of redundancy.* A statically determinate problem has no redundancy. In such cases no constraint can be removed without causing a loss of stability and maybe the collapse of the structure. Figure 14.2 shows beams supported in various ways, representing different degrees of redundancy. In order to obtain the reactions and to compute internal forces and moments, the number of independent deformation equations to be found must be equal to the number of the degrees of redundancy characterizing the structure.

In the case of statically indeterminate beams, the deformation equations are either those pertaining to rotation or those pertaining to deflection. Both of these were discussed in the previous chapter. For most cases there are several ways of solving the problem. Consider, for example, the beam shown in Figure 14.1. As it has one degree of redundancy, one deformation equation is needed to supplement the static equilibrium equations. The analyst may either remove one constraint from point A, turning the beam into a simple supported one, or

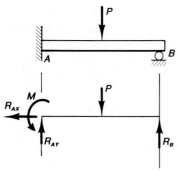

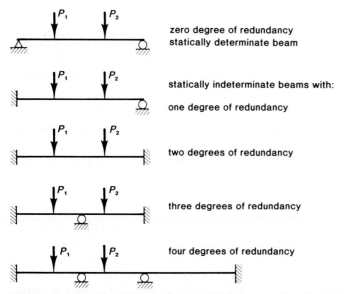

Figure 14.1 *Typical statically indeterminate problem.*

Figure 14.2 *The interpretation of "degrees of redundancy" of beams.*

remove the support from point *B,* turning the structure into a cantilever beam. In the first instance, the change of the fixed wall into a pinned support will allow the beam to rotate at point *A.* The amount of rotation can be computed by the use of the first and third moment area theorems shown in Example 13.2. By applying an equal but opposite rotation at point *A,* in addition to the load on the beam, the effect of the walled-in support can be recreated, and the statically indeterminate problem can be solved. The problem could also be solved by first considering the beam to be cantilevered at point *A.* In this case point *B* would deflect an amount that could be determined through the application of the second moment area theorem. If now a force acting as R_B is applied to the cantilever, causing a deflection equal in magnitude but opposite in direction to that caused by the force P, the configuration shown in Figure 14.1 is recreated and the problem is solved. Similar methods may be applied in the solution of problems having two or more degrees of redundancy. Of course, in such cases there are several elastic deformation expressions required to supplement the static equilibrium equations. Depending on the ingenuity of the analyst, the different possible paths of solution may lead to more or less complexity. Many techniques have been proposed; generally they are all based on the concepts outlined above. In the next section of this chapter, one of the many recommended techniques will be shown in detail: the one based on the use of the moment area theorems.

In practice, statically indeterminate beam problems are routinely encountered. For this reason the solution of common simple cases is available in tables that are usual features of technical handbooks. A typical table is included in Appendix B. More complicated structures of many degrees of redundancy are attacked by formal, *generalized solution techniques.* Two of these will be discussed in detail in this chapter. These are the *three-moment equation method* and the *moment distribution method.* The former involves simultaneous solution of a number of algebraic equations and the latter is an iterative technique. More

Example 14.1

Draw the approximated deflection and moment distribution curves for the structure shown in Figure 14.3.

Solution:

To construct the deflection diagram for the structure, one must first realize that there are three critical points at which the location of the deflected shape is known. First of all, as a result of the force P, the end of the cantilever must deflect a certain amount, designated z_A. The point is shown in an exaggerated manner so that the character of the deformed line can be visualized. It is also evident that both points B and C will remain on the line representing no deflection (i.e., $z = 0$). At the fixed end the slope of the line will be horizontal. However, at point B the elasticity of the $B-C$ span will cause the beam to rotate as shown. As a result of these we can now draw a smooth curve connecting points A, B, and C such that the line is tangent to the slopes shown at B and C. This can occur only if the curvature of the line changes its sign between B and C. Thus there must be a point of inflection in this region; Equation 13.11 shows that the moment will be zero at that point. To the left of the inflection point the center of curvature of the line is located below the curve. According to our sign convention, this condition is associated with negative bending moments. On the right side of the inflection point the moments are positive. Since the loading involves a concentrated force only, the moment diagram will be linear, as shown.

complex structural problems involving statically indeterminate beams are generally solved by electronic digital computers. The introduction of these methods are beyond the scope of this text.

In solving indeterminate structures a helpful first step is to construct an approximate *deflection diagram*. Once such a graph is drawn, a rough sketch showing the expected shape of the moment diagram may be constructed that will indicate portions of the beam where the moment is expected to be positive, negative, or zero. Using the sign convention of Figure 6.2, positive moments are to be plotted on the side of the beam where compressive stresses occur. At points of inflection of the deflection diagram, the bending moment is zero and is changing from positive to negative or vice versa. At pinned supports the moment will have extreme values unless they are at the end of the beam (in which case the moment is zero). The deflection curve is drawn as a smooth curve beginning from an exaggerated deflection point at the location of the dominant external loading within a span (or on a cantilever), through the support points, where there are no deflections. At a fixed support, the tangent of the deflection curve is known to be zero also. In essence, a deflection diagram gives helpful hints on the shape of the moment diagram. The typical procedure of drawing approximate deflection and moment distribution graphs is demonstrated in the following example.

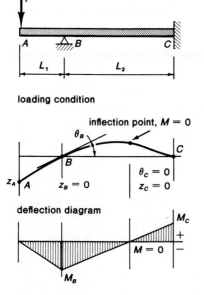

loading condition

deflection diagram

moment diagram

Figure 14.3 *Sketching deflection and moment diagrams.*

Example 14.2

A simply-supported beam is fixed at both ends and loaded with a constant spread load as shown in Figure 14.4. Construct the deflection and moment diagrams.

Solution:

As the beam as well as the loading is symmetrical about the center, one may realize that the largest deflection z_C will occur there. Also from symmetry we will note that the slope of the tangent to the graph of deflection at point C is zero. Both the deflection and the slope are zero at the two supports. We assign an arbitrary value to z_C and draw the elastic line through the three points, making its slope zero at each of them. Such a curve, tangent to two parallel lines, will have two points of inflection, as shown in the figure. These are points where the moment is zero. Closer to the ends than these inflection points, the moments are negative; between them it is positive. As the loading involves uniformly distributed forces, the bending moment diagram will be parabolic. The maximum negative moments occur at the wall.

A conclusion based on the first moment area theorem may be drawn here. As the slope is zero at the walls, the area of the moment diagram between the two walls must equal zero. The same statement is valid between points A and C as well. This indicates that the sum of the negative moment areas must equal the positive moment area in the diagram. This in itself would be sufficient to find a solution for this problem.

14.2 Superposition Techniques Using Moment Area Theorems

As suggested in the preceding section, one method to solve statically indeterminate structural problems is first to remove some of the constraints that cause the indeterminacy and then solve the remaining statically determinate beam, subject to all external loadings. As the next step, deformations (rotations and deflections) of this simplified structure are computed for the locations from which the constraints were removed. These deformations are those that would not have occurred if the constraints had been left in place. To regain the integrity of the original structure, appropriate forces and moments should be applied to the simplified structure such that their effects counteract these deformations. Initially these forces and/or moments are not known, but the way they relate to deflections and rotations of the simplified structure is given by the moment

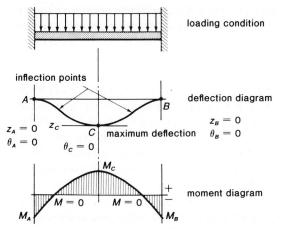

Figure 14.4 *Properties of the deflection and moment diagrams for a statically indeterminate beam.*

area theorems. In this manner one may derive appropriate equations to express the relationship between known deformations and unknown "external" loadings placed on the simplified structure. In essence the described procedure results in two solutions for the simplified statically determinate structure: one for the external loading for the simplified structure, another for the loadings placed upon it to recreate the effect of the constraints initially removed. The superposition of the two solutions will be the solution of the indeterminate problem. The following examples will demonstrate some typical superposition techniques.

Example 14.3

A simple beam with three supports is loaded by a single concentrated force P, as shown in Figure 14.5. Show two different ways the problem may be analyzed using the method of superposition.

Solution:

By *removing any one of the supports* the problem turns into a statically determinate one. As there are three supports, there could be three such approaches. For each, the deflection at the site of the missing support may be computed by the moment area method. Should the missing reaction force be placed on the structure, the corresponding equal but opposite deflection can be derived.

A fourth method of solution may be introduced by retaining all supports but *cutting the beam* over the middle one so that each span is independent of the other. As a result the slope of the beam at one side of the central support will be different from the slope of the other side. The difference of the slope at the cut may be expressed as an angle of rotation. By placing a moment over the center point, the sides of the cut may be rotated to recreate the effect of original material in contact over the continuous beam. Adding the reaction forces at the supports (balancing the superimposed moment) to the reactions computed for the two statically determinate beams (under the external loading) will solve the indeterminate structure. Of the four possible solutions, two will be demonstrated here. First, the problem will be solved by removing the left support. Next, the problem will be reworked by using the method of cutting the beam over the center support.

First Solution:

With the support at point A removed, the beam shown in Figure 14.5 becomes statically determinate. As the location of the external force P is symmetrical, the two reaction forces will be

$$R_B' = R_C' = P/2$$

The peak moment under P is

$$M = \frac{PL}{4}$$

Because of symmetry, the slope of the deflected beam at both supports will be the same. By the first moment diagram, the change of the slope from point B to point C will equal the area of the reduced moment diagram, that is

$$A^* = \frac{ML}{2\,EI} = \frac{PL^2}{8\,EI}$$

The angle α is therefore

$$\alpha = \frac{A^*}{2} = \frac{PL^2}{16\,EI} \text{ (in radians)}$$

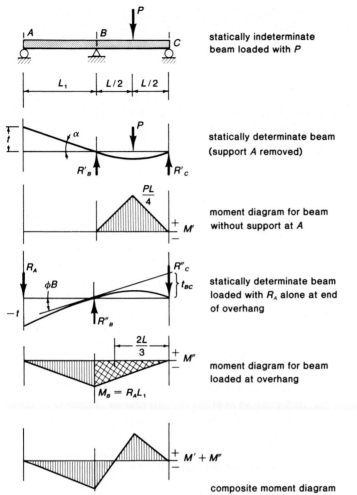

statically indeterminate
beam loaded with P

statically determinate beam
(support A removed)

moment diagram for beam
without support at A

statically determinate beam
loaded with R_A alone at end
of overhang

moment diagram for beam
loaded at overhang

composite moment diagram

Figure 14.5 *Statically indeterminate beam solved by superposition,*
support A removed.

Example 14.3 *(Continued)*

Accordingly, the deflection of point A due to the external force P is

$$t = \alpha L_1 = \left(\frac{PL^2}{16\,EI} \right) L_1$$

But in the original statically indeterminate beam, point A is held in place by a pinned support. To recreate this effect, the force R_A is to be applied at the end of the cantilever such that at that point the resulting deflection is equal to $-t$.

By the static equilibrium of forces, the two reactions resisting R_A at points B and C are such that

$$R_B{}'' = R_A + R_C{}''$$

Writing the equilibrium of moments at point B leads to

$$R_A L_1 = L\,R_C{}''$$

and therefore

$$R_C{}'' = \left(\frac{L_1}{L} \right) R_A$$

and

$$R_B{}'' = \left(1 + \frac{L_1}{L} \right) R_A$$

To determine the deflection of point A due to the force R_A, one first needs the tangential slope of the elastic beam at point B. This may be obtained by considering the cross-hatched area of the moment diagram between supports B and C.

The slope at point B may be defined by the second moment area theorem applied over the cross-hatched area of the reduced moment diagram, defining ϕ_B as

$$\phi_B = \frac{t_{CB}}{L} = \frac{1}{L} \left(\frac{2}{3} \right) L \left(\frac{R_A L_1 L}{2\,EI} \right)$$

$$= \frac{R_A L_1 L}{3\,EI}$$

Following the path demonstrated in Example 13.5, we can now determine the deflection under the force R_A. The first component is given by the angle of rotation at point B times the length L_1. The second component is given by the second moment area theorem, that is, the reduced moment area between points A and B times the distance of its centroid from point A. Hence, total deflection of point A due to R_A shall be

$$\phi_B L_1 + \frac{1}{2}\left(\frac{R_A L_1^2}{EI}\right)\left(\frac{2}{3}\right) L_1$$

To push point A back to where it belongs, this must equal $-t$, the deflection due to the external load P. The only unknown in this equation is R_A. Setting the two deflection equations equal to each other the unknown R_A can be expressed as

$$R_A = -\frac{3}{16}\left(\frac{P L^2}{L_1 L + L_1^2}\right)$$

Summing the reactions obtained for the two loadings and adding the two moment diagrams will complete the solution of this indeterminate problem. The resultant combined moment at support B will be

$$M_B = R_A L_1 = -\frac{3}{16}\left(\frac{P L^2}{L + L_1}\right)$$

Second Solution:

In this approach, all three supports will be left intact, but the beam will be assumed to be cut at support B. As a result the two spans will act independently of each other. The left span has no external loadings on it, and therefore its deformation will be zero. On the right span, the centrally located force P will result in a symmetric triangular moment diagram. Just as in the previous solution, the rotation of the right span at point B is

$$a = \frac{PL^2}{16\,EI}\ \text{(in radians)}$$

Since the left span is not loaded, there is no rotation on the left side of the cut beam. Hence the cutting of the beam will result in a separation of the ends equal to the angle a. To rejoin the parts of the structure, an equal but opposite rotation must be caused by a moment at point B. Initially we do not know the magnitude of this moment. It must be derived by the moment area theorems. Since at any one location in a beam there must be but a single internal moment, this same moment, M_B, should act on both spans of the cut beam. The static equilibrium conditions require that both simple spans resist the moment acting at support B by a couple. For the left span the reaction is

$$R_A = \frac{M_B}{L_1}$$

At the right span the reaction is (when the span is unloaded)

$$R_B{}'' = \frac{M_B}{L}$$

Example 14.3 *(Continued)*

The corresponding directions are noted in Figure 14.6. Later we will have to remember to add all of the appropriate reactions with careful observation of their respective signs (see Example 14.5).

By the use of the first and third moment area theorems we will find that the moment M_B will result in two angular deformations at the center support, one for each span. The first moment area theorem gives us the *change* of slope between points A and B as the reduced moment area in that span, that is,

$$\Delta\phi_{\text{left}} = \frac{M_B L_1}{2\, EI}$$

The third moment area theorem, as before, will give us the slope at point B as

$$\phi_{B,\text{ left}} = \frac{M_B L_1}{3\, EI}$$

Similarly, for the right side,

$$\phi_{B,\text{ right}} = \frac{M_B L}{3\, EI}$$

The sum of these two rotations must equal the angle α. Therefore,

$$\frac{P L^2}{16\, EI} = -M_B \left(\frac{L_1 + L}{3\, EI} \right)$$

and

$$M_B = -\frac{3}{16} \left(\frac{P L^2}{L_1 + L} \right)$$

This result is identical to the one obtained by the first method of solution. It is sufficient to plot the complete moment diagram for the statically indeterminate structure. From the foregoing the reactions can be determined also. From these, or directly from the moment diagram, the shear diagram can also be determined.

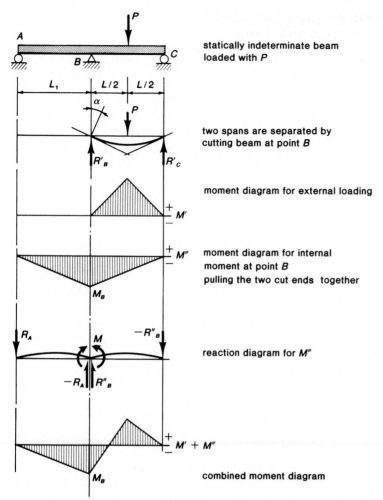

statically indeterminate beam
loaded with P

two spans are separated by
cutting beam at point B

moment diagram for external loading

moment diagram for internal
moment at point B
pulling the two cut ends together

reaction diagram for M''

combined moment diagram

Figure 14.6 *Statically indeterminate beam solved by superposition,
beam cut at center support.*

14.3 Simultaneous Solutions of Three Moment Equations

As the previously worked example demonstrates, the technique of superposition involves tedious and unimaginative work, even if the structure has no more than one or two degrees of redundancy. For more complex statically indeterminate structures the method of superposition becomes impractical. For instance, for a beam of four supports (one pinned and three on rollers) it would be necessary to first remove two constraints, determine two different deformations at each support removed, and finally compute and superimpose on the original solution two separate reaction forces or moments. The number of alternative paths to the solution is large, and it is easy to make errors that are hard to trace. In situations like this, the advantage of a standardized approach is obvious. One of such standardized techniques is the *method of three moments*, which is based on the superposition technique described in the previous section. Its fundamental approach involves the *cutting* of the statically indeterminate beam at *all internal supports*, as in the second method of solution of Example 14.3. Cutting the beam at these points converts the problem into a set of statically determinate beam problems in which each adjoining beam will require an internal bending moment to be superimposed at the cut to close the angular gap created by the separation of the beam. By taking only two of the adjoining beams at a time, one may write equations for these moments. Such equations will generally contain three bending moments: one on the left, one at the center where the two beams are joined, and one at the right. Figure 14.7 shows an example for this arrangement. With the five supports shown, one needs to write three separate equations, since to create statically determinate beams one needs to cut the structure at three points: *B, C,* and *D*. Therefore at the outset M_B, M_C, and M_D are not known. (Should there be only three supports, as in Example 14.3, the only cut needed would be at point *B,* and a single equation would be developed for M_B.) In the beam with five supports we need three equations. M_B and M_D both appear twice, and M_C appears three times. M_C is included as the right-end moment at the first equation, the center moment at the second equation, and the left-end moment for the third equation. For the case shown in Figure 14.7, M_A and M_E would be known at the beginning. Therefore we have three equations and three unknowns, a problem amenable to an algebraic solution.

The derivation of the *three-moment equation* is very much similar to the somewhat simpler problem shown in the second solution of Example 14.3. It will not be included here. The interested reader may consult the Bibliography for its mathematical derivation. The equation was originally derived in 1857 by B. P. E. Clapeyron in the general form

(14.1)
$$L_L M_L + 2\left(L_L + \frac{I_L}{I_R} L_R \right) M_C + \frac{I_L}{I_R} L_R M_R$$

$$= -6A_L \frac{x_L}{L_L} - 6A_R \frac{x_R}{L_R} \frac{I_L}{I_R}$$

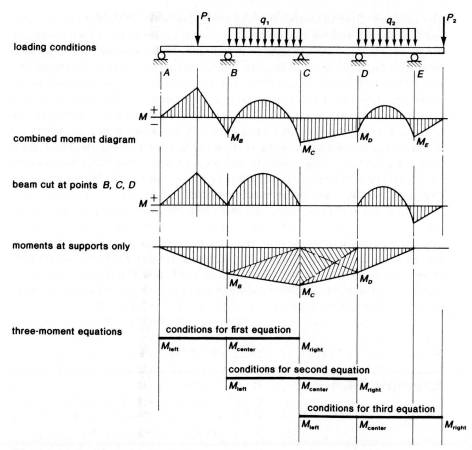

loading conditions

combined moment diagram

beam cut at points *B, C, D*

moments at supports only

three-moment equations

Figure 14.7 *Technique of solution using three-moment equation.*

The subscript L refers to the left of the cut, C to the site of the cut (the center), and R to the right of the cut. I refers to the moment of inertia of the beams on the two spans. A_L and A_R in this equation are the areas of the moment diagrams on the left and right spans. They are computed for the external loading on the statically determined (cut) spans and not reduced (not to be divided by EI). As shown in Figure 14.8 the values x_L and x_R are the centroidal distances of A_L and A_R measured from the left and right supports, respectively, and *not* from the center. In setting up the equation the initially unknown moments should be identified by the subscripts denoting their location in the whole structure, such as M_B, M_C, and M_D, rather than by their location in a particular three-moment equation. The subscripts L, C, and R are not to be carried into the equations, as they denote different moments in each equation.

As stated in the beginning, the three-moment equation deals with *simply supported beams* that are assumed to be cut at the support points that cause indeterminacy. If a structure contains fixed supports, they are first replaced by pinned ones. This is done by adding an additional span at the fixed end with zero length. Because of this additional span, the fixed end will add one more to the number of three-moment equations: the moment at the fixed end will be a central moment in the added equation.

Example 14.4

Solve the three-supported beam problem shown in Example 14.3 by the use of the three-moment equation.

Solution:
Since there is only one degree of redundancy in this problem, only one application of Equation 14.1 will be needed. By observing Figures 14.5 and 14.6 we find that many of the variables are zero:

$$M_L = A_L = x_L = 0$$
$$M_R = 0$$

Furthermore,

$$L_L = L_1; I_L = I_R = \text{constant}; L_R = L$$
$$x_R = L/2$$
$$A_R = PL^2/8$$

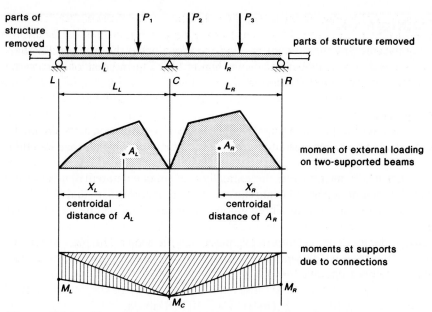

parts of structure removed

parts of structure removed

L L_L C L_R R

A_L A_R

moment of external loading on two-supported beams

X_L X_R

centroidal distance of A_L

centroidal distance of A_R

moments at supports due to connections

M_L M_C M_R

Figure 14.8   *Notations for three-moment equation, Equation 14.1.*

Substituting in Equation 14.1, one obtains

$$2(L_1 + L)M_B = -6\left(\frac{PL^2}{8}\right)\left(\frac{L}{2L}\right)$$

Rearranging, we have

$$M_B = -\frac{3}{16}\left(\frac{PL^2}{L + L_1}\right)$$

which is the same result that was obtained in Example 14.3.

Example 14.5

Using the three-moment equation, find the moments at the supports for the statically indeterminate beam shown in Figure 14.9. The moment of inertia for each span is shown in the drawing.

Solution:

First, the fixed support at point D is replaced by a combination of two pinned supports, D and E, zero distance apart. On the same drawing all span lengths are changed into inches.

The next drawing shows the statically determinate moment diagrams for the independent, separated beams formed by cuts at points B, C, and D. Below this is indicated the way the three independent equations will be set up. The moment diagrams for the two loaded spans are both symmetrical, owing to the fact that the loads are symmetrically placed on the spans. The peak moment for the first span is 144 kip inches. Therefore the area of the first moment diagram between points A and B is

$$A_{AB} = \frac{1}{2} (144)(144) = 10{,}368 \text{ kip} \cdot \text{in}^2$$

For the second span, between B and C, the peak moment is 120 kip inches. The area of this moment diagram of parabolic shape equals

$$A_{BC} = \frac{2}{3} (120)(120) = 9{,}600 \text{ kip} \cdot \text{in}^2$$

The first three-moment equation, comprising supports A, B, and C, is then

$$0 + 2 \left[144 + \left(\frac{180}{200} \right) 120 \right] M_B + \left(\frac{180}{200} \right) 120 \, M_C$$

$$= -6 (10{,}368) \left(\frac{72}{144} \right) - 6 (9{,}600) \left(\frac{60}{120} \right) \left(\frac{180}{200} \right)$$

Solving and rearranging this equation will lead to

$$M_B = -0.214 \, M_C - 113.14$$

The second three-moment equation, written for supports B, C, and D, will be

$$120 \, M_B + 2 \left[120 + \left(\frac{200}{210} \right) 120 \right] M_C + \left(\frac{200}{210} \right) 120 \, M_D$$

$$= -6 (9{,}600) \left(\frac{60}{120} \right)$$

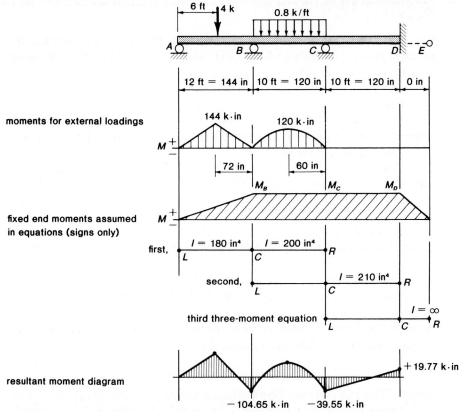

Figure 14.9 *Solution of a statically indeterminate beam of three degrees of freedom by simultaneous solution of three moment equations.*

After solving and rearranging this equation, we get

$$M_C = -0.256\,M_B - 0.244\,M_D - 61.46$$

The third three-moment equation, including supports C, D, and E, takes the form

$$120\,M_C + 2\,(120 + 0)\,M_D + 0 = 0$$

The right side of the equation is zero since there are no external loads present. Simplifying and rearranging will result in

$$M_C = -2\,M_D$$

Example 14.5 *(Continued)*

This equation indicates that the moment at the wall will be of a sign opposite that at support C. Observing the deformation of the span under the given loading will confirm this. Figure 14.3 sheds more light on this fact.

The next step in the computations is the simultaneous solution of the three resulting equations. This is where errors usually occur in practice: one must be careful with the arithmetic operations. Substitution of the results into the original equations to check one's computations is highly recommended.

The resulting moments are:

$$M_A = 0;$$
$$M_B = -104.65 \text{ kip} \cdot \text{in}$$
$$M_C = -39.55 \text{ kip} \cdot \text{in}$$
$$M_D = +19.77 \text{ kip} \cdot \text{in}$$

The shape of the moment diagram can now be shown in the drawing.

The last step in the analysis is the determination of the reactions at all supports. Here, again, the principle of superposition is applied. For example, at point A the reaction is composed of two parts: one is the reaction caused by the

14.4 The Iterative Technique of Moment Distribution

A most popular method of solving statically indeterminate beam problems is based on an iterative technique first proposed by H. Cross in 1936. This is a *method of relaxation,* in which initially assumed moments at the supports are successively corrected in a systematic manner until solution is found.

The initial values of the moments at the supports are assumed to be those corresponding to a set of two-supported beams with all ends fixed at the supports. Such solutions for the most common loading conditions are generally available in tables such as the one included in Appendix B. Other solutions, for more complicated loading conditions, may be obtained by using the principle of superposition in constructing composite moment diagrams from simpler tabulated data. Some spans where unusual external loadings are present may be solved by the techniques introduced in the previous sections.

When the moments at the supports at all initially fixed ends are determined, there will generally be differences in the magnitude of the moments on the two sides of each support. This means that if the supports are released, there is an unbalanced moment on most supports, causing joint rotation. Allowing for this, the method provides for an adjustment at each support, one by one, such

external load of the centrally located 4 kips, that is 2 kips, pointing upward. The second component of R_A is the couple resisting the moment M_B at support B. This moment results in a force that points downward to restrain the beam at point A. Its magnitude is equal to the moment divided by the span of the first beam, that is, 144 inches. The reaction at A therefore equals

$$R_A = +2 - \frac{104.65}{144} = 2 - 0.73 = +1.27 \text{ kips}$$

At point B the situation is more involved. The reaction is composed of the reactions to the external loads on both the first and the second spans. The sum

of these is $2 + 0.8(10)\frac{1}{2} = 6$ kips. Added to these, with care exercised in

the determination of the signs, are the force due to M_B acting on the left (0.73 kip, acting downwards at this point); the same moment resisted by the right span by the couple creating a force of $-104.65/120 = -0.87$ kip acting upward; and the effect of the moment over support C, which creates a force of $-39.55/120 = -0.33$ kip downward. As a result of all these, the reaction at support B is 5.81 kips upward. In case of doubt it is advisable to show each span in the form of a free body diagram, showing all external and internal (at the cuts) forces and moments acting on it.

that the moments on each side become equal. Once the end moments at all supports are equal on both sides there is no need to "freeze" (to fix) the supports, hence all can be considered pinned. Once all these balancing moments are found, the problem is solved.

If a particular fixed support is released, the difference in moment, ΔM, may be eliminated by applying an equal but opposite value. An example of applying a moment to a pinned support between two fixed ones is shown in Figure 14.10. Since the change of slope between the two fixed ends must be zero, the total area of the moment diagram must be zero as well. The angle of rotation caused by the adjusting moment ΔM at the pinned support in the middle equals, by the first moment area theorem,

(14.2)
$$\phi = \frac{\Delta M_1 L_1}{E_1 I_1} = \frac{\Delta M_2 L_2}{E_2 I_2}$$

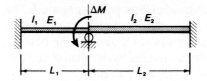

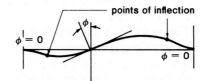

loading diagram

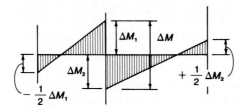

deformation diagram

moment diagram

Figure 14.10 *The distribution of the balancing moment between two adjoining spans that are fixed at the ends.*

Consequently the total adjusting moment must be split between the two adjoining beams as

(14.3)
$$\frac{\Delta M_1}{\Delta M_2} = \frac{E_1 I_1 / L_1}{E_2 I_2 / L_2} = \frac{(EI/L)_1}{(EI/L)_2}$$

Since

$$\Delta M_1 + \Delta M_2 = \Delta M$$

it follows that

(14.4)
$$\Delta M_1 = R_1 \, \Delta M$$

(14.5)
$$\Delta M_2 = R_2 \, \Delta M$$

where

(14.6)
$$R_1 = \frac{(E\,I/L)_1}{(E\,I/L)_1 + (E\,I/L)_2}$$

and

(14.7)
$$R_2 = \frac{(E\,I/L)_2}{(E\,I/L)_1 + (E\,I/L)_2}$$

The quantity R in these equations is called the *distribution factor*. For each joint the sum of the distribution factors must equal one. Based on the foregoing, the adjusting moment ΔM is distributed between the two sides of the released support according to the *relative stiffness* of the adjoining beams. The signs of the adjusting moments are such that, after the distributed adjusting moments are added to the fixed end moments, the sums are equal on the two sides of the support, that is, the moments at the support are balanced. Accordingly, the larger moment will be reduced and the smaller will be increased.

As indicated by Figure 14.5, applying a balancing moment at the released support will cause two moment-reactions in the fixed ends. It may be shown by the moment area theorem that the moments at the fixed supports shall equal

(14.8)
$$\Delta M_{\text{left}} = -\frac{1}{2} \Delta M_1$$

and

(14.9)
$$\Delta M_{\text{right}} = -\frac{1}{2} \Delta M_2$$

These equations give rise to the so-called *carry-over factor* of $-\frac{1}{2}$. If there were no carry-over due to the adjusting moments at each support, the method discussed here would involve one simple relaxation at each of the supports. However, after each relaxation the previously balanced neighboring support moments will be altered. For this reason the procedure is an iterative one. The successive balancing of moments and the carrying over of the appropriate balancing moments to the neighboring support is continued until the magnitude of the carry-overs is negligible.

There are two special cases of the balancing of supports at the end of the structure. One occurs when the last support of the beam is initially a fixed one. Fixed supports need not be relaxed, therefore any initial fixed end moment and all successive carry-over moments will be absorbed by them. At such supports all incoming moments are merely summed. The other case occurs when the last support is a pinned one. Rollers or pinned supports cannot carry moments. Therefore an initial fixed moment should be balanced by an equal and opposite balancing moment, which then should be carried over (by multiplying by $-\frac{1}{2}$) to the neighboring support. Incoming carry-over moment should be returned likewise to the neighboring support. In the case of such pinned end supports, it is advantageous to consider the loading conditions tabulated (see Appendix B) for one end fixed and the other pinned. This will eliminate the need for the initial relaxation of the pinned end. In the case of cantilevered ends the situation is similar. The loading of the cantilever will define the moment over the first pinned support, and successive relaxations will not alter this value.

The methodical procedure of the moment relaxation technique is shown in the following example.

Example 14.6

Solve the statically indeterminate beam problem presented in Example 14.5 by the moment distribution method.

Solution:

The problem and its solution are shown in Figure 14.11. First, the fixed end moments are determined for the external loadings, using the solutions catalogued in Appendix B. The initial unadjusted differential moments are shown at each support. To distribute these between the adjoining beams, the stiffness factor for each span is computed. Since the material of all beams is the same, the Young's modulus does not appear in the computations (E = constant). Using the adjacent stiffness factors, the distribution factors are calculated for each support points. Note that at the left end its value is -1, and on the right end it is zero. The following line shows the calculated fixed end moments.

The sequence of successive relaxation of the supports is up to the analyst. Here, support A and support C are started on the first step. At support A the fixed end moment of -72 kip inches is balanced by $+72$. Multiplying by the carry-over factor of $-\frac{1}{2}$, a moment of -36 is carried over to the left of support B. The carry-over steps are indicated by arrows in the drawing.

At point C the difference of moments on the two sides is 80 kip inches. Multiplying this by the distribution factors will result in 39.2 on the left side. Since the fixed end moment on this side is -80 and on the other side is zero, the adjusting moment will have a positive sign on the left and a negative one on the right. Adding the resulting moments on both sides one will see that the support is balanced. Multiplying both balancing moments by the carry-over factor ($-\frac{1}{2}$), the right side of support B will have an additional moment placed on it that equals -19.6.

In the next step, support B is analyzed. Adding the moments on each side, the difference between the sums on the left and right sides will be 8.4. This is the amount that should be distributed (with careful attention to the signs). The procedure from here on is self-explanatory.

When the carry-over moments shrink to negligible values, the computations can be terminated. The resulting moments at the supports are shown in the last row of the computations. These are equal to the moments obtained using the three-moment equations in Example 14.5.

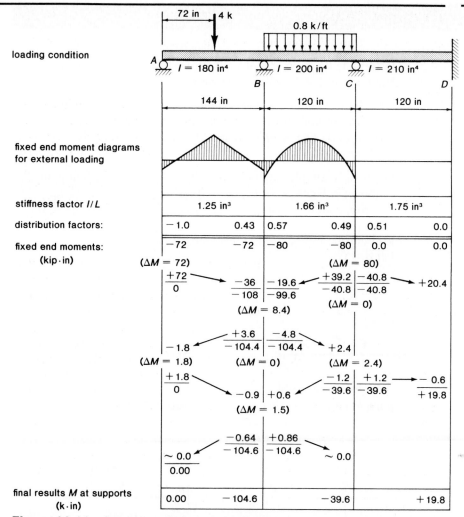

loading condition

fixed end moment diagrams
for external loading

stiffness factor I/L

distribution factors:

fixed end moments:
(kip·in)

final results M at supports
(k·in)

Figure 14.11 *Iterative solution of statically indeterminate beam problem
using moment distribution.*

Problems

14.1–
14.5   Determine the moments at supports for the beams shown in Figures P 14.1 to P 14.5, respectively.

14.6   Find the maximum downward deflection of the small aluminum beam shown in Figure P 14.6 due to an applied force P of 400 newtons. The beam's constant flexural rigidity, EI, is $60 \text{ N} \cdot \text{m}^2$.

14.7   Beam AC is simply supported at its two ends and at its midpoint B, as shown in Figure P 14.7. Before the 10-kip load P is applied, however, contact is made between support B and the beam so that support B carries part of the load. Determine the amount of the load carried by each of the three supports, A, B, and C. Let $\delta = 0.5$ in.

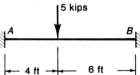

Figure P 14.1

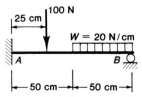

Figure P 14.2

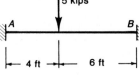

Figure P 14.3

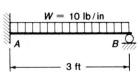

Figure P 14.4

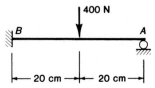

Figure P 14.5

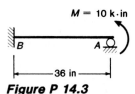

Figure P 14.6

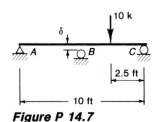

Figure P 14.7

14.8 Find the maximum downward deflection of the beam shown in Figure P 14.8. Let $EI = 1.9$ kN·mm².

14.9–

14.16 Find the reactions at the supports by using the three-moment equation method for the structures shown in Figures P 14.9 to P 14.16, respectively.

EI = constant

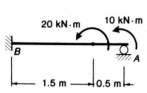

Figure P 14.8

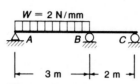

Figure P 14.9

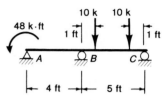

Figure P 14.10

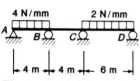

Figure P 14.11

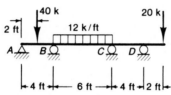

Figure P 14.12

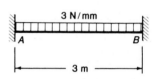

Figure P 14.13

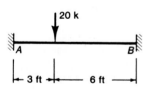

Figure P 14.14

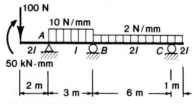

Figure P 14.15

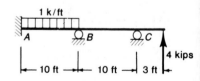

Figure P 14.16

14.17–
14.20 Rework Problem 14.9, 14.11, 14.15, and 14.16 by using the moment distribution method.

14.21 For the statically indeterminate beam shown in Figure P 14.21, the loading and spanning are changeable for x from 0 to 180. Find M_B in newton millimeters for different given values of x. (Answers for each unit increment of x are available in instructor's manual.)

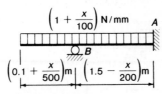

Figure P 14.21

Column Design

15

1. To understand that members loaded in axial compression are subject to the possibility of buckling.

2. To be able to use Euler's buckling formula, modified as necessary for differing end restraints.

3. To understand the limitations of Euler's buckling formula.

4. To be able to check a column design.

Objectives

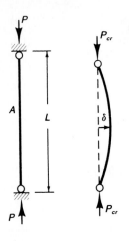

Figure 15.1
Pin-ended column.

(a)   (b)

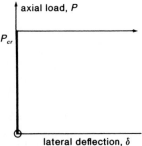

Figure 15.2
**Load-deflection behavior
of a perfect column.**

15.1 The Basis of Column Buckling

In Chapter 8 the stresses in an axially loaded member were found to be distributed across the section in a uniform manner. Thus the stress was given by

(15.1)
$$\sigma = \frac{P}{A}$$

where P is the axial force and A is the cross-sectional area of the member. An axially loaded member in which the force is causing compression stress in the member is called a *column*. Such a member is illustrated is Figure 15.1a. Unfortunately, it is not sufficient just to find the stress in a column from Equation 15.1 because a column may also be subject to *buckling*. As is illustrated in Figure 15.1b, a column may be acted upon by an axial force large enough to cause the column to collapse. The smallest force required to cause the column to collapse is called the buckling load or the critical load and is given the symbol P_{cr}. Column buckling is a particularly dangerous phenomenon in that it may occur completely without warning while the axial stress in the member is still well within the elastic range of material behavior. It is characterized by large deflections of the column (as shown in Figure 15.1b) and a loss of load-carrying capacity of the member. For these reasons it is important that some reliable method be developed for the prediction of buckling in members loaded in axial compression.

15.2 Euler's Buckling Formula

Some very important early investigations of column buckling were carried out by Leonard Euler, a Swiss mathematician, in the mid-eighteenth century. He showed mathematically that the buckling of a column could be predicted, provided certain assumptions were made. These assumptions were:

a. The column is initially perfectly straight and has pinned ends, as shown in Figure 15.1a
b. The material behaves elastically throughout its loading;
c. As the column is loaded the lateral deflection, δ, at midheight of the column may be graphed as shown in Figure 15.2. The column, if initially perfectly straight, has no lateral deflection at all until the axial load reaches the critical buckling load, at which point the lateral deflection increases without measure.

On these assumptions Euler proposed a column buckling formula given by

(15.2)
$$P_{cr} = \frac{\pi^2 EI}{L^2}$$

where P_{cr} is the critical buckling load, E is the modulus of elasticity of the material, and I is the moment of inertia of the section.

Equation 15.2 involves the moment of inertia of the section, and it is important that the correct moment of inertia be used. Since a column will buckle at the lowest possible axial load, the minimum moment of inertia of the section must be used. (Care must be exercised with this principle, as will be demonstrated in the following sections.)

Equation 15.2 may also be written in terms of stress, since the cross-sectional area of the member is available if the moment of inertia is available. Thus the critical buckling stress, σ_{cr}, is

(15.3)
$$\sigma_{cr} = \frac{P_{cr}}{A} = \frac{\pi^2 EI}{AL^2}$$

15.3 Radius of Gyration and Slenderness Ratio

In discussion of column buckling a very useful property of a cross section is the *radius of gyration, r,* which was defined in Chapter 7 by the relation

(15.4)
$$r = \sqrt{\frac{I}{A}}$$

and has the units of length (e.g., inches or millimeters).

It is now recalled that in Appendix A, usually two values of moment of inertia are tabulated, and there are therefore usually two different values of radius of gyration also. These two values represent the radii of inertia about the x- and y-axes and are defined by the equations

(15.5)
$$r_x = \sqrt{\frac{I_x}{A}}$$

and

(15.6)
$$r_y = \sqrt{\frac{I_y}{A}}$$

Equations 15.3 and 15.4 may be combined to give

(15.7)
$$\sigma_{cr} = \frac{\pi^2 E}{\left(\dfrac{L}{r}\right)^2}$$

The denominator of this equation is the square of the quantity L/r, which, for convenience, is called the *slenderness ratio.* The slenderness ratio is a dimensionless number that, for most columns, is in the range of 0 to 200. Notice that it includes the effects of a number of things on the buckling behavior of the column. Not only does it include the effects of the cross section being used (the values of I and A involved in the calculation of r), but it also includes the physical length of the column being investigated.

Example 15.1

A pin-ended column 24 feet long is to be built from a wide-flange section W 12 × 45. The column is to be made of steel with a modulus of elasticity of 30,000 kips per square inch and a yield stress of 36 kips per square inch. Find the minimum axial force that will cause the column to buckle and the critical buckling stress.

Solution:

In Appendix A the following properties of a W 12 × 45 section are found:

$$A = 13.2 \text{ in}^2$$

$$I_x = 350 \text{ in}^4$$

$$I_y = 50 \text{ in}^4$$

Using these values, the critical buckling force P_{cr} may be found from Equation 15.2. Since it is unknown which moment of inertia to use, Equation 15.2 will be applied twice, giving two values of $P_{cr}-P_{crx}$ representing buckling about the x-axis and P_{cry} representing buckling about the y-axis:

$$P_{crx} = \frac{\pi^2 EI_x}{L^2} = \frac{\pi^2 (30 \times 10^3) \, 350}{(24 \times 12)^2} = 1{,}249 \text{ kips}$$

$$P_{cry} = \frac{\pi^2 EI_y}{L^2} = \frac{\pi^2 (30 \times 10^3) \, 50}{(24 \times 12)^2} = 178 \text{ kips}$$

Since the value of P_{cry} calculated is less than the value of P_{crx}, the column will buckle about its y-axis at an axial load of 178 kips. In this instance, this could have been foreseen as the buckling must take place about the axis with the smallest moment of inertia. This rule does not apply universally, and so it is always a good idea to calculate the buckling force about both axes.

Calculation of the critical buckling stress is a relatively easy matter, now that the buckling load is known. From Equation 15.1, the critical buckling stress is

$$\sigma_{cr} = \frac{P_{cr}}{A} = \frac{178}{13.2} = 13.5 \text{ ksi}$$

The critical buckling stress may also be calculated directly from Equation 15.7. If it is assumed that the axis of buckling is at present unknown, the equation may be applied for buckling about both axes:

$$r_x = \sqrt{\frac{I_x}{A}} = \sqrt{\frac{350}{13.2}} = 5.15 \, \text{in}$$

Thus

$$\sigma_{crx} = \frac{\pi^2 E}{\left(\dfrac{L}{r_x}\right)^2} = \frac{\pi^2(30 \times 10^3)}{\left(\dfrac{24 \times 12}{5.15}\right)^2} = 94.6 \, \text{ksi}$$

Similarly,

$$r_y = \sqrt{\frac{I_y}{A}} = \sqrt{\frac{50}{13.2}} = 1.94 \, \text{in}$$

Thus

$$\sigma_{cry} = \frac{\pi^2 E}{\left(\dfrac{L}{r_y}\right)^2} = \frac{\pi^2(30 \times 10^3)}{\left(\dfrac{24 \times 12}{1.94}\right)^2} = 13.5 \, \text{ksi}$$

Since σ_{cry} is the smaller of the two stresses, the critical buckling stress is 13.5 kips per square inch, and it applies to buckling about the y-axis of the section.

It is noted that the critical buckling stress of the column is much less than the yield stress of the material and so the column will buckle before the material yields. Thus the material will perform elastically for all loads less than the buckling load.

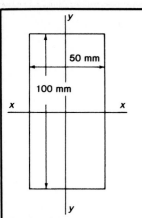

Figure 15.3
Cross section of column for Example 15.2.

Example 15.2

A column is to be constructed from a piece of wood 3 meters long with a rectangular cross section having the dimensions 50 by 100 millimeters. If the modulus of elasticity of the wood is 8.0 gigapascals, and the ends of the column are pinned, find the critical buckling stress of the column.

Solution:

For convenience x- and y-axes are established as shown in Figure 15.3. The moments of inertia of the section about the two axes are

$$I_x = \frac{bd^3}{12} = \frac{(50 \times 10^{-3})(100 \times 10^{-3})^3}{12} = 4.167 \times 10^{-6} \text{ m}^4$$

and

$$I_y = \frac{db^3}{12} = \frac{(100 \times 10^{-3})(50 \times 10^{-3})^3}{12} = 1.042 \times 10^{-6} \text{ m}^4$$

Also the cross-sectional area is

$$A = (50 \times 10^{-3})(100 \times 10^{-3}) = 5 \times 10^{-3} \text{ m}^2$$

Thus the radii of gyration are

Example 15.3

The structure shown in Figure 15.4a is comprised of pin-ended members as shown. Rod AB is a solid cylindrical shaft of 3 inches outside diameter, and rod AC is a solid cylindrical shaft of 4 inches outside diameter. Both rods are made of steel ($E = 30 \times 10^6$ psi). It is desired to have a factor of safety of two against buckling of the rods. Find the maximum allowable force P that can be applied to the structure. Assume the material is performing elastically throughout the loading.

Solution:

Before conducting the buckling analysis, the forces in the members in terms of the force P must be found. This may be achieved by isolating the joint A in exactly the same manner as was done when analyzing trusses by the method of joints (see Chapter 4). A free body diagram of joint A is shown in Figure 15.4b.

$$r_x = \sqrt{\frac{I_x}{A}} = \sqrt{\frac{4.167 \times 10^{-6}}{5 \times 10^{-3}}} = 0.0289 \text{ m}$$

and

$$r_y = \sqrt{\frac{I_y}{A}} = \sqrt{\frac{1.042 \times 10^{-6}}{5 \times 10^{-3}}} = 0.01443 \text{ m}$$

The respective slenderness ratios are therefore

$$\frac{L}{r_x} = \frac{3.0}{0.0289} = 103.8$$

$$\frac{L}{r_y} = \frac{3.0}{0.01443} = 207.9$$

Choosing the larger slenderness ratio, since it gives the lower critical stress, the critical buckling stress of the column is given by

$$\sigma_{cr} = \frac{\pi^2 E}{\left(\dfrac{L}{r}\right)^2} = \frac{\pi^2 (8 \times 10^9)}{(207.9)^2} = 1.83 \times 10^6 \text{ Pa} = 1.83 \text{ MPa}$$

At this stress the column buckles about its y-axis.

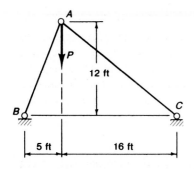

(a)

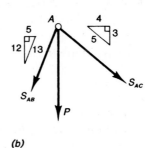

(b)

Figure 15.4 *Frame for Example 15.3.*

Example 15.3 (Continued)

Horizontal and vertical equilibrium equations are now used to find the unknown member forces S_{AB} and S_{AC}

$$\Sigma F_H = 0 \overset{+}{\rightarrow}$$

$$S_{AC} \cdot \left(\frac{4}{5}\right) - S_{AB} \cdot \left(\frac{5}{13}\right) = 0$$

$$\Sigma F_V = 0 \uparrow +$$

$$S_{AB} \cdot \left(\frac{12}{13}\right) + P + S_{AC} \cdot \left(\frac{3}{5}\right) = 0$$

Solving these gives

$$S_{AB} = -0.825 \, P$$

$$S_{AC} = -0.397 \, P$$

The negative signs on these forces indicates that, provided the force P is positive in the direction shown, both rods will be in compression and will therefore be subject to the possibility of buckling. The moment of inertia of each rod is now required:

$$I_{AB} = \frac{\pi r^4}{4} = \frac{\pi \times 1.5^4}{4} = 3.98 \text{ in}^4$$

$$I_{BC} = \frac{\pi r^4}{4} = \frac{\pi \times 2^4}{4} = 12.57 \text{ in}^4$$

Since the rods are both of circular cross section, the moment of inertia of the rods is the same about any axis that passes through the centroid of the section. Equation 15.2 is now applied to find the axial force that would cause each rod to buckle. The units used are kips and inches.

$$P_{\text{cr}AB} = \frac{\pi^2 EI_{AB}}{L_{AB}^2} = \frac{\pi^2 (30 \times 10^3) \, 3.98}{(13 \times 12)^2} = 48.4 \text{ kips}$$

$$P_{\text{cr}AC} = \frac{\pi^2 EI_{AC}}{L_{AC}^2} = \frac{\pi^2 (30 \times 10^3) \, 12.57}{(20 \times 12)^2} = 64.6 \text{ kips}$$

Since a factor of safety of two is specified, the maximum allowable forces in the members are given by

$$S_{AB_{all}} = \frac{P_{cr_{AB}}}{2} = -\frac{48.4}{2} = -24.2 \text{ kips}$$

and

$$S_{AC_{all}} = \frac{P_{cr_{AC}}}{2} = -\frac{64.6}{2} = -32.3 \text{ kips}$$

The negative signs indicate compressive forces, since buckling occurs only when members are in compression.

If the force in the member AB has reached its allowable value, the value of the force P is given by

$$S_{AB_{all}} = -0.824 \, P = -24.2 \text{ kips}$$

i.e.,

$$P = 29.3 \text{ kips}$$

Similarly, if the force in the rod AC has reached its maximum value, the value of the force P is given by

$$S_{AC_{all}} = -0.397 \, P = -32.3 \text{ kips}$$

i.e.,

$$P = 81.4 \text{ kips}$$

The lower of these two values is chosen, and so the maximum allowable force P that can be applied to the structure is 29.3 kips.

15.4 Varying End Restraint and Effective Length Factor

The columns considered so far have all been pin-ended columns, but columns with other end conditions often occur. To account for this an *effective slenderness ratio* is defined as kL/r, where k is an *effective length factor*. Equation 15.7 is then modified to have the form

(15.8)
$$\sigma_{cr} = \frac{\pi^2 E}{\left(\dfrac{kL}{r}\right)^2}$$

Each possible end restraint is now specified in terms of an effective column length. It may be easier to consider the column to have an effective length L_e given by

(15.9)
$$L_e = k \cdot L$$

where L is still the nominal length of the column. The effective length of a pin-ended column is equal to the true length of the column and thus $k = 1.0$. It is convenient to consider the effective length of a column as the distance between points of zero moment on the column (whether pin supports or points of inflection in the deflected shape of the buckled column). In each of the end restraints considered in Figure 15.5, the distance L is the nominal length of the column and the distance L_e is the effective length of the column. The diagrams show several common end restraints for columns, but these are not the only possible end restraints. For easy reference the buckled shape of the column is shown by the dotted line. Note that the effective length factor varies between 0.5 and 2.0 for the cases cited.

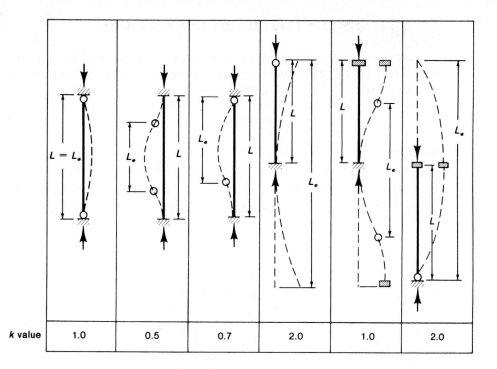

| k value | 1.0 | 0.5 | 0.7 | 2.0 | 1.0 | 2.0 |

end restraints

⊥ end fixed against translation and rotation

⊥ end free to rotate but unable to move (pinned joint)

end free to move laterally but unable to rotate

end free to translate and rotate

Figure 15.5 Approximate effective length factors.

Example 15.4

A column 40 feet long is to be constructed from a steel wide-flange section W 14 $\times$ 82. The steel has a yield stress of 50 kips per square inch and a modulus of elasticity of 30,000 kips per square inch. If the end restraints are as shown in Figure 15.6, find the axial load that will cause buckling of the column. Assume that Euler's buckling formula is valid.

Solution:

From the tables of Appendix A the following properties of a W 14 $\times$ 82 section are extracted or calculated

$$A = 24.1 \text{ in}^2$$

$$I_x = 882 \text{ in}^4$$

$$I_y = 148 \text{ in}^4$$

$$r_x = \sqrt{\frac{I_x}{A}} = 6.05 \text{ in}$$

$$r_y = \sqrt{\frac{I_y}{A}} = 2.48 \text{ in}$$

Figure 15.6 indicates that for buckling about the strong axis, i.e., the x-axis, the column is pin-ended and thus the effective length factor for buckling about this axis is $k_x = 1.0$. However, for buckling about the weak axis, i.e., the y-axis, the column is fixed at both ends. Reference to Figure 15.5 shows that the effective length factor for this case is 0.5, and thus $k_y = 0.5$. The slenderness ratios for buckling about both axes may now be found:

$$\frac{k_x L}{r_x} = \frac{1.0 \times (40 \times 12)}{6.05} = 79.3$$

$$\frac{k_y L}{r_y} = \frac{0.5 \times (40 \times 12)}{2.48} = 96.8$$

Reference to Equation 15.8 shows that the larger the effective slenderness ratio, the lower the critical buckling stress. Thus the larger of the two slenderness

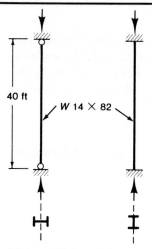

Figure 15.6 Column of Example 15.4.

ratios calculated is chosen. This value, 96.8, corresponds to buckling about the weak axis or the y-axis of the section. Using Equation 15.8 the critical buckling stress is found:

$$\sigma_{cr} = \frac{\pi^2 E}{\left(\dfrac{kL}{r}\right)^2} = \frac{\pi^2 \times 30{,}000}{(96.8)^2} = 31.6 \text{ ksi}$$

Thus, the applied axial force that will cause the column to buckle is given by Equation 15.1:

$$\begin{aligned} P_{cr} &= \sigma_{cr} \cdot A \\ &= 31.6 \times 24.1 \\ &= 762 \text{ kips} \end{aligned}$$

It is noted that the critical buckling stress is below the yield strength of the material.

Example 15.5

A column is to be constructed from a piece of wood with a 50 by 150 millimeter cross section as shown in Figure 15.7a. The column is to be 4.0 meters long, and the end restraints offered to buckling of the column are shown in Figure 15.7b. If the modulus of elasticity of the wood is 8.0 gigapascals, find the buckling load of the column.

Solution:

Before the buckling analysis can be carried out, the properties of the section must be determined:

$$A = 50 \times 150 = 7{,}500 \text{ mm}^2 = 7.5 \times 10^{-3} \text{ m}^2$$

$$I_x = \frac{bd^3}{12} = \frac{50 \times (100)^3}{12} = 1.406 \times 10^7 \text{ mm}^4$$

$$= 1.406 \times 10^{-5} \text{ m}^4$$

$$I_y = \frac{db^3}{12} = \frac{150 \times (50)^3}{12} = 1.563 \times 10^6 \text{ mm}^4$$

$$= 1.563 \times 10^{-6} \text{ m}^4$$

Thus,

$$r_x = \sqrt{\frac{I_x}{A}} = \sqrt{\frac{1.406 \times 10^7}{7{,}500}} = 43.30 \text{ mm}$$

and

$$r_y = \sqrt{\frac{I_y}{A}} = \sqrt{\frac{1.563 \times 10^6}{7{,}500}} = 14.42 \text{ mm}$$

The effective length factors for buckling about the two axes are found from Figure 15.5 to be

$$k_x = 0.7$$

and

$$k_y = 0.5$$

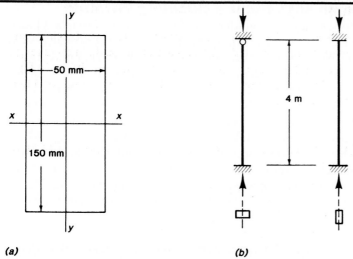

(a)   (b)

Figure 15.7   Column of Example 15.5.

Thus the effective slenderness ratios are

$$\frac{k_x \cdot L}{r_x} = \frac{0.7 \times (4.0 \times 10^3)}{43.3} = 64.7$$

$$\frac{k_y L}{r_y} = \frac{0.5 \times (4.0 \times 10^3)}{14.42} = 138.7$$

Since the larger of these gives the smaller critical buckling stress by Euler's buckling formula, it is concluded that buckling takes place about the y-axis. The critical buckling stress is found using Equation 15.8:

$$\sigma_{cr} = \frac{\pi^2 E}{\left(\dfrac{kL}{r}\right)^2} = \frac{\pi^2 \times 8.0 \times 10^9}{(138.7)^2} = 4.10 \times 10^6 \, \text{Pa} = 4.10 \, \text{MPa}$$

Thus the buckling load may be found using Equation 15.1:

$$P_{cr} = \sigma_{cr} \cdot A$$
$$= (4.10 \times 10^6) \times (7.5 \times 10^{-3})$$
$$= 3.08 \times 10^4 \, \text{N}$$
$$= 30.8 \, \text{kN}$$

15.5 Problems with Euler's Buckling Formula

In Section 15.2 the assumptions on which the Euler buckling formula is based were enumerated. Unfortunately, these assumptions are never completely valid for real columns. There are a number of reasons that the assumptions are not entirely sound, but a brief examination of two of the assumptions will convey some appreciation of the difficulties of column design. The two problems considered are the problem of material behavior and that of column out-of-straightness. There are significant other problems that still are unsolved by engineers.

The problem of material behavior is illustrated by reference to Figure 15.8. Euler's buckling formula has the form of a hyperbolic parabola, and thus, as the column length decreases, the critical buckling stress increases. In the limit, the critical buckling stress would approach infinity for a very short column. Clearly this situation is not governed by Euler's formula because the material would yield in compression before the column buckled. It is reasonable, then, to set the yield stress of the material as the upper limit of the critical buckling stress. Thus in Figure 15.8, the solid horizontal line at the yield stress σ_y is an upper bound on the critical buckling stress.

Even this modification of Euler's buckling formula is not satisfactory. It can be shown experimentally that long columns tend to buckle at an axial stress closely approximating the value predicted by Euler's buckling formula. Very short columns, on the other hand, tend to buckle at a stress closely approximated by the yield stress of the material. There is, however, a whole group of intermediate column lengths for which the buckling stress cannot be adequately approximated either by Euler's buckling formula or by the material yield stress. The experimental results tend to show that the critical buckling stress for such columns tends to follow the curve shown dotted in Figure 15.8. It is important that technologists be able to predict the strength as intermediate columns. The exact form of the dotted line of Figure 15.8 is still in doubt, and recent evidence indicates that it may vary with the shape of the section being investigated.

The problem of column out-of-straightness arises because a real column is never perfectly straight, although Euler's buckling formula assumes that this is so. Most design codes specify a maximum permissible out-of-straightness for columns, and it is usual to allow an out-of-straightness of about 1/1,000 of the column length. (Thus a column 10 feet long may be up to 1/8 inch out of straight at its center.) To see the difference this makes to column behavior, reference should be made to Figure 15.9, which is really a modified version of Figure 15.2. The solid line in Figure 15.9b indicates the ideal behavior of a column that is initially perfectly straight. The dotted line shows the real behavior of a column with an initial mid-height deflection (or out-of-straightness) of δ_o. As indicated, the column tends to deflect as the load is increased, rather than suddenly as was the case for ideal columns. The important concern, however, is that a real column tends to buckle at a lower critical stress than an ideal column of the same length. This is indicated by the fact that the peak of the curve for the real column is lower than the value of P_{cr} derived for an ideal column.

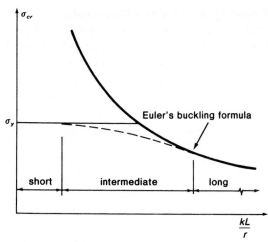

Figure 15.8 *Euler's buckling curve.*

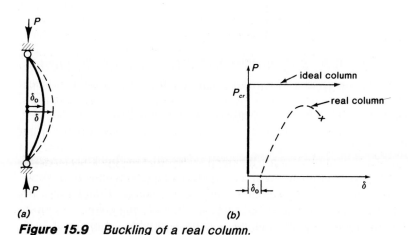

(a)   *(b)*

Figure 15.9 *Buckling of a real column.*

15.6 Column Design Formula

In Section 15.5 some of the problems associated with column design were discussed. However, since columns are very common in engineering and technology, some method of designing them must be established. The American Institute of Steel Construction publishes a "Specification for the Design, Fabrication and Erection of Structural Steel for Buildings," which is regularly updated. For many years the specification (along with others) has presented the same approach to column design, and so this approach is now briefly discussed.

In Figure 15.8, the general shape of the column buckling curve is given. To assist in column design, equations for the various portions of this curve are given. This curve is redrawn in two portions in Figure 15.10. The transition between the two curves A and B occurs at a value given by

$$\frac{kL}{r} = C_c = \sqrt{\frac{2\pi^2 E}{\sigma_y}}$$

(15.10)

The curve B to the right of the point C_c is the familiar Euler's buckling curve of Equation 15.7; i.e.,

$$\sigma_{cr} = \frac{\pi^2 E}{\left(\dfrac{kL}{r}\right)^2}$$

(15.11)

The curve A to the left of the point C_c is assumed to have a parabolic shape given by

$$\sigma_{cr} = \left[1 - \frac{\left(\dfrac{kL}{r}\right)^2}{2C_c^2}\right]\sigma_y$$

(15.12)

Clearly, when $\frac{kL}{r} = 0$, $\sigma_{cr} = \sigma_y$. In other words, the buckling stress of a short column is equal to the yield strength of the material. If the value of $\frac{kL}{r} = C_c$ is substituted into Equation 15.12, it is readily seen that the critical buckling stress is given by $\sigma_{cr} = \frac{\sigma_y}{2}$.

Equations 15.11 and 15.12 are assumed to describe the buckling stress of columns, and they generally fit observed experimental results quite well. However, for design use they are insufficient, since some factor of safety must be introduced. It is generally considered that buckling of long columns is more critical than buckling of short columns, and so a variable factor of safety is assumed. Equations for curves A and B are divided by the appropriate factors of safety to produce the design curves C and D, respectively, as shown in Figure 15.11. Curves C and D represent the maximum allowable stresses in a column if there is to be an adequate factor of safety. Once again, the transition between the two curves is taken at $\frac{kL}{r} = C_c$, where C_c has the value given in Equation 15.10.

Curve D is derived from curve B by introducing a constant factor of safety of 23/12; that is, the allowable stress is defined as 12/23 of the buckling stress. If σ_a is the allowable axial stress permitted on the column, then curve D is given by

$$(\sigma_a)_{\text{curve } D} = \frac{(\sigma_{cr})_{\text{curve } B}}{\text{factor of safety}}$$

(15.13)

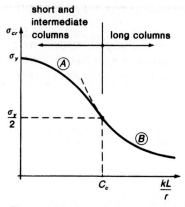

Figure 15.10 Column buckling curve.

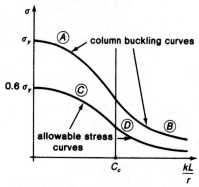

Figure 15.11 Allowable stress curves for columns.

$$\sigma_a = \frac{\left[\dfrac{\pi^2 E}{\left(\dfrac{kL}{r}\right)^2}\right]}{\left(\dfrac{23}{12}\right)}$$

or

$$\sigma_a = \frac{12\pi^2 E}{23\left(\dfrac{kL}{r}\right)^2} \tag{15.14}$$

In the derivation of curve C, however, a variable factor of safety is used; the equation of curve C is

$$(\sigma_a)_{\text{curve } C} = \frac{(\sigma_{\text{cr}})_{\text{curve } A}}{\text{factor of safety}} \tag{15.15}$$

$$\sigma_a = \frac{\left[1 - \dfrac{\left(\dfrac{kL}{r}\right)^2}{2\,C_c^2}\right]\sigma_y}{\left[\dfrac{5}{3} + \dfrac{3\left(\dfrac{kL}{r}\right)}{8\,C_c} - \dfrac{\left(\dfrac{kL}{r}\right)^3}{8\,C_c^3}\right]} \tag{15.16}$$

The denominator of Equation 15.16 represents the factor of safety, which varies from 5/3 when $\dfrac{kL}{r} = 0$ to $\dfrac{23}{12}$ when $\dfrac{kL}{r} = C_c$.

Example 15.6

A wide-flange steel section W 12 × 45 is to be used as a column. The steel used has a yield strength of 36 kips per square inch (and $E = 30 \times 10^3$ ksi). Find the maximum allowable compressive force that may be applied to the column under the following conditions:

    a. The column is 25 feet long and is pinned at both ends for buckling about both axes.

    b. The column is 30 feet long and is fixed at both ends for buckling about both axes.

Solution:

The required properties of a W 12 × 45 section are found from Appendix A:

$$A = 13.2 \text{ in}^2$$
$$r_x = 5.15 \text{ in}$$

and

$$r_y = 1.94 \text{ in}$$

The transition value between design curves is given by Equation 15.10 as

$$C_c = \sqrt{\frac{2\pi^2 E}{\sigma_y}} = \sqrt{\frac{2\pi^2 \cdot 30{,}000}{36}} = 128.3$$

Part a. The value of $\dfrac{kL}{r}$ for this column must first be found. From Figure 15.5, $k = 1.0$ for a column with pinned ends. Thus

$$\frac{kL}{r_x} = \frac{1.0 \times (25 \times 12)}{5.15} = 58.3$$

and

$$\frac{kL}{r_y} = \frac{1.0 \times (25 \times 12)}{1.94} = 154.6$$

Since the larger of these values controls the design, the critical value of $\dfrac{kL}{r}$ is 154.6. This is larger than the value of C_c, and thus curve D (in Figure 15.11) determines the maximum allowable stress permitted in the column. Using Equation 15.14 for curve D,

$$\sigma_a = \frac{12\,\pi^2 E}{23\left(\dfrac{kL}{r}\right)^2} = \frac{12 \cdot \pi^2 \cdot 30{,}000}{23 \cdot (154.6)^2} = 6.46 \text{ ksi}$$

The maximum axial load, P_a, that may be applied to the members is given by

$$P_a = \sigma_a \cdot A$$

i.e.,

$$P_a = 6.46 \times 13.2 = 85.3 \text{ kips}$$

Part b. From Figure 15.5, the effective length factor k for a column fixed at both ends is found to be 0.5. Thus

$$\frac{kL}{r_x} = \frac{0.5\,(30 \times 12)}{5.15} = 35.0$$

$$\frac{kL}{r_y} = \frac{0.5\,(30 \times 12)}{1.94} = 92.8$$

Once again $\dfrac{kL}{r_y} = 92.8$ controls the design, but now it is less than the calculated value of C_c, and so curve C (of Figure 15.11) must be used to calculate the allowable stress. Using Equation 15.16,

$$\sigma_a = \frac{\left[1 - \dfrac{\left(\dfrac{kL}{r}\right)^2}{2\,C_c^2}\right]\sigma_y}{\left[\dfrac{5}{3} + \dfrac{3\left(\dfrac{kL}{r}\right)}{8\,C_c} - \dfrac{\left(\dfrac{kL}{r}\right)^3}{8\,C_c^3}\right]} = \frac{\left[1 - \dfrac{(92.8)^2}{2(128.3)^2}\right]36}{\left[\dfrac{5}{3} + \dfrac{3(92.8)}{8(128.3)} - \dfrac{(92.8)^3}{8(128.3)^3}\right]} = 14.06 \text{ ksi}$$

Thus the maximum axial load that may be applied to this column is

$$P_a = \sigma_a \cdot A = 14.06 \times 13.2 = 185.6 \text{ kips}$$

15.1 A 10-foot-long column with the cross section shown in Figure P 15.1 is simply supported on both ends. Determine a) the direction of buckling, b) the buckling stress. ($E = 29 \times 10^6$ psi.)

15.2 Find the shortest length L for a fixed-ended steel column, having a cross-sectional area of 50 by 80 millimeters, for which the elastic Euler formula applies. ($E = 200$ GPa; the proportional limit is 20 MPa.)

15.3 A rectangular steel tube 10 feet long with both ends pinned serves as a column. The cross section of the column is shown in Figure P 15.3. Find a) the critical axial load P_{cr} of the column; b) the critical stress in the column. ($E = 29 \times 10^6$ psi.)

15.4 A square steel tube, 4 by 4 centimeters in outside dimensions with 1/2-centimeter-thick walls and 2 meters long, is to be used as a column with one end fixed, the other free. If the column design load P is one-third of the critical load P_{cr}, determine a) the design load P; b) the maximum stress in the column when the load P is applied. ($E = 200$ GPa.)

15.5 A round steel bar 1.5 inches in diameter and 5 feet long is shown in Figure P 15.5. If cables and connections are properly designed, what force F may be applied to the assembly before the bar buckles? Use Euler's formula. ($E = 29 \times 10^6$ psi.)

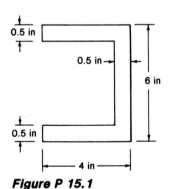

Figure P 15.1

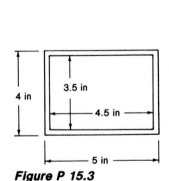

Figure P 15.3

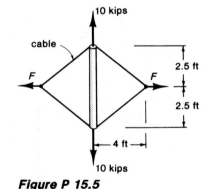

Figure P 15.5

15.6   For the solid steel bar and cable assembly shown in Figure P 15.6, determine the diameter of the steel bar such that the factor of safety is 2.5. ($E = 200$ GPa.)

15.7   In Figure P 15.7 the square solid bar is made of steel ($E = 29 \times 10^6$ psi). Find the size of the bar if the factor of safety is 2.0.

15.8   For the two sections in Figure P 15.8, find $\dfrac{\text{area}_A}{\text{area}_B}$ and $\dfrac{P_{cr_A}}{P_{cr_B}}$.

15.9   The pin-connected aluminum-alloy frame shown in Figure P 15.9 carries concentrated loads F horizontally and $2F$ vertically. Assuming buckling can occur only in the plane of the frame, determine the value of F that will cause an unstable condition. Use the Euler formula as a criterion for member buckling. Both sections are 40 by 40 millimeters. ($E = 70$ GPa for the alloy.)

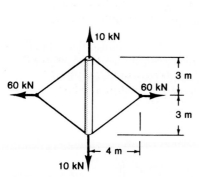

Figure P 15.6

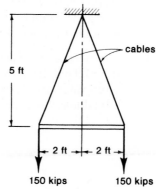

Figure P 15.7

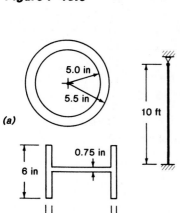

Figure P 15.8

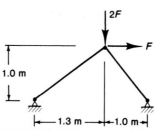

Figure P 15.9

Column Design   455

15.10 For a given cross-sectional area, compare the buckling load of two solid bars acting as pin-ended columns, one with a circular section and the other with a square section.

15.11 For the structure shown in Figure P 15.11, if the cross section of the solid member AB is circular, what must be the diameter in order to prevent buckling? Use the Euler formula with a factor of safety 2.0. ($E = 70$ GPa.)

15.12 A structure made from 2-inch standard steel pipes is to be used for lifting loads vertically with a pulley at A, as shown in Figure P 15.12. What load rating may be assigned to this structure? Use the Euler formula and a factor of safety 2.0. Consider all joints pinned, and assume that the connection details are adequate. ($E = 29 \times 10^3$ ksi.)

15.13 An allowable axial load for a 4-meter-long pin-ended column of a certain linearly elastic material is 20 kilonewtons. Five different columns made of the same material and having the same cross section have the supporting conditions shown in Figure P 15.13. Using the column capacity for the 4-meter column as the criterion, what are the allowable loads for the five columns?

15.14 For the structure shown in Figure P 15.14, x is changeable from 0 to 180. Find the allowable force F applied at the end for different given values of x. (Answers for each unit increment of x are available in the instructor's manual.) ($EI = 70$ GPa.)

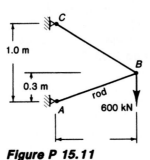

Figure P 15.11

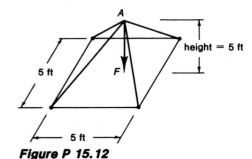

Figure P 15.12

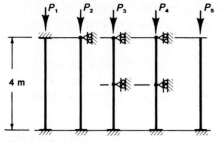

Figure P 15.13

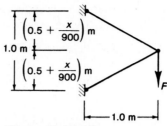

Figure P 15.14

Solutions
for Selected
Problems

Chapter 2

2.5 $M_A = 21.59 \text{ kN} \cdot \text{m} \;\circlearrowleft$

      $M_D = 1.59 \text{ kN} \cdot \text{m} \;\circlearrowleft$

2.8 $M = 4.39 \text{ kN} \cdot \text{m}$

2.12   Moment about the origin: $M = 9.7.0 \text{ N} \cdot \text{m}$

      Moment about point A: $M = 81.8 \text{ N} \cdot \text{m}$

2.19 $M_{Ox} = -40 \text{ kip} \cdot \text{ft}$

      $M_{Oy} = 70 \text{ kip} \cdot \text{ft}$

      $M_{Oz} = -195 \text{ kip} \cdot \text{ft}$

2.22 $M_{Ox} = 0$

      $M_{Oy} = -3 \text{ N} \cdot \text{m}$

      $M_{Oz} = -109 \text{ N} \cdot \text{m}$

Chapter 3

3.11 $B_x = 375 \text{ lb} \;\nwarrow$

      $A_x = 25 \text{ lb} \;\rightarrow$

      $A_y = 200 \text{ lb} \downarrow$

3.14 $A_x = 2 \text{ kN} \leftarrow$

      $A_y = 1 \text{ kN} \downarrow$

      $B_y = 9 \text{ kN} \downarrow$

3.17 $B_x = 75 \text{ lb} \leftarrow$

      $B_y = 204.9 \text{ lb} \uparrow$

      $M_{ZB} = 1{,}620.8 \text{ lb} \cdot \text{ft} \;\circlearrowright$

3.21 $A_x = 80 \text{ k} \leftarrow$

      $B_x = 0$

      $C_x = 70 \text{ k} \leftarrow$

      $D_x = 70 \text{ k} \leftarrow$

      $E_x = 70 \text{ k} \rightarrow$

$A_y = 80 \text{ k} \uparrow$

$B_y = 120 \text{ k} \uparrow$

$C_y = 100 \text{ k} \uparrow$

$D_y = 100 \text{ k} \downarrow$

$E_y = 20 \text{ k} \downarrow$

3.24 $A_x = 565.3 \text{ N} \leftarrow$

$A_y = 4{,}230.78 \text{ N} \uparrow$

$A_z = 269.3 \text{ N} \nearrow$

Chapter 4

4.2 $AB, BC, AC, AJ, EF, EG, OI, EH, IJ$

4.7 $F_{AF} = 5/3 \text{ k } (T)$

$F_{AB} = 4/3 \text{ k } (C)$

$F_{BF} = 0$

$F_{BC} = -4/3 \text{ k } (C)$

$F_{FC} = -5/3 \text{ k } (C)$

4.11 $A_x = 8.24 \text{ k} \leftarrow$

$B_x = 1.76 \text{ k} \leftarrow$

$B_y = 5 \text{ k} \uparrow$

$FB = 10.35 \text{ k } (T)$
$GC = -2.02 \text{ k } (C)$
$ED = 2.40 \text{ k } (T)$
$BD = -2.65 \text{ k } (C)$
$EF = 2.62 \text{ k } (T)$
$EG = -2.19 \text{ k } (C)$
$GF = 1.86 \text{ k } (T)$

4.14 $AB = -13.26 \text{ kN } (C)$
$BC = -26.65 \text{ kN } (C)$
$CD = 9.88 \text{ kN } (T)$
$DE = 6.63 \text{ kN } (T)$
$CE = 26.52 \text{ kN } (C)$
$BE = 26.52 \text{ kN } (T)$
$AE = 33.15 \text{ kN } (C)$

4.24 $BA = 20 \text{ kN } (T)$

Chapter 7

7.3 $\bar{x} = 0, \bar{y} = 11.05 \text{ in}$

7.6 $\bar{x} = 0, \bar{y} = 157.78 \text{ mm}$

7.11 $I_{xx} = 8{,}007.09 \text{ in}^4, I_{yy} = 1{,}714.52 \text{ in}^4$

$\bar{I}_{xx} = 2{,}536.11 \text{ in}^4, \bar{I}_{yy} = 1{,}714.52 \text{ in}^4$

7.14 $\bar{I}_{xx} = 1{,}067{,}200{,}000 \text{ mm}^4$

$\bar{I}_{xx} = 395{,}047{,}733.2 \text{ mm}^4$

7.18 $I_{xx} = 585.41 \text{ m}^4, I_{yy} = 3{,}651.66 \text{ m}^4$

$\bar{I}_{xx} = 585.41 \text{ m}^4, \bar{I}_{yy} = 1{,}942.74 \text{ m}^4$

Chapter 8

8.5 $\Delta L_T = 7.425 \times 10^{-3} \text{ in longer}$

8.11   2

8.14 $\sigma_{net} = 450 \text{ kN/m}^2$

8.18 $P = 2.5 \text{ N}$

8.23 $\sigma_{max} = 7.9 \text{ ksi}$

Chapter 9

9.1   25%

9.6   23.9 MPa

9.9   3 in

9.11 $\tau = 7.37 \text{ ksi at 1 ft from the wall}$

9.16 $-2.70°$

Chapter 10

10.3 $M = 90 \text{ k} \cdot \text{ft at point } A$

10.9 $M = 8 \text{ k} \cdot \text{ft at point } B$

10.17 $\sigma_{max} = \pm 272.13 \text{ k/ft}^2 \text{ at point } B$

10.22 $\sigma_{max} = \pm 2{,}073.2 \text{ kN/m}^2 \text{ at point } A$

10.24 $M_{all} = 0.90 \text{ N} \cdot \text{m}$

Chapter 11

11.2 $\tau_{max} = 18.78$ MPa at N.A.

      $\tau_{min} = 0$ on top and bottom surfaces.

11.3 $\tau_1 = 9.15$ ksi

      $\tau_2 = 0$

11.5 $\tau_{max} = 11.08$ ksi at support A

11.7 $\tau_1 = 11$ ksi

      $\tau_2 = 9.43$ ksi

      $\tau_3 = 0.31$ ksi

11.10 $\tau_{max} = 69.79$ MPa

Chapter 12

12.2 $\sigma_{max}(T) = 11{,}375$ kN/m² on bottom of beam

      $\sigma_{max}(C) = -11{,}125$ kN/m² on top of beam

12.5 $\tau_{max} = 18{,}859$ ksi $\uparrow$ at point D

12.8 $\sigma_{max} = 381.94$ kN/cm² (C)

      $\tau = 133.25$ kN/cm² $\rightarrow$ and 0.25 kN/cm²$\downarrow$

12.13 $\sigma = 21.19$ ksi (T)

12.16

      $\sigma = 6.82$ MPa

      $\tau = 31.82$ MPa

      25° $\sigma = -56.82$ MPa

      $\tau = 10.61$ MPa

Chapter 13

13.7 $v_{max} = \dfrac{WL^4}{8EI}$

13.8 $v_{max} = \dfrac{PL^3}{96EI} + \dfrac{35}{5{,}184}\dfrac{WL^4}{EI}$

13.12 $v_A = -PL^3/8$

13.16 $v_A = \dfrac{-11}{24}\dfrac{ML^2}{EI}$

13.17 $v_A = \dfrac{ML^2}{18EI}$

      $\theta_A = \dfrac{ML}{6EI}$

Chapter 14

14.2 $M_A = 26{,}640$ N·mm

14.6 $v_{max} = -3.98$ mm

14.9 $R_C = 675$ N

      $R_B = 4{,}125$ N

      $R_A = 2{,}550$ N

14.13 $M_A = 2.25 \times 10^6$ N·mm $\circlearrowleft$

      $M_B = 2.25 \times 10^6$ N·mm $\circlearrowright$

14.16 $V_A = 5.2$ kips$\uparrow$

      $V_B = 6.7$ kips $\uparrow$

      $V_C = 5.9$ kips $\downarrow$

Chapter 15

15.1 Buckling will occur about vertical axis

$\sigma = 31.06$ ksi

15.5 $F = 46.2$ kips

15.7 $b = 1.85$ in

15.8 $\dfrac{\text{Area}_A}{\text{Area}_B} = 1.0$

$\dfrac{P_{crA}}{P_{crB}} = 8.31$

15.14 $R = 1.26$ mm

$d = 2.52$ mm

Properties of Common Steel Engineering Sections

Design data: W shapes

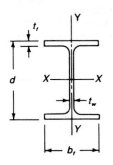

Designation (New series)	Area A (in^2)	Depth d (in)	Flange Width b_f (in)	Flange Thickness t_f (in)	Web thickness t_w (in)
W 36 × 300	88.3	36.74	16.655	1.680	0.945
W 36 × 210	61.8	36.69	12.180	1.360	0.830
W 33 × 152	44.7	33.49	11.565	1.055	0.635
W 30 × 211	62.0	30.94	15.105	1.315	0.775
W 30 × 132	38.9	30.31	10.545	1.000	0.615
W 27 × 178	52.3	27.81	14.085	1.190	0.725
W 27 × 114	33.5	27.29	10.070	0.930	0.570
W 24 × 162	47.7	25.00	12.955	1.220	0.705
W 24 × 94	27.7	24.31	9.065	0.875	0.515
W 21 × 147	43.2	22.06	12.510	1.150	0.720
W 21 × 93	27.3	21.62	8.420	0.930	0.580
W 21 × 57	16.7	21.06	6.555	0.650	0.405
W 18 × 119	35.1	18.97	11.265	1.060	0.655
W 18 × 71	20.8	18.47	7.635	0.810	0.495
W 18 × 50	14.7	17.99	7.495	0.570	0.355
W 16 × 100	29.4	16.97	10.425	0.985	0.585
W 16 × 57	16.8	16.43	7.120	0.715	0.430
W 14 × 426	125	14.67	16.695	3.035	1.875
W 14 × 211	62.0	14.72	15.800	1.560	0.980
W 14 × 82	24.1	14.31	10.130	0.855	0.510
W 14 × 53	15.6	13.92	8.060	0.660	0.370
W 14 × 38	11.2	14.10	6.770	0.515	0.310
W 12 × 152	44.7	13.71	12.480	1.400	0.870
W 12 × 45	13.2	12.06	8.045	0.575	0.335
W 12 × 22	6.48	12.31	4.030	0.425	0.260
W 10 × 112	32.9	11.36	10.415	1.250	0.755
W 10 × 54	15.8	10.09	10.030	0.615	0.370
W 10 × 45	13.3	10.10	8.020	0.620	0.350
W 10 × 19	5.62	10.24	4.020	0.395	0.250
W 8 × 31	9.13	8.00	7.995	0.435	0.285
W 6 × 25	7.34	6.38	6.080	0.455	0.320
W 5 × 19	5.54	5.15	5.030	0.430	0.270
W 4 × 13	3.83	4.16	4.060	0.345	0.280

I = Moment of inertia Z = Section modulus r = Radius of gyration

| Elastic properties | | | | | | Torsional constant |
| Axis X – X | | | Axis Y – Y | | | |
I (in^4)	Z (in^3)	r (in)	I (in^4)	Z (in^3)	r (in)	J (in^4)
20,300	1,110	15.2	1,300	156	3.83	64.2
13,200	719	14.6	411	67.5	2.58	28.0
8,160	487	13.5	273	47.2	2.47	12.4
10,300	663	12.9	757	100	3.49	27.9
5,770	380	12.2	196	37.2	2.25	9.72
6,990	502	11.6	555	78.8	3.26	19.5
4,090	299	11.0	159	31.5	2.18	7.33
5,170	414	10.4	443	68.4	3.05	18.5
2,700	222	9.87	109	24.0	1.98	5.26
3,630	329	9.17	376	60.1	2.95	15.4
2,070	192	8.70	92.9	22.1	1.84	6.03
1,170	111	8.36	30.6	9.35	1.35	1.77
2,190	231	7.90	253	44.9	2.69	10.6
1,170	127	7.50	60.3	15.8	1.70	3.48
800	88.9	7.38	40.1	10.7	1.65	1.24
1,490	175	7.10	186	35.7	2.52	7.73
758	92.2	6.72	43.1	12.1	1.60	2.22
6,600	707	7.26	2,360	283	4.34	331
2,660	338	6.55	1,030	130	4.07	44.6
882	123	6.05	148	29.3	2.48	5.08
541	77.8	5.89	57.7	14.3	1.92	1.94
385	54.6	5.88	26.7	7.88	1.55	0.80
1,430	209	5.66	454	72.8	3.19	25.8
350	58.1	5.15	50.0	12.4	1.94	1.31
156	25.4	4.91	4.66	2.31	0.848	0.29
716	126	4.66	236	45.3	2.68	15.1
303	60.0	4.37	103	20.6	2.56	1.82
248	49.1	4.33	53.4	13.3	2.01	1.51
96.3	18.8	4.14	4.29	2.14	0.874	0.23
110.	27.5	3.47	37.1	9.27	2.02	0.54
53.4	16.7	2.70	17.1	5.61	1.52	0.46
26.2	10.2	2.17	9.13	3.63	1.28	0.31
11.3	5.46	1.72	3.86	1.90	1.00	0.15

Design data: HP shapes

| | | | Flange | | Web |
Designation (New series)	Area A (in^2)	Depth d (in)	Width b_f (in)	Thickness t_f (in)	thickness t_w (in)
HP 14 × 117	34.4	14.21	14.885	0.805	0.805
HP 13 × 100	29.4	13.15	13.205	0.765	0.765
HP 12 × 84	24.6	12.28	12.295	0.685	0.685
HP 10 × 57	16.8	9.99	10.225	0.565	0.565
HP 8 × 36	10.6	8.02	8.155	0.445	0.445

I = Moment of inertia Z = Section modulus r = Radius of gyration

Design data: Structural tees cut from W shapes

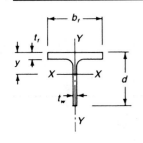

| | | | Flange | | Stem | |
Designation (New series)	Area A (in^2)	Depth d (in)	Width b_f (in)	Thickness t_f (in)	thickness t_w (in)	d/t_w
WT 18 × 150	44.1	18.370	16.655	1.680	0.945	19.4
WT 18 × 105	30.9	18.345	12.180	1.360	0.830	22.1
WT 16.5 × 120.5	35.4	17.090	15.860	1.400	0.830	20.6
WT 16.5 × 76	22.4	16.745	11.565	1.055	0.635	26.4
WT 15 × 105.5	31.0	15.470	15.105	1.315	0.775	20.0
WT 13.5 × 89	26.1	13.905	14.085	1.190	0.725	19.2
WT 10.5 × 73.5	21.6	11.030	12.510	1.150	0.720	15.3
WT 10.5 × 28.5	8.37	10.530	6.555	0.650	0.405	26.0
WT 9 × 59.5	17.5	9.485	11.265	1.060	0.655	14.5
WT 9 × 35.5	10.4	9.235	7.635	0.810	0.495	18.7
WT 8 × 50	14.7	8.485	10.425	0.985	0.585	14.5
WT 8 × 28.5	8.38	8.215	7.120	0.715	0.430	19.1
WT 8 × 15.5	4.56	7.940	5.525	0.440	0.275	28.9

I = Moment of inertia Z = Section modulus r = Radius of gyration

| Elastic properties | | | | | | Torsional constant |
| Axis X – X | | | Axis Y – Y | | | |
I (in⁴)	Z (in³)	r (in)	I (in⁴)	Z (in³)	r (in)	J (in⁴)
1,220	172	5.96	443	59.5	3.59	8.02
886	135	5.49	294	44.5	3.16	6.25
650	106	5.14	213	34.6	2.94	4.24
294	58.8	4.18	101	19.7	2.45	1.97
119	29.8	3.36	40.3	9.88	1.95	0.77

| Axis X – X | | | | Axis Y – Y | | |
I (in⁴)	Z (in³)	r (in)	y (in)	I (in⁴)	Z (in³)	r (in)
1,230	86.1	5.27	4.13	648	77.8	3.83
985	73.1	5.65	4.87	206	33.8	2.58
871	65.8	4.96	3.85	466	58.8	3.63
592	47.4	5.14	4.26	136	23.6	2.47
610	50.5	4.43	3.40	378	50.1	3.49
414	38.2	3.98	3.05	278	39.4	3.26
204	23.7	3.08	2.39	188	30.0	2.95
90.4	11.8	3.29	2.85	15.3	4.67	1.35
119	15.9	2.60	2.03	126	22.5	2.69
78.2	11.2	2.74	2.26	30.1	7.89	1.70
76.8	11.4	2.28	1.76	93.1	17.9	2.52
48.7	7.77	2.41	1.94	21.6	6.06	1.60
27.4	4.64	2.45	2.02	6.20	2.24	1.17

Design data: Angles, equal legs and unequal legs

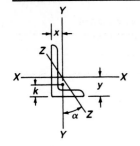

Size and thickness		Weight per foot	Area	Axis X − X		
(in)	k (in)	(lb)	(in²)	I (in⁴)	Z (in³)	r (in)
L8 × 8 × 1⅛	1¾	56.9	16.7	98.0	17.5	2.42
½	1⅛	26.4	7.75	48.6	8.36	2.50
L8 × 6 × 1	1½	44.2	13.0	80.8	15.1	2.49
½	1	23.0	6.75	44.3	8.02	2.56
L8 × 4 × 1	1½	37.4	11.0	69.6	14.1	2.52
½	1	19.6	5.75	38.5	7.49	2.59
L7 × 4 × ¾	1¼	26.2	7.69	37.8	8.42	2.22
½	1	17.9	5.25	26.7	5.81	2.25
L6 × 6 × 1	1½	37.4	11.0	35.5	8.57	1.80
¾	1¼	28.7	8.44	28.2	6.66	1.83
½	1	19.6	5.75	19.9	4.61	1.86
⅜	⅞	14.9	4.36	15.4	3.53	1.88
L6 × 4 × ¾	1¼	23.6	6.94	24.5	6.25	1.88
½	1	16.2	4.75	17.4	4.33	1.91
⅜	⅞	12.3	3.61	13.5	3.32	1.93
L5 × 5 × ¾	1¼	23.6	6.94	15.7	4.53	1.51
½	1	16.2	4.75	11.3	3.16	1.54
⅜	⅞	12.3	3.61	8.74	2.42	1.56
L3 × 2½ × ⅜	¾	6.6	1.92	1.66	0.810	0.928
¼	⅝	4.5	1.31	1.17	0.561	0.945
L3 × 2 × ⅜	¹¹/₁₆	5.9	1.73	1.53	0.781	0.940
¼	⁹/₁₆	4.1	1.19	1.09	0.542	0.957
³/₁₆	½	3.07	0.902	0.842	0.415	0.966
L2 × 2 × ⅜	¹¹/₁₆	4.7	1.36	0.479	0.351	0.594
¼	⁹/₁₆	3.19	0.938	0.348	0.247	0.609
⅛	⁷/₁₆	1.65	0.484	0.190	0.131	0.626

I = Moment of inertia   Z = Section modulus   r = Radius of gyration

	Axis Y−Y			Axis Z−Z		
y (in)	I (in⁴)	Z (in³)	r (in)	x (in)	r (in)	tan α
2.41	98.0	17.5	2.42	2.41	1.56	1.000
2.19	48.6	8.36	2.50	2.19	1.59	1.000
2.65	38.8	8.92	1.73	1.65	1.28	0.543
2.47	21.7	4.79	1.79	1.47	1.30	0.558
3.05	11.6	3.94	1.03	1.05	0.846	0.247
2.86	6.74	2.15	1.08	0.859	0.865	0.267
2.51	9.05	3.03	1.09	1.01	0.860	0.324
2.42	6.53	2.12	1.11	0.917	0.872	0.335
1.86	35.5	8.57	1.80	1.86	1.17	1.000
1.78	28.2	6.66	1.83	1.78	1.17	1.000
1.68	19.9	4.61	1.86	1.68	1.18	1.000
1.64	15.4	3.53	1.88	1.64	1.19	1.000
2.08	8.68	2.97	1.12	1.08	0.860	0.428
1.99	6.27	2.08	1.15	0.987	0.870	0.440
1.94	4.90	1.60	1.17	0.941	0.877	0.446
1.52	15.7	4.53	1.51	1.52	0.975	1.000
1.43	11.3	3.16	1.54	1.43	0.983	1.000
1.39	8.74	2.42	1.56	1.39	0.990	1.000
0.956	1.04	0.581	0.736	0.706	0.522	0.676
0.911	0.743	0.404	0.753	0.661	0.528	0.684
1.04	0.543	0.371	0.559	0.539	0.430	0.428
0.993	0.392	0.260	0.574	0.493	0.435	0.440
0.970	0.307	0.200	0.583	0.470	0.439	0.446
0.636	0.479	0.351	0.594	0.636	0.389	1.000
0.592	0.348	0.247	0.609	0.592	0.391	1.000
0.546	0.190	0.131	0.626	0.546	0.398	1.000

Design data: Rectangular tubing

Size	Gauge	Weight	Area	Axis X–X Moment of inertia, I	Section modulus, Z
(in)		(lb/ft)	(in²)	(in⁴)	(in³)
2 × 3	.250	7.10	2.09	2.20	1.469
2 × 4	.250	8.80	2.59	4.69	2.345
2 × 6	.250	12.02	3.53	13.30	4.434
	.375	16.84	4.95	16.75	5.582
2 × 8	.250	15.42	4.53	29.22	7.306
	.375	21.94	6.45	38.12	9.530
2 × 9	.250	17.12	5.03	40.45	8.990
	.375	24.49	7.20	53.46	11.880
3 × 4	.250	10.50	3.09	6.45	3.225
	.313	12.69	3.56	7.30	3.649
3 × 6	.250	13.72	4.03	17.44	5.813
	.375	19.39	5.70	22.61	7.537
3 × 8	.250	17.12	5.03	36.72	9.180
	.375	24.49	7.20	48.95	12.240
3 × 10	.250	20.34	5.98	64.72	12.940
	.375	29.19	8.58	86.70	17.340
4 × 6	.250	15.42	4.54	21.6	7.19
	.375	21.94	6.45	28.6	9.52
	.500	27.68	8.14	33.4	11.07
4 × 8	.250	18.82	5.54	44.2	11.06
	.375	27.04	7.95	59.9	14.97
	.500	34.48	10.14	71.6	17.87
4 × 10	.250	22.04	6.48	76.6	15.32
	.375	31.73	9.33	104.1	20.82
	.500	40.55	11.90	125.0	24.90
4 × 12	.250	25.44	7.48	122.8	20.47
	.375	36.83	10.83	168.9	28.14
	.500	47.35	13.90	205.0	34.20
5 × 6	.250	17.12	5.03	25.7	8.57
	.375	24.49	7.20	34.6	11.52
5 × 9	.250	22.04	6.48	68.1	15.13
	.375	31.73	9.33	92.8	20.63
	.500	40.55	11.90	112.0	24.80
5 × 10	.250	23.74	6.98	88.4	17.70
	.375	34.27	10.08	121.5	24.29
6 × 8	.250	22.04	6.48	58.4	14.59
	.375	31.73	9.33	79.7	19.91
	.500	40.55	11.93	96.2	23.98
6 × 10	.250	25.44	7.48	100.3	20.07
	.375	36.83	10.83	138.7	27.74
	.500	47.35	13.93	169.5	33.90
6 × 12	.250	28.83	8.48	157.3	26.22
	.375	41.93	12.33	219.4	36.57
	.500	54.15	15.90	271.0	45.20

Radius of gyration, r	Moment of inertia, I	Section modulus, Z	Radius of gyration, r
(in)	(in⁴)	(in³)	(in)
1.03	1.15	1.147	0.74
1.35	1.53	1.532	0.77
1.94	2.24	2.244	0.80
1.84	2.74	2.741	0.74
2.54	3.03	3.029	0.82
2.43	3.75	3.749	0.76
2.83	3.41	3.415	0.82
2.72	4.25	4.253	0.77
1.45	4.10	2.733	1.15
1.43	4.64	3.093	1.14
2.08	5.87	3.912	1.21
1.99	7.46	4.971	1.14
2.70	7.76	5.176	1.24
2.61	10.06	6.705	1.18
3.29	9.55	6.367	1.26
3.18	12.47	8.312	1.21
2.18	11.51	5.754	1.59
2.10	15.10	7.549	1.53
2.02	17.40	8.700	1.46
2.83	15.03	7.515	1.65
2.74	20.04	10.021	1.59
2.65	23.57	11.784	1.53
3.44	18.35	9.17	1.68
3.34	24.58	12.30	1.62
3.23	29.00	14.50	1.56
4.05	21.87	10.93	1.71
3.95	29.52	14.76	1.65
3.84	35.20	17.60	1.59
2.26	19.43	7.77	1.96
2.19	26.04	10.42	1.90
3.24	27.56	11.02	2.06
3.15	37.33	14.93	2.00
3.06	44.60	17.80	1.93
3.56	30.38	12.15	2.09
3.47	41.35	16.54	2.03
3.00	37.61	12.54	2.41
2.92	51.14	17.05	2.34
2.84	61.37	20.46	2.27
3.66	45.88	15.29	2.48
3.58	63.03	21.01	2.41
3.49	76.54	25.51	2.34
4.31	54.15	18.05	2.53
4.22	74.91	24.97	2.46
4.13	92.00	30.70	2.40

Design data: Rectangular tubing (Continued)

| Size | Gauge | Weight | Area | Axis X−X | |
				Moment of inertia, I	Section modulus, Z
(in)		*(lb/ft)*	*(in²)*	*(in⁴)*	*(in³)*
6 × 14	.250	32.23	9.48	231.0	33.00
	.375	47.03	13.80	325.0	46.40
	.500	60.95	17.90	404.0	57.80
8 × 10	.250	28.83	8.48	124	24.8
	.375	41.93	12.33	174	34.7
	.500	54.15	15.90	215	43.0
8 × 12	.250	32.23	9.5	192	32.0
	.375	47.03	13.8	270	45.0
	.500	60.95	17.9	337	56.2
8 × 16	.250	39.03	11.5	386	48.3
	.375	57.23	16.8	550	68.7
	.500	74.54	21.9	694	86.7
8 × 20	.250	45.83	13.5	672	67.2
	.375	67.43	19.8	964	96.4
	.500	88.14	25.9	1,230	123.0
10 × 14	.250	39.03	11.5	326	46.5
	.375	57.23	16.8	464	66.3
	.500	74.54	21.9	587	83.8
12 × 16	.250	45.83	13.5	510	63.8
	.375	67.43	19.8	733	91.6
	.500	88.14	25.9	934	117.0
12 × 20	.250	52.63	15.5	867	86.7
	.375	77.63	22.8	1,250	125.0
	.500	101.74	29.9	1,610	161.0

Axis Y—Y

Radius of gyration, r	Moment of inertia, I	Section modulus, Z	Radius of gyration, r
(in)	(in⁴)	(in³)	(in)
4.94	62.40	20.80	2.57
4.85	86.90	29.00	2.51
4.75	107.00	35.70	2.45
3.83	88.4	22.1	3.23
3.75	123.4	30.9	3.16
3.67	153.0	38.1	3.10
4.50	103	25.9	3.30
4.42	145	36.3	3.24
4.34	181	45.2	3.18
5.80	133	33.4	3.41
5.71	189	47.2	3.35
5.62	237	59.3	3.29
7.06	164	40.9	3.48
6.97	232	58.1	3.42
6.88	293	73.4	3.36
5.33	195	39.1	4.13
5.25	278	55.6	4.06
5.17	350	70.1	4.00
6.15	330	55.0	4.95
6.08	473	78.8	4.88
6.00	602	100.0	4.82
7.48	399	66.5	5.08
7.41	574	95.7	5.02
7.33	734	122.0	4.95

Design data: Square tubing

Size	Gauge	Weight	Area	Moment of inertia, I	Section modulus, Z	Radius of gyration, r
(in)		(lb/ft)	(in²)	(in⁴)	(in³)	(in)
2×2	.188	4.31	1.27	0.67	0.667	0.72
	.250	5.40	1.59	0.77	0.768	0.70
$2\frac{1}{2} \times 2\frac{1}{2}$	.188	5.59	1.64	1.42	1.137	0.93
	.250	7.10	2.09	1.68	1.348	0.90
3×3	.250	8.80	2.59	3.15	2.101	1.10
	.313	10.00	2.94	3.51	2.341	1.09
$3\frac{1}{2} \times 3\frac{1}{2}$	.250	10.50	3.09	5.28	3.020	1.31
	.313	12.12	3.56	5.99	3.425	1.29
4×4	.250	12.02	3.54	7.99	3.99	1.50
	.375	16.84	4.95	10.15	5.08	1.43
	.500	20.88	6.14	11.23	5.62	1.35
6×6	.250	18.82	5.54	29.8	9.95	2.32
	.375	27.04	7.95	40.4	13.48	2.25
	.500	34.48	10.14	48.4	16.13	2.18
8×8	.250	25.44	7.48	73.4	18.4	3.13
	.375	36.83	10.83	101.5	25.4	3.06
	.500	47.35	13.93	124.1	31.0	2.98
10×10	.250	32.23	9.48	148.	29.6	3.95
	.375	47.03	13.83	208	41.6	3.88
	.500	60.95	17.93	260	52.0	3.81
12×12	.250	39.03	11.5	261	43.5	4.77
	.375	57.23	16.8	372	61.9	4.70
	.500	74.54	21.9	470	78.3	4.63
14×14	.250	45.83	13.5	420	60.0	5.58
	.375	67.43	19.8	603	86.2	5.52
	.500	88.14	25.9	769	110.0	5.45

Shear, Moment, and Deflection Formulas for Beams

Shear, Moment, and Deflection Formulas for Beams

(Notation: W = load; w = unit load; M is positive when clockwise; V is positive when upward; y is positive when upward. Constraining moments, applied couples, loads, and reactions are positive when acting as shown.)

Statically Determinate Cases

Loading, support, and reference number	Reactions R_1 and R_2, vertical shear V	Bending moment M and maximum bending moment
1. Cantilever, end load 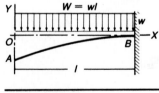	$R_2 = +W$ $V = -W$	$M = -Wx$ Max $M = -Wl$ at B
2. Cantilever, intermediate load 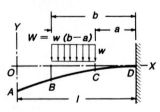	$R_2 = +W$ $(A$ to $B)\ V = 0$ $(B$ to $C)\ V = -W$	$(A$ to $B)\ M = 0$ $(B$ to $C)\ M = -W(x-b)$ Max $M = -Wa$ at C
3. Cantilever, uniform load	$R_2 = +W$ $V = -\dfrac{W}{l}x$	$M = -\dfrac{1}{2}\dfrac{W}{l}x^2$ Max $M = -\tfrac{1}{2}Wl$ at B
4. Cantilever, partial uniform load	$R_2 = +W$ $(A$ to $B)\ V = 0$ $(B$ to $C)\ V = -\dfrac{W}{b-a}(x-l+b)$ $(C$ to $D)\ V = -W$	$(A$ to $B)\ M = 0$ $(B$ to $C)\ M = -\dfrac{1}{2}\dfrac{W}{b-a}(x-l+b)^2$ $(C$ to $D)\ M = -\tfrac{1}{2}W(2x - 2l + a + b)$ Max $M = -\tfrac{1}{2}W(a+b)$ at D

Deflection y, maximum deflection, and end slope θ

$$y = -\frac{1}{6}\frac{W}{EI}(x^3 - 3l^2x + 2l^3)$$

$$\text{Max } y = -\frac{1}{3}\frac{Wl^3}{EI} \text{ at } A$$

$$\theta = +\frac{1}{2}\frac{Wl^2}{EI} \text{ at } A$$

$$(A \text{ to } B) \; y = -\frac{1}{6}\frac{W}{EI}(-a^3 + 3a^2l - 3a^2x)$$

$$(B \text{ to } C) \; y = -\frac{1}{6}\frac{W}{EI}[(x - b)^3 - 3a^2(x - b) + 2a^3]$$

$$\text{Max } y = -\frac{1}{6}\frac{W}{EI}(3a^2l - a^3)$$

$$\theta = +\frac{1}{2}\frac{Wa^2}{EI} \; (A \text{ to } B)$$

$$y = -\frac{1}{24}\frac{W}{EIl}(x^4 - 4l^3x + 3l^4)$$

$$\text{Max } y = -\frac{1}{8}\frac{Wl^3}{EI}$$

$$\theta = +\frac{1}{6}\frac{Wl^2}{EI} \text{ at } A$$

$$(A \text{ to } B) \; y = -\frac{1}{24}\frac{W}{EI}[4(a^2 + ab + b^2)(l - x) - a^3 - ab^2 - a^2b - b^3]$$

$$(B \text{ to } C) \; y = -\frac{1}{24}\frac{W}{EI}\left[6(a + b)(l - x)^2 - 4(l - x)^3 + \frac{(l - x - a)^4}{b - a} \right]$$

$$(C \text{ to } D) \; y = -\frac{1}{12}\frac{W}{EI}[3(a + b)(l - x)^2 - 2(l - x)^3]$$

$$\text{Max } y = -\frac{1}{24}\frac{W}{EI}[4(a^2 + ab + b^2)l - a^3 - ab^2 - a^2b - b^3] \text{ at } A$$

$$\theta = +\frac{1}{6}\frac{W}{EI}[a^2 + ab + b^2] \; (A \text{ to } B)$$

Source: Roark, Raymond J. *Formulas for Stress and Strain.* 3d ed. New York: McGraw-Hill, 1954. Reproduced by permission from McGraw-Hill Book Co.

Shear, Moment, and Deflection Formulas for Beams (Continued)

Loading, support, and reference number	Reactions R_1 and R_2, vertical shear V	Bending moment M and maximum bending moment
5. Cantilever, triangular load 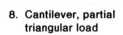 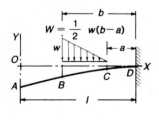	$R_2 = +W$ $V = -\dfrac{W}{l^2}x^2$	$M = -\dfrac{1}{3}\dfrac{W}{l^2}x^3$ Max $M = -\frac{1}{3}Wl$ at B
6. Cantilever, partial triangular load	$R_2 = +W$ (A to B) $V = 0$ (B to C) $V = -\dfrac{W(x-l+b)^2}{(b-a)^2}$ (C to D) $V = -W$	(A to B) $M = 0$ (B to C) $M = -\dfrac{1}{3}\dfrac{W(x-l+b)^3}{(b-a)^2}$ (C to D) $M = -\frac{1}{3}W(3x - 3l + b + 2a)$ Max $M = -\frac{1}{3}W(b + 2a)$ at D
7. Cantilever, triangular load	$R_2 = +W$ $V = -W\left(\dfrac{2lx - x^2}{l^2}\right)$	$M = -\dfrac{1}{3}\dfrac{W}{l^2}(3lx^2 - x^3)$ Max $M = -\frac{2}{3}Wl$ at B
8. Cantilever, partial triangular load	$R_2 = +W$ (A to B) $V = 0$ (B to C) $V = -W\left[1 - \dfrac{(l-a-x)^2}{(b-a)^2}\right]$ (C to D) $V = -W$	(A to B) $M = 0$ (B to C) $M =$ $-\dfrac{1}{3}W\left[\dfrac{3(x-l+b)^2}{b-a} - \dfrac{(x-l+b)^3}{(b-a)^2}\right]$ (C to D) $M = -\frac{1}{3}W(-3l + 3x + 2b + a)$ Max $M = -\frac{1}{3}W(2b + a)$ at D

$$y = -\frac{1}{60}\frac{W}{EIl^2}(x^5 - 5l^4x + 4l^5)$$

$$\text{Max } y = -\frac{1}{15}\frac{Wl^3}{EI} \text{ at } A$$

$$\theta = +\frac{1}{12}\frac{Wl^2}{EI} \text{ at } A$$

$$(A \text{ to } B) \; y = -\frac{1}{60}\frac{W}{EI}[(5b^2 + 10ba + 15a^2)(l - x) - 4a^3 - 2ab^2 - 3a^2b - b^3]$$

$$(B \text{ to } C) \; y = -\frac{1}{60}\frac{W}{EI}\left[(20a + 10b)(l - x)^2 - 10(l - x)^3 + 5\frac{(l - x - a)^4}{b - a} - \frac{(l - x - a)^5}{(b - a)^2} \right]$$

$$(C \text{ to } D) \; y = -\frac{1}{6}\frac{W}{EI}[(2a + b)(l - x)^2 - (l - x)^3]$$

$$\text{Max } y = -\frac{1}{60}\frac{W}{EI}[(5b^2 + 10ba + 15a^2)l - 4a^3 - 2ab^2 - 3a^2b - b^3] \text{ at } A$$

$$\theta = +\frac{1}{12}\frac{W}{EI}(3a^2 + 2ab + b^2) \; (A \text{ to } B)$$

$$y = -\frac{1}{60}\frac{W}{EIl^2}(-x^5 - 15l^4x + 5lx^4 + 11l^5)$$

$$\text{Max } y = -\frac{11}{60}\frac{Wl^3}{EI} \text{ at } A$$

$$\theta = +\frac{1}{4}\frac{W}{EI}l^2 \text{ at } A$$

$$(A \text{ to } B) \; y = -\frac{1}{60}\frac{W}{EI}[(5a^2 + 10ab + 15b^2)(l - x) - a^3 - 2a^2b - 3ab^2 - 4b^3]$$

$$(B \text{ to } C) \; y = -\frac{1}{60}\frac{W}{EI}\left[\frac{(l - x - a)^5}{(b - a)^2} - 10(l - x)^3 + (10a + 20b)(l - x)^2 \right]$$

$$(C \text{ to } D) \; y = -\frac{1}{6}\frac{W}{EI}[(a + 2b)(l - x)^2 - (l - x)^3]$$

$$\text{Max } y = -\frac{1}{60}\frac{W}{EI}[(5a^2 + 10ab + 15b^2)l - a^3 - 2a^2b - 3ab^2 - 4b^3] \text{ at } A$$

$$\theta = +\frac{1}{12}\frac{W}{EI}(a^2 + 2ab + 3b^2) \; (A \text{ to } B)$$

Shear, Moment, and Deflection Formulas for Beams (Continued)

Loading, support, and reference number	Reactions R_1 and R_2, vertical shear V	Bending moment M and maximum bending moment
9. Cantilever, end couple	$R_2 = 0$ $V = 0$	$M = M_0$ Max $M = M_0(A$ to $B)$
10. Cantilever, intermediate couple	$R_2 = 0$ $V = 0$	$(A$ to $B)$ $M = 0$ $(B$ to $C)$ $M = M_0$ Max $M = M_0$ $(B$ to $C)$
11. End supports, center load	$R_1 = +\frac{1}{2}W \quad R_2 = +\frac{1}{2}W$ $(A$ to $B)$ $V = +\frac{1}{2}W$ $(B$ to $C)$ $V = -\frac{1}{2}W$	$(A$ to $B)$ $M = +\frac{1}{2}Wx$ $(B$ to $C)$ $M = +\frac{1}{2}W(l - x)$ Max $M = +\frac{1}{4}Wl$ at B
12. End supports, intermediate load	$R_1 = +W\frac{b}{l} \quad R_2 = +W\frac{a}{l}$ $(A$ to $B)$ $V = +W\frac{b}{l}$ $(B$ to $C)$ $V = -W\frac{a}{l}$	$(A$ to $B)$ $M = +W\frac{b}{l}x$ $(B$ to $C)$ $M = +W\frac{a}{l}(l - x)$ Max $M = +W\frac{ab}{l}$ at B
13. End supports, uniform load	$R_1 = +\frac{1}{2}W \quad R_2 = +\frac{1}{2}W$ $V = \frac{1}{2}W\left(1 - \frac{2x}{l}\right)$	$M = \frac{1}{2}W\left(x - \frac{x^2}{l}\right)$ Max $M = +\frac{1}{8}Wl$ at $x = \frac{1}{2}l$

Deflection y, maximum deflection, and end slope θ

$y = \dfrac{1}{2}\dfrac{M_0}{EI}(l^2 - 2lx + x^2)$

$\text{Max } y = +\dfrac{1}{2}\dfrac{M_0 l^2}{EI} \text{ at } A$

$\theta = -\dfrac{M_0 l}{EI} \text{ at } A$

$(A \text{ to } B) \; y = \dfrac{M_0 a}{EI}\left(l - \dfrac{1}{2}a - x\right)$

$(B \text{ to } C) \; y = \dfrac{1}{2}\dfrac{M_0}{EI}[(x - l + a)^2 - 2a(x - l + a) + a^2]$

$\text{Max } y = \dfrac{M_0 a}{EI}\left(l - \dfrac{1}{2}a\right) \text{ at } A$

$\theta = -\dfrac{M_0 a}{EI} \; (A \text{ to } B)$

$(A \text{ to } B) \; y = -\dfrac{1}{48}\dfrac{W}{EI}(3l^2 x - 4x^3)$

$\text{Max } y = -\dfrac{1}{48}\dfrac{W l^3}{EI} \text{ at } B$

$\theta = -\dfrac{1}{16}\dfrac{W l^2}{EI} \text{ at } A, \qquad \theta = +\dfrac{1}{16}\dfrac{W l^2}{EI} \text{ at } C$

$(A \text{ to } B) \; y = -\dfrac{Wbx}{6EIl}[2l(l - x) - b^2 - (l - x)^2]$

$(B \text{ to } C) \; y = -\dfrac{Wa(l - x)}{6EIl}[2lb - b^2 - (l - x)^2]$

$\text{Max } y = -\dfrac{Wab}{27EIl}(a + 2b)\sqrt{3a(a + 2b)} \text{ at } x = \sqrt{\dfrac{1}{3}a(a + 2b)} \text{ when } a > b$

$\theta = -\dfrac{1}{6}\dfrac{W}{EI}\left(bl - \dfrac{b^2}{l}\right) \text{ at } A; \qquad \theta = +\dfrac{1}{6}\dfrac{W}{EI}\left(2bl + \dfrac{b^3}{l} - 3b^2\right) \text{ at } C$

$y = -\dfrac{1}{24}\dfrac{Wx}{EIl}(l^3 - 2lx^2 + x^3)$

$\text{Max } y = -\dfrac{5}{384}\dfrac{W l^3}{EI} \text{ at } x = \dfrac{1}{2}l$

$\theta = -\dfrac{1}{24}\dfrac{W l^2}{EI} \text{ at } A \qquad \theta = +\dfrac{1}{24}\dfrac{W l^2}{EI} \text{ at } B$

Shear, Moment, and Deflection Formulas for Beams (Continued)

Loading, support, and reference number	Reactions R_1 and R_2, vertical shear V	Bending moment M and maximum bending moment
14. End supports, partial uniform load $d = l - \frac{1}{2}b - \frac{1}{2}a$ 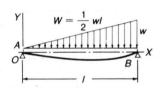	$R_1 = W\dfrac{d}{l}$ $R_2 = \dfrac{W}{l}\left(a + \dfrac{1}{2}c\right)$ $(A \text{ to } B)\ V = R_1$ $(B \text{ to } C)\ V = R_1 - W\dfrac{x-a}{c}$ $(C \text{ to } D)\ V = R_1 - W$	$(A \text{ to } B)\ M = R_1 x$ $(B \text{ to } C)\ M = R_1 x - W\dfrac{(x-a)^2}{2c}$ $(C \text{ to } D)\ M = R_1 x - W(x - \tfrac{1}{2}a - \tfrac{1}{2}b)$ $\text{Max } M = W\dfrac{d}{l}\left(a + \dfrac{cd}{2l}\right)$ at $x = a + \dfrac{cd}{l}$
15. End supports, triangular load $W = \frac{1}{2}wl$ 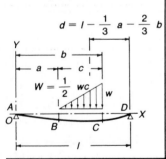	$R_1 = \tfrac{1}{3}W$ $R_2 = \tfrac{2}{3}W$ $V = W\left(\dfrac{1}{3} - \dfrac{x^2}{l^2}\right)$	$M = \dfrac{1}{3}W\left(x - \dfrac{x^3}{l^2}\right)$ $\text{Max } M = 0.128Wl$ at $x = l\left(\dfrac{\sqrt{3}}{3}\right) = 0.5774l$
16. End supports, partial triangular load $d = l - \frac{1}{3}a - \frac{2}{3}b$ $W = \frac{1}{2}wc$	$R_1 = W\dfrac{d}{l}$ $R_2 = W\dfrac{l-d}{l}$ $(A \text{ to } B)\ V = +R_1$ $(B \text{ to } C)\ V = R_1 - \left(\dfrac{x-a}{c}\right)^2 W$ $(C \text{ to } D)\ V = R_1 - W$	$(A \text{ to } B)\ M = R_1 x$ $(B \text{ to } C)\ M = R_1 x - W\dfrac{(x-a)^3}{3c^2}$ $(C \text{ to } D)\ M = R_1 x - \tfrac{1}{3}W(3x - a - 2b)$ $\text{Max. } M = W\dfrac{d}{l}\left(a + \dfrac{2}{3}c\sqrt{\dfrac{d}{l}}\right)$ at $x = a + c\sqrt{\dfrac{d}{l}}$
17. End supports, triangular load $W = \frac{1}{2}wl$ 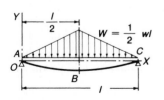	$R = \tfrac{1}{2}W$ $R_2 = \tfrac{1}{2}W$ $(A \text{ to } B)\ V = \dfrac{1}{2}W\left(1 - 4\dfrac{x^2}{l^2}\right)$ $(B \text{ to } C)\ V = -\dfrac{1}{2}W\left(1 - 4\dfrac{(l-x)^2}{l^2}\right)$	$(A \text{ to } B)\ M = \dfrac{1}{6}W\left(3x - 4\dfrac{x^3}{l^2}\right)$ $(B \text{ to } C)\ M = \dfrac{1}{6}W\left[3(l-x) - 4\dfrac{(l-x)^3}{l^2}\right]$ $\text{Max } M = \tfrac{1}{6}Wl$ at B

Deflection y, maximum deflection, and end slope θ

$(A \text{ to } B)\ y = \frac{1}{48EI}\left\{ 8R_1(x^3 - l^2x) + Wx\left[\frac{8d^3}{l} - \frac{2bc^2}{l} + \frac{c^3}{l} + 2c^2 \right] \right\}$

$(B \text{ to } C)\ y = \frac{1}{48EI}\left\{ 8R_1(x^3 - l^2x) + Wx\left[\frac{8d^3}{l} - \frac{2bc^2}{l} + \frac{c^3}{l} + 2c^2 \right] - 2W\frac{(x-a)^4}{c} \right\}$

$(C \text{ to } D)\ y = \frac{1}{48EI}\left\{ 8R_1(x^3 - l^2x) + Wx\left[\frac{8d^3}{l} - \frac{2bc^2}{l} + \frac{c^3}{l} \right] - 8W(x - \frac{1}{2}a - \frac{1}{2}b)^3 + W(2bc^2 - c^3) \right\}$

$\theta = \frac{1}{48EI}\left[-8R_1l^2 + W\left(\frac{8d^3}{l} - \frac{2bc^2}{l} + \frac{c^3}{l} + 2c^2 \right) \right]$ at A;

$\theta = \frac{1}{48EI}\left[16R_1l^2 - W\left(24d^2 - \frac{8d^3}{l} + \frac{2bc^2}{l} - \frac{c^3}{l} \right) \right]$ at B

$y = -\frac{1}{180}\frac{Wx}{EIl^2}(3x^4 - 10l^2x^2 + 7l^4)$

Max $y = -0.01304\frac{Wl^3}{EI}$ at $x = 0.519l$

$\theta = -\frac{7}{180}\frac{Wl^2}{EI}$ at A; $\quad \theta = +\frac{8}{180}\frac{Wl^2}{EI}$ at B.

$(A \text{ to } B)\ y = \frac{1}{6EI}\left\{ R_1(x^3 - l^2x) + Wx\left[\frac{d^3}{l} + \frac{1}{6}c^2\left(1 - \frac{b}{l} \right) + \frac{17}{270}\frac{c^3}{l} \right] \right\}$

$(B \text{ to } C)\ y = \frac{1}{6EI}\left[R_1(x^3 - l^2x) - \frac{1}{10}W\frac{(x-a)^5}{c^2} + Wx\left(\frac{d^3}{l} + \frac{1}{6}c^2 - \frac{1}{6}c^2\frac{b}{l} + \frac{17}{270}\frac{c^3}{l} \right) \right]$

$(C \text{ to } D)\ y = \frac{1}{6EI}\left\{ R_1(x^3 - l^2x) - W\left[d^3 - d^3\frac{x}{l} - \frac{1}{6}bc^2\left(1 - \frac{x}{l} \right) + \frac{17}{270}c^3\left(1 - \frac{x}{l} \right) \right] \right\}$

$\theta = \frac{1}{6EI}\left[-R_1l^2 + W\left(\frac{d^3}{l} + \frac{1}{6}c^2 + \frac{17}{270}\frac{c^3}{l} - \frac{1}{6}\frac{c^2b}{l} \right) \right]$ at A

$\theta = \frac{1}{6EI}\left[2R_1l^2 + W\left(\frac{d^3}{l} + \frac{17}{270}\frac{c^3}{l} - \frac{1}{6}\frac{c^2b}{l} - 3d^2 \right) \right]$ at D

$(A \text{ to } B)\ y = \frac{1}{6}\frac{Wx}{EIl^2}\left(\frac{1}{2}l^2x^2 - \frac{1}{5}x^4 - \frac{5}{16}l^4 \right)$

Max $y = -\frac{1}{60}\frac{Wl^3}{EI}$ at B

$\theta = -\frac{5}{96}\frac{Wl^2}{EI}$ at A; $\quad \theta = +\frac{5}{96}\frac{Wl^2}{EI}$ at C

Shear, Moment, and Deflection Formulas for Beams (Continued)

Loading, support, and reference number	Reactions R_1 and R_2, vertical shear V	Bending moment M and maximum bending moment
18. End supports, triangular load	$R_1 = \frac{1}{2}W$ $R_2 = \frac{1}{2}W$ $(A \text{ to } B)\ V = \frac{1}{2}W\left(\frac{l-2x}{l}\right)^2$ $(B \text{ to } C)\ V = -\frac{1}{2}W\left(\frac{2x-l}{l}\right)^2$	$(A \text{ to } B)\ M = \frac{1}{2}W\left(x - 2\frac{x^2}{l} + \frac{4}{3}\frac{x^3}{l^2}\right)$ $(B \text{ to } C)\ M = \frac{1}{2}W\left[(l-x) - 2\frac{(l-x)^2}{l}\right.$ $\left. + \frac{4}{3}\frac{(l-x)^3}{l^2}\right]$ $\text{Max } M = \frac{1}{12}Wl \text{ at } B$
19. End supports, end couple	$R_1 = -\frac{M_0}{l}$ $R_2 = +\frac{M_0}{l}$ $V = R_1$	$M = M_0 + R_1 x$ $\text{Max } M = M_0 \text{ at } A$
20. End supports, intermediate couple	$R_1 = -\frac{M_0}{l}$ $R_2 = +\frac{M_0}{l}$ $(A \text{ to } C)\ V = R_1$	$(A \text{ to } B)\ M = R_1 x$ $(B \text{ to } C)\ M = R_1 x + M_0$ $\text{Max } -M = R_1 a \text{ just left of } B$ $\text{Max } +M = R_1 a + M_0 \text{ just right of } B$

Deflection y, maximum deflection, and end slope θ

$(A \text{ to } B)\ y = \dfrac{1}{12}\dfrac{W}{EI}\left(x^3 - \dfrac{x^4}{l} + \dfrac{2}{5}\dfrac{x^5}{l^2} - \dfrac{3}{8}l^2 x\right)$

$\text{Max } y = -\dfrac{3}{320}\dfrac{Wl^3}{EI} \text{ at } B$

$\theta = -\dfrac{1}{32}\dfrac{Wl^2}{EI} \text{ at } A;\qquad \theta = +\dfrac{1}{32}\dfrac{Wl^2}{EI} \text{ at } B$

$y = \dfrac{1}{6}\dfrac{M_0}{EI}\left(3x^2 - \dfrac{x^3}{l} - 2lx\right)$

$\text{Max } y = -0.0642\dfrac{M_0 l^2}{EI} \text{ at } x = 0.422l$

$\theta = -\dfrac{1}{3}\dfrac{M_0 l}{EI} \text{ at } A \qquad \theta = +\dfrac{1}{6}\dfrac{M_0 l}{EI} \text{ at } B$

$(A \text{ to } B)\ y = \dfrac{1}{6}\dfrac{M_0}{EI}\left[\left(6a - 3\dfrac{a^2}{l} - 2l\right)x - \dfrac{x^3}{l}\right]$

$(B \text{ to } C)\ y = \dfrac{1}{6}\dfrac{M_0}{EI}\left[3a^2 + 3x^2 - \dfrac{x^3}{l} - \left(2l + 3\dfrac{a^2}{l}\right)x\right]$

$\theta = -\dfrac{1}{6}\dfrac{M_0}{EI}\left(2l - 6a + 3\dfrac{a^2}{l}\right) \text{ at } A;\qquad \theta = +\dfrac{1}{6}\dfrac{M_0}{EI}\left(l - 3\dfrac{a^2}{l}\right) \text{ at } C$

$\theta = \dfrac{M_0}{EI}\left(a - \dfrac{a^2}{l} - \dfrac{1}{3}l\right) \text{ at } B$

Statically Indeterminate Cases

Loading, support, and reference number	Reactions R_1 and R_2, constraining moments M_1 and M_2, and vertical shear V	Bending moment M and maximum positive and negative bending moments
21. One end fixed, one end supported Center load 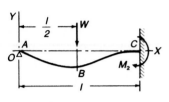	$R_1 = \tfrac{5}{16}W \quad R_2 = \tfrac{11}{16}W$ $M_2 = \tfrac{3}{16}Wl$ $(A \text{ to } B) \; V = +\tfrac{5}{16}W$ $(B \text{ to } C) \; V = -\tfrac{11}{16}W$	$(A \text{ to } B) \; M = \tfrac{5}{16}Wx$ $(B \text{ to } C) \; M = W(\tfrac{1}{2}l - \tfrac{11}{16}x)$ Max $+M = \tfrac{5}{32}Wl$ at B Max $-M = -\tfrac{3}{16}Wl$ at C
22. One end fixed, one end supported Intermediate load 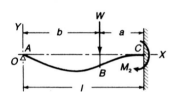	$R_1 = \dfrac{1}{2}W\left(\dfrac{3a^2l - a^3}{l^3}\right) \quad R_2 = W - R_1$ $M_2 = \dfrac{1}{2}W\left(\dfrac{a^3 + 2al^2 - 3a^2l}{l^2}\right)$ $(A \text{ to } B) \; V = +R_1$ $(B \text{ to } C) \; V = R_1 - W$	$(A \text{ to } B) \; M = R_1 x$ $(B \text{ to } C) \; M = R_1 x - W(x - l + a)$ Max $+M = R_1(l - a)$ at B; max possible value $= 0.174 \, Wl$ when $a = 0.634l$ Max $-M = -M_2$ at C; max possible value $= -0.1927 \, Wl$ when $a = 0.4227l$
23. One end fixed, one end supported Uniform load 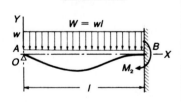	$R_1 = \tfrac{3}{8}W \quad R_2 = \tfrac{5}{8}W$ $M_2 = \tfrac{1}{8}Wl$ $V = W\left(\dfrac{3}{8} - \dfrac{x}{l}\right)$	$M = W\left(\dfrac{3}{8}x - \dfrac{1}{2}\dfrac{x^2}{l}\right)$ Max $+M = \tfrac{9}{128}Wl$ at $x = \tfrac{3}{8}l$ Max $-M = -\tfrac{1}{8}Wl$ at B

Deflection y, maximum deflection, and end slope θ

$(A \text{ to } B)\ y = \dfrac{1}{96}\dfrac{W}{EI}(5x^3 - 3l^2x)$

$(B \text{ to } C)\ y = \dfrac{1}{96}\dfrac{W}{EI}\left[5x^3 - 16\left(x - \dfrac{l}{2}\right)^3 - 3l^2x \right]$

$\text{Max } y = -0.00932\dfrac{Wl^3}{EI} \text{ at } x = 0.4472l$

$\theta = -\dfrac{1}{32}\dfrac{Wl^2}{EI} \text{ at } A$

$(A \text{ to } B)\ y = \dfrac{1}{6EI}[R_1(x^3 - 3l^2x) + 3Wa^2x]$

$(B \text{ to } C)\ y = \dfrac{1}{6EI}\{R_1(x^3 - 3l^2x) + W[3a^2x - (x - b)^3]\}$

If $a < 0.586l$, max y is between A and B at: $x = l\sqrt{1 - \dfrac{2l}{3l-a}}$

If $a > 0.586l$, max y is at: $x = \dfrac{l(l^2 + b^2)}{3l^2 - b^2}$

If $a = 0.586l$, max y is at B and $= -0.0098\dfrac{Wl^3}{EI}$, max possible deflection

$\theta = \dfrac{1}{4}\dfrac{W}{EI}\left(\dfrac{a^3}{l} - a^2\right) \text{ at } A$

$y = \dfrac{1}{48}\dfrac{W}{EIl}(3lx^3 - 2x^4 - l^3x)$

$\text{Max } y = -0.0054\dfrac{Wl^3}{EI} \text{ at } x = 0.4215l$

$\theta = -\dfrac{1}{48}\dfrac{Wl^2}{EI} \text{ at } A$

Shear, Moment, and Deflection Formulas for Beams (Continued)

Loading, support, and reference number	Reactions R_1 and R_2, constraining moments M_1 and M_2, and vertical shear V	Bending moment M and maximum positive and negative bending moments
24. One end fixed, one end supported Partial uniform load 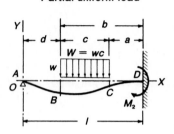	$R_1 = \dfrac{1}{8}\dfrac{W}{l^3}[4l(a^2 + ab + b^2)$ $\quad - a^3 - ab^2 - a^2b - b^3]$ $R_2 = W - R_1$ $M_2 = -R_1 l + \frac{1}{2}W(a + b)$ $(A \text{ to } B)\ V = +R_1$ $(B \text{ to } C)\ V = R_1 - W\left(\dfrac{x - d}{c}\right)$ $(C \text{ to } D)\ V = R_1 - W$	$(A \text{ to } B)\ M = R_1 x$ $(B \text{ to } C)\ M = R_1 x - W\dfrac{(x - d)^2}{2c}$ $(C \text{ to } D)\ M = R_1 x - W(x - d - \frac{1}{2}c)$ $\text{Max} +M = R_1\left(d + \dfrac{1}{2}\dfrac{R_1}{W}c\right)$ $\quad \text{at } x = d + \dfrac{R_1}{W}c$ $\text{Max} -M = -M_2$
25. One end fixed, one end supported Triangular load 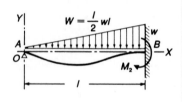	$R_1 = \frac{1}{5}W \qquad R_2 = \frac{4}{5}W$ $M_2 = \frac{2}{15}Wl$ $V = W\left(\dfrac{1}{5} - \dfrac{x^2}{l^2}\right)$	$M = W\left(\dfrac{1}{5}x - \dfrac{1}{3}\dfrac{x^3}{l^2}\right)$ $\text{Max} +M = 0.06Wl \text{ at } x = 0.4474l$ $\text{Max} -M = -M_2$

Deflection y, maximum deflection, and end slope θ

$(A \text{ to } B)\ y = \dfrac{1}{EI}\left[R_1\left(\dfrac{1}{6}x^3 - \dfrac{1}{2}l^2 x\right) + Wx(\tfrac{1}{3}a^2 + \tfrac{1}{2}ac + \tfrac{1}{8}c^2)\right]$

$(B \text{ to } C)\ y = \dfrac{1}{EI}\left[R_1\left(\dfrac{1}{6}x^3 - \dfrac{1}{2}l^2 x\right) + Wx\left(\dfrac{1}{2}a^2 + \dfrac{1}{2}ac + \dfrac{1}{6}c^2\right) - W\dfrac{(x-d)^4}{24c}\right]$

$(C \text{ to } D)\ y = \dfrac{1}{EI}\left\{R_1\left(\dfrac{1}{6}x^3 - \dfrac{1}{2}l^2 x + \dfrac{1}{3}l^3\right) + W[\tfrac{1}{3}(a + \tfrac{1}{2}c)^3 - \tfrac{1}{2}(a + \tfrac{1}{2}c)^2 l - \tfrac{1}{8}(x - d - \tfrac{1}{2}c)^3 + \tfrac{1}{2}(a + \tfrac{1}{2}c)^2 x]\right\}$

$\theta = -\dfrac{1}{EI}\left[\dfrac{1}{2}R_1 l^2 - W\left(\dfrac{1}{2}a^2 + \dfrac{1}{2}ac + \dfrac{1}{6}c^2\right)\right]$ at A

$y = \dfrac{1}{60}\dfrac{W}{EIl}\left(2lx^3 - l^3 x - \dfrac{x^5}{l}\right)$

$\text{Max } y = -0.00477\dfrac{Wl^3}{EI}$ at $x = l\sqrt{\dfrac{1}{5}}$

$\theta = -\dfrac{1}{60}\dfrac{Wl^2}{EI}$ at A

Shear, Moment, and Deflection Formulas for Beams (Continued)

Loading, support, and reference number	Reactions R_1 and R_2, constraining moments M_1 and M_2, and vertical shear V	Bending moment M and maximum positive and negative bending moments
26. One end fixed, one end supported Partial triangular load 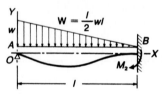	$R_1 = \dfrac{1}{20}\dfrac{W}{l^2}[(10ab + 15a^2 + 5b^2)l$ $\quad - 4a^3 - 2ab^2 - 3a^2b - b^3]$ $R_2 = W - R_1$ $M_2 = -R_1 l + \tfrac{1}{3}W(2a + b)$ $(A \text{ to } B)\ V = R_1$ $(B \text{ to } C)\ V = R_1 - W\left(\dfrac{x-d}{c}\right)^2$ $(C \text{ to } D)\ V = R_1 - W$	$(A \text{ to } B)\ M = R_1 x$ $(B \text{ to } C)\ M = R_1 x - \dfrac{1}{3}W\dfrac{(x-d)^3}{c^2}$ $(C \text{ to } D)\ M = R_1 x - W(x - d - \tfrac{2}{3}c)$ $\text{Max} +M = R_1\left(d + \tfrac{2}{3}c\sqrt{\dfrac{R_1}{W}}\right)$ $\quad \text{at } x = d + c\sqrt{\dfrac{R_1}{W}}$ $\text{Max} -M = -M_2$
27. One end fixed, one end supported Triangular load	$R_1 = \tfrac{11}{20}W \qquad R_2 = \tfrac{9}{20}W$ $M_2 = \tfrac{7}{60}Wl$ $V = W\left(\dfrac{11}{20} - \dfrac{2x}{l} + \dfrac{x^2}{l^2}\right)$	$M = W\left(\dfrac{11}{20}x - \dfrac{x^2}{l} + \dfrac{1}{3}\dfrac{x^3}{l^2}\right)$ $\text{Max} +M = 0.0846Wl \text{ at } x = 0.329l$ $\text{Max} -M = -\tfrac{7}{60}Wl \text{ at } B$

Deflection y, maximum deflection, and end slope θ

$(A \text{ to } B) \; y = \dfrac{1}{EI}\left[R_1\left(\dfrac{1}{6}x^3 - \dfrac{1}{2}l^2x\right) + Wx(\tfrac{1}{2}a^2 + \tfrac{1}{3}ac + \tfrac{1}{12}c^2) \right]$

$(B \text{ to } C) \; y = \dfrac{1}{EI}\left[R_1\left(\dfrac{1}{6}x^3 - \dfrac{1}{2}l^2x\right) + Wx\left(\dfrac{1}{2}a^2 + \dfrac{1}{3}ac + \dfrac{1}{12}c^2\right) - W\dfrac{(x-d)^5}{60c^2} \right]$

$(C \text{ to } D) \; y = \dfrac{1}{EI}\left\{ R_1\left(\dfrac{1}{6}x^3 - \dfrac{1}{2}l^2x + \dfrac{1}{3}l^3\right) + W[\tfrac{1}{2}(a + \tfrac{1}{2}c)^2x - \tfrac{1}{6}(x - d - \tfrac{2}{3}c)^3 + \tfrac{1}{6}(a + \tfrac{1}{2}c)^3 - \tfrac{1}{2}(a + \tfrac{1}{2}c)^2l] \right\}$

$\theta = -\dfrac{1}{EI}\left[\dfrac{1}{2}R_1l^2 - W\left(\dfrac{1}{12}c^2 + \dfrac{1}{3}ac + \dfrac{1}{2}a^2\right) \right]$ at A

$y = \dfrac{1}{120}\dfrac{W}{EIl}\left(11lx^3 - 3l^3x - 10x^4 + 2\dfrac{x^5}{l} \right)$

$\text{Max } y = -0.00609\dfrac{Wl^3}{EI}$ at $x = 0.402l$

$\theta = -\dfrac{1}{40}\dfrac{Wl^2}{EI}$ at A

Shear, Moment, and Deflection Formulas for Beams (Continued)

Loading, support, and reference number	Reactions R_1 and R_2, constraining moments M_1 and M_2, and vertical shear V	Bending moment M and maximum positive and negative bending moments
28. One end fixed, one end supported Partial triangular load $W = \frac{1}{2}wc$	$R_1 = \frac{1}{20}\frac{W}{l^3}[(10ab + 5a^2 + 15b^2)l$ $- a^3 - 2a^2b - 3ab^2 - 4b^3]$, $R_2 = W - R_1$ $M_2 = -R_1l + \frac{1}{3}W(a + 2b)$ $(A \text{ to } B)\ V = R_1$ $(B \text{ to } C)\ V = R_1 - \frac{2(x-d)}{c}W$ $+ \frac{(x-d)^2}{c^2}W$ $(C \text{ to } D)\ V = R_1 - W$	$(A \text{ to } B)\ M = R_1x$ $(B \text{ to } C)\ M = R_1x - \frac{(x-d)^2}{c}W + \frac{(x-d)^3}{3c^2}W$ $(C \text{ to } D)\ M = R_1x - W(x - d - \frac{1}{3}c)$ $\text{Max} +M = R_1\left(d + c - c\sqrt{1 - \frac{R_1}{W}}\right)$ $\quad - \frac{1}{3}Wc\left(1 - \sqrt{1 - \frac{R_1}{W}}\right)^2\left(2 + \sqrt{1 - \frac{R_1}{W}}\right)$ $\text{at } x = d + c\left(1 - \sqrt{1 - \frac{R_1}{W}}\right)$ $\text{Max} -M = -M_2 \text{ at } D$
29. One end fixed, one end supported End couple 	$R_1 = -\frac{3}{2}\frac{M_0}{l} \qquad R_2 = +\frac{3}{2}\frac{M_0}{l}$ $M_2 = \frac{1}{2}M_0$ $V = -\frac{3}{2}\frac{M_0}{l}$	$M = \frac{1}{2}M_0\left(2 - 3\frac{x}{l}\right)$ $\text{Max} +M = M_0 \text{ at } A$ $\text{Max} -M = -\frac{1}{2}M_0 \text{ at } B$

Deflection y, maximum deflection, and end slope θ

$(A \text{ to } B) \quad y = \frac{1}{EI}\left\{ R_1\left(\frac{1}{6}x^3 - \frac{1}{2}l^2x\right) + Wx(\frac{1}{2}a^2 + \frac{2}{3}ac + \frac{1}{4}c^2) \right\}$

$(B \text{ to } C) \quad y = \frac{1}{EI}\left\{ R_1\left(\frac{1}{6}x^3 - \frac{1}{2}l^2x\right) + W\left[\frac{1}{60}\frac{(x-d)^5}{c^2} - \frac{1}{12}\frac{(x-d)^4}{c} + (\frac{1}{2}a^2 + \frac{2}{3}ac + \frac{1}{4}c^2)x\right]\right\}$

$(C \text{ to } D) \quad y = \frac{1}{EI}\left\{ R_1\left(\frac{1}{6}x^3 - \frac{1}{2}l^2x + \frac{1}{3}l^3\right) - \frac{1}{2}W(l-x)^2\left[\left(a + \frac{2}{3}c\right) - \frac{l-x}{3}\right]\right\}$

$\theta = -\frac{1}{EI}\left[\frac{1}{2}R_1l^2 - W\left(\frac{1}{2}a^2 + \frac{2}{3}ac + \frac{1}{4}c^2\right)\right]$ at A

$y = \frac{1}{4}\frac{M_0}{EI}\left(2x^2 - \frac{x^3}{l} - lx\right)$

Max $y = -\frac{1}{27}\frac{M_0l^2}{EI}$ at $x = \frac{1}{3}l$

$\theta = -\frac{1}{4}\frac{M_0l}{EI}$ at A

Shear, Moment, and Deflection Formulas for Beams (Continued)

Loading, support, and reference number	Reactions R_1 and R_2, constraining moments M_1 and M_2, and vertical shear V	Bending moment M and maximum positive and negative bending moments
30. One end fixed, one end supported Intermediate couple 	$R_1 = -\dfrac{3}{2}\dfrac{M_0}{l}\left(\dfrac{l^2 - a^2}{l^2}\right)$ $R_2 = +\dfrac{3}{2}\dfrac{M_0}{l}\left(\dfrac{l^2 - a^2}{l^2}\right)$ $M_2 = \dfrac{1}{2}M_0\left(1 - 3\dfrac{a^2}{l^2}\right)$ $(A \text{ to } B)\ V = R_1$ $(B \text{ to } C)\ V = R_1$	$(A \text{ to } B)\ M = R_1 x$ $(B \text{ to } C)\ M = R_1 x + M_0$ $\text{Max} +M = M_0\left[1 - \dfrac{3a(l^2 - a^2)}{2l^3}\right]$ at B (to right) $\text{Max} -M = -M_2$ at C (when $a < 0.275l$) $\text{Max} -M = R_1 a$ at B (to left) (when $a > 0.275l$)
31. Both ends fixed. Center load 	$R_1 = \tfrac{1}{2}W \qquad R_2 = \tfrac{1}{2}W$ $M_1 = \tfrac{1}{8}Wl \qquad M_2 = \tfrac{1}{8}Wl$ $(A \text{ to } B)\ V = +\tfrac{1}{2}W$ $(B \text{ to } C)\ V = -\tfrac{1}{2}W$	$(A \text{ to } B)\ M = \tfrac{1}{8}W(4x - l)$ $(B \text{ to } C)\ M = \tfrac{1}{8}W(3l - 4x)$ $\text{Max} +M = \tfrac{1}{8}Wl$ at B $\text{Max} -M = -\tfrac{1}{8}Wl$ at A and C
32. Both ends fixed. Intermediate load 	$R_1 = \dfrac{Wb^2}{l^3}(3a + b)$ $R_2 = \dfrac{Wa^2}{l^3}(3b + a)$ $M_1 = W\dfrac{ab^2}{l^2} \qquad M_2 = W\dfrac{a^2 b}{l^2}$ $(A \text{ to } B)\ V = R_1$ $(B \text{ to } C)\ V = R_1 - W$	$(A \text{ to } B)\ M = -W\dfrac{ab^2}{l^2} + R_1 x$ $(B \text{ to } C)\ M = -W\dfrac{ab^2}{l^2} + R_1 x - W(x - a)$ $\text{Max} +M = -W\dfrac{ab^2}{l^2} + R_1 a$ at B; max possible value $= \tfrac{1}{8}Wl$ when $a = \tfrac{1}{2}l$ $\text{Max} -M = -M_1$ when $a < b$; max possible value $= -0.1481Wl$ when $a = \tfrac{1}{3}l$ $\text{Max} -M = -M_2$ when $a > b$; max possible value $= -0.1481Wl$ when $a = \tfrac{2}{3}l$

Deflection y, maximum deflection, and end slope θ

$(A \text{ to } B)\ y = \dfrac{M_0}{EI}\left[\dfrac{l^2 - a^2}{4l^3}(3l^2x - x^3) - (l - a)x\right]$

$(B \text{ to } C)\ y = \dfrac{M_0}{EI}\left[\dfrac{l^2 - a^2}{4l^3}(3l^2x - x^3) - lx + \dfrac{1}{2}(x^2 + a^2)\right]$

$\theta = \dfrac{M_0}{EI}\left(a - \dfrac{1}{4}l - \dfrac{3}{4}\dfrac{a^2}{l}\right)$ at A

$(A \text{ to } B)\ y = -\dfrac{1}{48}\dfrac{W}{EI}(3lx^2 - 4x^3)$

$\text{Max } y = -\dfrac{1}{192}\dfrac{Wl^3}{EI}$ at B

$(A \text{ to } B)\ y = \dfrac{1}{6}\dfrac{Wb^2x^2}{EIl^3}(3ax + bx - 3al)$

$(B \text{ to } C)\ y = \dfrac{1}{6}\dfrac{Wa^2(l - x)^2}{EIl^3}[(3b + a)(l - x) - 3bl]$

$\text{Max } y = -\dfrac{2}{3}\dfrac{W}{EI}\dfrac{a^3b^2}{(3a + b)^2}$ at $x = \dfrac{2al}{3a + b}$ if $a > b$

$\text{Max } y = \dfrac{2}{3}\dfrac{W}{EI}\dfrac{a^2b^3}{(3b + a)^2}$ at $x = l - \dfrac{2bl}{3b + a}$ if $a < b$

Shear, Moment, and Deflection Formulas for Beams (Continued)

Loading, support, and reference number	Reactions R_1 and R_2, constraining moments M_1 and M_2, and vertical shear V	Bending moment M and maximum positive and negative bending moments
33. Both ends fixed. Uniform load 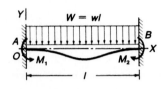	$R_1 = \tfrac{1}{2}W \quad R_2 = \tfrac{1}{2}W$ $M_1 = \tfrac{1}{12}Wl \quad M_2 = \tfrac{1}{12}Wl$ $V = \dfrac{1}{2}W\left(1 - \dfrac{2x}{l}\right)$	$M = \dfrac{1}{2}W\left(x - \dfrac{x^2}{l} - \dfrac{1}{6}l\right)$ Max $+M = \tfrac{1}{24}Wl$ at $x = \tfrac{1}{2}l$ Max $-M = -\tfrac{1}{12}Wl$ at A and B
34. Both ends fixed. Partial uniform load $d = l - \dfrac{1}{2}a - \dfrac{1}{2}b$ 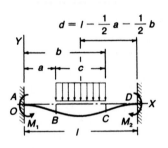	$R_1 = \dfrac{1}{4}\dfrac{W}{l^2}\left(12d^2 - 8\dfrac{d^3}{l} + 2\dfrac{bc^2}{l} - \dfrac{c^3}{l} - c^2\right)$ $R_2 = W - R_1$ $M_1 = -\dfrac{1}{24}\dfrac{W}{l}\left(24\dfrac{d^3}{l} - 6\dfrac{bc^2}{l} + 3\dfrac{c^3}{l} + 4c^2 - 24d^2\right)$ $M_2 = \dfrac{1}{24}\dfrac{W}{l}\left(24\dfrac{d^3}{l} - 6\dfrac{bc^2}{l} + 3\dfrac{c^3}{l} + 2c^2 - 48d^2 + 24dl\right)$ $(A \text{ to } B) \ V = R_1$ $(B \text{ to } C) \ V = R_1 - W\dfrac{x-a}{c}$ $(C \text{ to } D) \ V = R_1 - W$	$(A \text{ to } B) \ M = -M_1 + R_1 x$ $(B \text{ to } C) \ M = -M_1 + R_1 x - \dfrac{1}{2}W\dfrac{(x-a)^2}{c}$ $(C \text{ to } D) \ M = -M_1 + R_1 x - W(x - l + d)$ Max $+M$ is between B and C at $x = a + \dfrac{R_1}{W}c$ Max $-M = -M_1$ when $a < l - b$ Max $-M = -M_2$ when $a > l - b$
35. Both ends fixed Triangular load 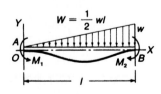	$R_1 = \tfrac{3}{10}W \quad R_2 = \tfrac{7}{10}W$ $M_1 = \tfrac{1}{15}Wl$ $M_2 = \tfrac{1}{10}Wl$ $V = W\left(\dfrac{3}{10} - \dfrac{x^2}{l^2}\right)$	$M = W\left(\dfrac{3}{10}x - \dfrac{1}{15}l - \dfrac{1}{3}\dfrac{x^3}{l^2}\right)$ Max $+M = 0.043\,Wl$ at $x = 0.548l$ Max $-M = -\tfrac{1}{10}Wl$ at B

$$y = \frac{1}{24}\frac{Wx^2}{EIl}(2lx - l^2 - x^2)$$

$$\text{Max } y = -\frac{1}{384}\frac{Wl^3}{EI} \text{ at } x = \frac{1}{2}l$$

$$(A \text{ to } B)\ y = \frac{1}{6EI}(R_1x^3 - 3M_1x^2)$$

$$(B \text{ to } C)\ y = \frac{1}{6EI}\left(R_1x^3 - 3M_1x^2 - \frac{1}{4}W\frac{(x-a)^4}{c} \right)$$

$$(C \text{ to } D)\ y = \frac{1}{6EI}[R_2(l-x)^3 - 3M_2(l-x)^2]$$

$$y = \frac{1}{60}\frac{W}{EI}\left(3x^3 - 2lx^2 - \frac{x^5}{2} \right)$$

$$\text{Max } y = -0.002617\frac{Wl^3}{EI} \text{ at } x = 0.525l$$

Loading, support, and reference number	Reactions R_1 and R_2, constraining moments M_1 and M_2, and vertical shear V	Bending moment M and maximum positive and negative bending moments
36. Both ends fixed **Partial triangular load** $d = l - \dfrac{1}{3}\,a - \dfrac{2}{3}\,b$ $W = \dfrac{1}{2}\,wc$	$R_1 = \dfrac{W}{l^2}\left(3d^2 - \dfrac{1}{6}c^2 + \dfrac{1}{3}\dfrac{bc^2}{l}\right.$ $\left. - \dfrac{17}{135}\dfrac{c^3}{l} - 2\dfrac{d^3}{l}\right)$ $R_2 = W - R_1$ $M_1 = -\dfrac{W}{l}\left(\dfrac{d^3}{l} + \dfrac{1}{9}c^2 + \dfrac{51}{810}\dfrac{c^3}{l}\right.$ $\left. - \dfrac{1}{6}\dfrac{c^2 b}{l} - d^2\right)$ $M_2 = \dfrac{W}{l}\left(\dfrac{d^3}{l} + \dfrac{1}{18}c^2 + \dfrac{51}{810}\dfrac{c^3}{l}\right.$ $\left. - \dfrac{1}{6}\dfrac{c^2 b}{l} - 2d^2 + dl\right)$ $(A\text{ to }B)\ V = R_1$ $(B\text{ to }C)\ V = R_1 - W\dfrac{(x-a)^2}{c^2}$ $(C\text{ to }D)\ V = R_1 - W$	$(A\text{ to }B)\ M = -M_1 + R_1 x$ $(B\text{ to }C)\ M = -M_1 + R_1 x - W\dfrac{(x-a)^3}{3c^2}$ $(C\text{ to }D)\ M = -M_1 + R_1 x - W(x - l + d)$ $\text{Max} +M$ at $x = a + c\sqrt{\dfrac{R_1}{W}}$ $\text{Max} -M = M_1$ when $d > \dfrac{l}{2}$, $-M_2$ when $d < \dfrac{l}{2}$
37. Both ends fixed **Intermediate couple**	$R_1 = -6\dfrac{M_0}{l^3}(al - a^2)$ $R_2 = 6\dfrac{M_0}{l^3}(al - a^2)$ $M_1 = -\dfrac{M_0}{l^2}(4la - 3a^2 - l^2)$ $M_2 = \dfrac{M_0}{l^2}(2la - 3a^2)$ $V = R_1$	$(A\text{ to }B)\ M = -M_1 + R_1 x$ $(B\text{ to }C)\ M = -M_1 + R_1 x + M_0$ $\text{Max} +M = M_0\left(4\dfrac{a}{l} - 9\dfrac{a^2}{l^2} + 6\dfrac{a^3}{l^3}\right)$ just right of B $\text{Max} -M = M_0\left(4\dfrac{a}{l} - 9\dfrac{a^2}{l^2} + 6\dfrac{a^3}{l^3} - 1\right)$ just left of B

Deflection y, maximum deflection, and end slope θ

$(A \text{ to } B)$ $y = \dfrac{1}{6EI}(R_1 x^3 - 3M_1 x^2)$

$(B \text{ to } C)$ $y = \dfrac{1}{EI}\left(\dfrac{1}{6}R_1 x^3 - \dfrac{1}{2}M_1 x^2 - \dfrac{1}{60}W\dfrac{(x-a)^5}{c^2}\right)$

$(C \text{ to } D)$ $y = \dfrac{1}{6EI}\{R_1(x^3 - 3l^2 x + 2l^3) - 3M_1(l-x)^2 + W[3d^2 x + d^3 - 3d^2 l - (x-l+d)^3]\}$

$(A \text{ to } B)$ $y = -\dfrac{1}{6EI}(3M_1 x^2 - R_1 x^3)$

$(B \text{ to } C)$ $y = \dfrac{1}{6EI}[(M_0 - M_1)(3x^2 - 6lx + 3l^2) - R_1(3l^2 x - x^3 - 2l^3)]$

Max $+y$ at $x = \dfrac{2M_1}{R_1}$ if $a > \tfrac{2}{3}l$

Max $-y$ at $x = l - \dfrac{2M_2}{R_2}$ if $a < \tfrac{2}{3}l$

Material Properties

Appendix

Metals	ρ Density (lb/in³)	σ_y Yield strength (psi)	σ_t Tension strength (psi)	E Modulus of elasticity (psi)	α Thermal expansion @ 68°F (in/in/°F)
Aluminum (99.9 +)	0.098	5,000	13,000	10×10^6	12.5×10^{-6}
Aluminum (6061-T6)	0.098	40,000	45,000	10×10^6	12.0×10^{-6}
Copper (99.9−)	0.323	—	37,000	16×10^6	9.0×10^{-6}
Brass (cold rolled)	0.316	60,000	75,000	15×10^6	9.8×10^{-6}
Bronze (cold rolled)	0.320	75,000	100,000	15×10^6	9.4×10^{-6}
Steel (A36) (structural)	0.284	36,000	58,000	30×10^6	6.6×10^{-6}
Steel (0.4% C hot rolled)	0.284	53,000	84,000	30×10^6	6.6×10^{-6}
Steel (0.8% C hot rolled)	0.284	76,000	122,000	30×10^6	6.6×10^{-6}
Stainless steel (18-8 annealed)	0.286	36,000	85,000	30×10^6	9.6×10^{-6}
Stainless steel (18-8 cold rolled)	0.286	165,000	190,000	30×10^6	9.6×10^{-6}
Cast iron-gray	0.260	Brittle	25,000	15×10^6	6.6×10^{-6}
Magnesium (99.9 +)	0.063	14,000	27,000	6.5×10^6	14×10^{-6}
Silver (99.9 +)	0.380	—	41,000	11×10^6	10×10^{-6}

Polymers	ρ Density (lb/in³)	σ_t Tensile strength (psi)	E Modulus of elasticity (psi)	α Thermal Expansion @ 68°F (in/in/°F)
Cellulose acetate (hard)	1.3	$(4.6\text{-}8.5) \times 10^6$	$(0.190\text{-}0.400) \times 10^6$	$(44\text{-}88) \times 10^{-6}$
Nylon 6/6	1.15	$(9.0\text{-}12.0) \times 10^6$	0.40×10^6	55×10^{-6}
Polyethylene	0.9	$(1.2\text{-}3.5) \times 10^6$	0.035×10^6	100×10^{-6}
Polystyrene	1.05	$(5.0\text{-}10.0) \times 10^6$	$(0.4\text{-}0.6) \times 10^6$	35×10^{-6}
PVC (rigid)	1.4	$(5.0\text{-}17.0) \times 10^6$	$(0.2\text{-}0.6) \times 10^6$	55×10^{-6}
Phenol-formaldehyde	1.3	$(6.0\text{-}9.0) \times 10^6$	0.5×10^6	40×10^{-6}
Rubber (synthetic)	1.5	$(1.0\text{-}4.0) \times 10^6$	$(0.005\text{-}0.010) \times 10^6$	—
Rubber (vulcanized)	1.2	$(1.0\text{-}4.0) \times 10^6$	0.5×10^6	45×10^{-6}
Wood (white pine)	0.5	$(7\text{-}12) \times 10^6$	$(1.4\text{-}1.6) \times 10^6$	—

Ceramics	ρ Density (lb/in³)	σ_t Tensile strength (psi)	σ_c Compressive strength (psi)	E Modulus of elasticity (psi)	α Thermal expansion @ 68°F (in/in/°F)
Alumina ($Al_2 O_3$)	4.000	35.8×10^3	412.5×10^3	52.4×10^6	5.0×10^{-6}
Silicon carbide (SiC)	3.170	—	—	—	2.5×10^{-6}
Plate glass	2.500	$(6\text{-}6.5) \times 10^3$	$(90.0\text{-}180.0) \times 10^3$	10.0×10^6	4.9×10^{-6}
Borosilicate glass	2.400	—	—	10.0×10^6	1.5×10^{-6}
Glass wool	0.024	—	—	—	—
Porcelain	2.530	2.5×10^3	49.1×10^3	—	2.6×10^{-6}
Quartz	2.650	7×10^3	160.0×10^3	10.0×10^6	7.0×10^{-6}
Common brick	1.750	—	$(10.0\text{-}20.0) \times 10^3$	—	5.0×10^{-6}
Firebrick	2.100	—	—	—	2.5×10^{-6}
Concrete	1.400-2.200	—	6.0×10^3	2.0×10^6	7.0×10^{-6}
Ice	0.910	—	—	1.3×10^6	28.0×10^{-6}

Note: Ceramic properties are average values.

Applications of Calculus in Strength of Materials

Many of the basic equations presented in this text are commonly derived with the aid of the fundamental concepts of differential and integral calculus.

Differential calculus is built on the concepts of *limits*. To understand the *limiting process,* consider a continuous function in a general form, $y = f(x)$. Selecting two values of the independent variable x_1 and x_2, we may evaluate our function to obtain the corresponding values y_1 and y_2. The difference between the selected x values may be written as

$$x_2 - x_1 = \Delta x \tag{1}$$

Likewise the difference between the two computed y values may be denoted by

$$y_2 - y_1 = \Delta y \tag{2}$$

The limiting process involves the ratio of these differences. The limit of the ratio $\Delta y / \Delta x$ when Δx is made smaller and smaller so that it approaches zero is called the *first derivative* of the function y at x_1. In mathematical notation it is written as

$$\lim_{x \to 0} \frac{\Delta y}{\Delta x} = \frac{dy}{dx} = f'(x) = y' \tag{3}$$

It may be shown that dy/dx represents the slope of the function $y = f(x)$ at the value x where the limiting process is performed. In other words, the function dy/dx represents the rate of change of the function along x.

Inspecting Figure 5.9, which shows graphs of the internal shear force V for various loading distributions $w(x)$ on simple beams, one may see that for any location x along the beam, the shear force relates to the loading as

$$\frac{dV}{dx} = w(x) \tag{4}$$

This means that the slope of the shear force diagram at any location x equals the value of the loading w at that point. Similarly, as indicated by Figure 6.13, the relationship between the shear force and the moment diagrams is

$$\frac{dM}{dx} = V(x) \tag{5}$$

expressing that the rate of change, or slope, of the moment diagram equals the value of the shear force at any location x.

By substituting Equation 5 into Equation 4 one obtains

$$\frac{d}{dx}\left(\frac{dM}{dx}\right) = \frac{d^2 M}{dx^2} = w(x) \tag{6}$$

This equation states that the *second derivative* of the mathematical function describing the moment diagram results in the loading distribution $w(x)$. Figure 6.10 shows some examples for this.

Reviewing Example 6.8 may be helpful in understanding these relationships. Successive cuts in the beam to isolate free bodies may be made at smaller and smaller intervals. The distance between successive cuts may be expressed as Δx. If the limiting process described by Equation 3 is used, it will lead to the results shown above.

In the geometry of the deflected beam, similar relationships hold true. Rearranging Equation 13.7 and changing deltas into d's, to signify the use of the limiting process, one obtains

$$(7) \qquad \frac{dz}{dx} = \phi$$

meaning that the rate of change of the deflection of the beam equals the change in slope. Likewise, from Equation 13.4 we find that the rate of change of the slope of the deflected beam equals the curvature; that is,

$$(8) \qquad \frac{d\phi}{dx} = \frac{1}{\rho}$$

Just as was done in the case of Equation 6, one may combine these two formulas into

$$(9) \qquad \frac{d}{dx}\left(\frac{dz}{dx}\right) = \frac{d^2 z}{dx^2} = \frac{1}{\rho}$$

Furthermore, from elastic considerations we have found that the curvature of a deflected beam relates to the bending moment diagram according to Equation 13.11:

$$(10) \qquad \frac{1}{\rho} = \frac{M(x)}{EI}$$

Substituting Equation 10 into Equation 8 leads to

$$(11) \qquad \frac{d\phi}{dx} = \frac{M(x)}{EI}$$

which states that the rate of change of the slope equals the value of the reduced moment at any point x along the beam. Furthermore, substituting Equation 7 into this formula results in

$$(12) \qquad \frac{M}{EI} = \frac{d^2 z}{dx^2}$$

indicating that the second derivative of the function representing the deflection of the beam equals the value of the reduced moment.

In conclusion, we may relate the loading diagram $w(x)$ to the deflection by substituting Equation 6 into Equation 12,

$$(13) \qquad w(x) = \frac{1}{EI}\frac{d^4 z}{dx^4} = \frac{1}{EI} z(x)''''$$

which shows that the loading diagram relates to the fourth derivative of the deflection diagram.

Integral calculus is also based on the concept of limits, in this case the limit of sums. Consider, for example, Equation 7.3, stating that the total weight of a body is equal to the sum of the weights of its elements. In this equation, the number of the individual elements is arbitrary. In the process of *integration,* however, the size of the individual elements is taken to be approaching zero, while, of course, their number increases toward infinity. At the limit one reaches the *integral* of the function, written as

(14)
$$W = \lim_{A_i \to 0} \sum_{i=1}^{\infty} \rho \quad t \varDelta A_i = \int_A \rho t \, dA$$

For a precise mathematical explanation of this concept see Earl W. Swokowski, *Calculus with Analytic Geometry,* 2d ed. (Boston: Prindle, Weber & Schmidt, 1979).

Applying the concept of integration to the definition of the moment of inertia, for instance, would require that Equation 7.10 be written in the form of

(15)
$$I = \int y^2 \, dA$$

The derivations of the torsion formula (Equation 9.9), the bending stress formula (Equation 10.5), and the formula for shear in bending (Equation 11.6) all make use of this concept.

In particular, the derivation of the moment area theorems serves as a good example for the use of integration. As shown in Figure 13.5, the change of slope between two points along a deflected beam is given by the area under the curvature diagram between these points. Equation 13.5 describes this as

(16)
$$\varDelta \phi_{AB} = \sum_{A}^{B} \left(\frac{1}{\rho} \right) \varDelta x$$

Applying the limiting process—reducing the size of the $\varDelta x$ value while increasing the number of slices into which the area under the curve is divided up—will lead, at the limit of $\varDelta x \to 0$, to

(17)
$$\int d \phi_{AB} = \int_A^B \frac{1}{\rho} \, dx$$

Substituting Equation 11 results in

(18)
$$\phi_{AB} = \int_A^B \frac{M(x)}{EI} \, dx$$

which is the statement of the first moment area theorem in integral form.

When the two boundaries of the integration, A and B, are given, we speak of a *definite integral*. Conversely, an integral formula without both limits given is called an *indefinite integral*. To evaluate such an integral one needs to know the value of the function at one point. This is expressed by an integration constant C such that

(19)
$$F = \int dF + C$$

If Equation 18, for instance, had been written in the indefinite form, one would need to define the initial slope ϕ_A at point A as the *boundary condition* of the function. The value of C is then obtained, after the integration is performed, by substituting this value ϕ_A at $x = A$. Detailing the procedures used in integrating various types of functions is beyond the scope of this text.

The functional relationship between differentiation and integration appears obvious when considering Equations 11 and 17. By elimination of the integral signs in Equation 17 and dividing by dx, one obtains Equation 11.

The mathematical relationships between loading, shear, moment, slope and deflection are shown below:

(20)
$$\int w(x)\, dx + C_1 = V(x) = EI\, z'''$$

(21)
$$\int\int w(x)\, dx\, dx + x\, C_1 + C_2 = M(x) = EI\, z''$$

(22)
$$\int\int\int w(x)\, dx\, dx\, dx + \frac{x^2}{2} C_1 + x\, C_2 + C_3 = EI\, \phi(x) = EI\, z'$$

(23)
$$\int\int\int\int w(x)\, dx\, dx\, dx\, dx + \frac{x^3}{6} C_1 + \frac{x^2}{2} C_2 + x\, C_3 + C_4 = EI\, z(x)$$

Multiple integral signs represent successive integrations of the loading function $w(x)$. Multiple primes represent successive differentiation of the deflection function $z(x)$. The constants C stand for the various boundary conditions associated with the shear, moment, slope angle, and deflection of the beam. These equations can be simply derived from Equation 13, which expresses the relationship between loading and deflection. Consider a free body diagram of a beam cut at $x = a$ and $x = b$. The internal shear force on the cut at $x = b$ equals the shear at $x = a$ plus the sum of the external loading between a and b. Integrating Equation 13 results in

(24)
$$\int w(x)\, dx + C_1 = EI \int_a^b \frac{d^4 z}{dx^4}\, dx = EI \frac{d^3 z}{dx^3} = V(b)$$

The integration constant C_1 here is the shear force at $x = a$, or $V(a)$. Equations 21, 22, and 23 may be derived in a similar manner.

Bibliography

Bassin, Milton G.; Brodsky, S. M.; and Wolkoff, H. 1979. *Statics and Strength of Materials,* 3d ed. New York: McGraw-Hill.

Beer, F. P., and Johnston, E. R. 1956. *Mechanics for Engineers: Statics.* 3d ed. New York: McGraw-Hill.

Eisenberg, Martin A. 1980. *Introduction to the Mechanics of Solids.* Reading, Mass.: Addison-Wesley.

Halperin, Don A. 1976. *Statics and Strength of Materials for Technology.* New York: John Wiley.

Muvdi, B. B., and McNabb, J. W. 1980. *Engineering Mechanics of Materials.* New York: Macmillan.

Popov, E. P. 1976. *Mechanics of Materials.* 2d ed. Englewood Cliffs, N.J.: Prentice-Hall.

Roark, Raymond J. and Young, Warren C. 1975. *Formulas for Stress and Strain.* 5th ed. New York: McGraw-Hill.

Shanley, F. R. 1957. *Strength of Materials.* New York: McGraw-Hill.

Shelley, Joseph F. 1980. *Engineering Mechanics* (Statics and dynamics). New York: McGraw-Hill.

Thrower, James R. 1980. *Technical Statics and Strength of Materials.* New York: Macmillan.

Index